KT-465-225

Game Theory with Applications to Economics

Game Theory with Applications to Economics

Second Edition

JAMES W. FRIEDMAN

University of North Carolina

New York Oxford

OXFORD UNIVERSITY PRESS

Oxford University Press

Oxford New York Toronto
Delhi Bombay Calcutta Madras Karachi
Petaling Jaya Singapore Hong Kong Tokyo
Nairobi Dar es Salaam Cape Town
Melbourne Auckland

and associated companies in
Berlin Ibadan

First published in 1986 by Oxford University Press, Inc.,
200 Madison Avenue, New York, New York 10016

First issued as an Oxford University Press paperback, 1991.

Oxford is a registered trademark of Oxford University Press.

Library of Congress Cataloging-in-Publication Data
Friedman, James W.
Game theory with applications to economics/James W. Friedman.—2nd ed.
p. cm. Includes bibliographical references.
ISBN 0-19-506355-4
ISBN 0-19-507053-4 (pbk)
1. Game theory. 2. Economics, Mathematical. I. Title.
HB144.F75 1990 330'.01'593—dc20 89-39691

9 8 7 6 5 4 3 2 1

Printed in the United States of America

To Stewart

Preface

Game theory is moving ahead very rapidly now and the breadth and depth of its application to economics, political science, and other areas is spectacular. One book cannot cover everything, and mine does not do so. I have tried to cover the basics plus a few specialized topics. The choice of specialized topics is governed by several considerations, only one of which is its intrinsic merit. While I do believe those advanced topics I have singled out are important, other topics of equal merit have been passed by. Why? This book gains if I stay closer to the advanced topics which I am in a relatively stronger position to treat. I hope that game theory books will flourish and that more will be written at various levels and with various emphases. We will all be the richer in knowledge.

This new edition differs from the previous one mainly in the coverage of noncooperative games. Chapter 2 is entirely new; it covers finite games and some of the better known refinements of noncooperative equilibrium. These topics had only cursory coverage in the first edition. The remaining chapters are revisions of the earlier edition. The new Chapter 3 is the old Chapter 2, and so forth to the end. The most extensive revisions appear in the new Chapters 4 and 5 where there is much more material on extensions of the "folk theorem for repeated games." The chapters on cooperative games, 6, 7, and 8, remain largely as they were.

In preparing this new edition, I have benefitted from comments on the first edition by a number of people, including Mark Flood, Ad Pikkemaat, Bob Rawthorn, Larry Samuelson, John Seater, and Jörgen Weibull. Takashi Yanagawa, Ngina Chiteji, Robert Boggs, and William Winfrey provided helpful comments on a draft of the new edition. I owe an especial debt to Larry Samuelson who gave me very thoughful, helpful, and detailed comments on the first edition and excellent suggestions for revision. Herb Addison, at the Oxford University Press, has been supportive, helpful, and cooperative both with the present and earlier editions of this book. Both versions are much better for his efforts. None of these people should be held accountable for good advice that I forgot or ignored, nor for any remaining faults. These are my lone responsibility.

Preface

Preface to the first edition

In writing this book, I have tried to reach an honors undergraduate or first-year graduate school audience in economics, having a moderate background in mathematics. The book discusses game theory, with examples from economics and sometimes from politics. It is, to quote a source I cannot now recall, "introductory but not elementary." It presumes no prior knowledge of game theory, but many topics are handled in considerable depth. There are two major respects in which the book is aimed at an economics audience. First, the selection of topics is influenced by my views of what is particularly fruitful for economics and, second, the examples are drawn mainly from economics in a way intended to illustrate the breadth and depth of game theoretic influence on that discipline. In addition to serving as a text for courses in game theory, I hope that this book will prove helpful to economists who have not specialized in game theory, and who wish to gain a knowledge of useful developments in the field.

The book can be read on any of several levels, depending on the mathematical background of the reader and on her or his willingness to work hard. A reader with only a working problem-solving knowledge of calculus, and with the ability to accept prose occasionally littered with mathematical symbols, should be able to understand all the examples and to follow everything except the proofs of theorems and lemmas. The proofs themselves vary greatly in their length and difficulty. Undoubtedly, many readers will find some proofs easy and other impenetrable. Although I intended to keep the mathematical depth uniformly modest, I soon found that doing so would force certain topics to be either omitted entirely or treated with insufficient completeness. Consequently, I compromised by including topics of varied technical difficulty. The most mathematically difficult portions are explained in words that give the reader an intuitive grasp.

Many writers have commented on their great intellectual debt to others, and my debt must be as large as most. In addition to the influence over the years of many teachers, colleagues, and students, there are two people whom I particularly want to single out: the late William Fellner and Martin Shubik. Willy taught the first-year graduate theory course at Yale when I entered the graduate program there. He was a man of great thoughtfulness and subtlety: in that course came my first introduction to a game-theoretic topic when we studied oligopoly. That provided a glimpse into an interesting area that shortly became a fascination. The following year, Martin Shubik was a visiting professor who taught a game theory/oligopoly course in which his knowledge, point of view, insight, and enthusiasm influenced my main lines of interest in economics.

Several people have read parts of an earlier draft of this book and have given me helpful comments, including Catherine Eckel, Nicholas Economides, Val Lambson, Douglas McManus, John McMillan, David Salant, Patricia Smith, Richard Steinberg, Manolis Tsiritakis, Chang-Chen Yang, and Allan Young. They eliminated typographical errors, improved the clarity of exposition, and corrected my errors. Robert Rosenthal read most of the previous draft and provided comments that helped me eliminate much nonsense and much murky prose. At many points in the text he has saved me from serious error. Some typing was done by Barbara Barker, Irene Dowdy, and Wadine Williams. I am grateful to them for pitching in when there was some time pressure. Most of the typing was done, with great skill, speed and good humor, by Vickie Carrol and Jay Willard. Much of what is good in the following pages results from the help of all these people. Any blame, of course, for any remaining mistakes and inaccuracies rests with me.

Chapel Hill, N.C.
June 1985 J. W. F.

A note on other books on game theory

There are many books on game theory from which a reader might benefit greatly. These include Luce and Raiffa (1957), Owen (1982), Shubik (1982, 1984), Harsanyi and Selten (1988), Moulin (1988), Roth (1979), and van Damme (1987). Luce and Raiffa have written the sort of classic book to which many of us must aspire. Although much out of date due to the developments of more than 30 years, it is a superb source for much of the central material of the field. The writing is lucid, the intuition supporting various models and the criticisms of them are insightful and illuminating, and the technical demands are never more than the minimum necessary for the subject. Owen's book is a fine modern text that is particularly strong on cooperative game theory. Shubik's two volumes, totalling over 1,200 pages, cover immense ground in both game theory proper and applications of game theory to social science disciplines. They surely provide the most comprehensive coverage of game theory and applications by a single hand and will be an invaluable source to serious students of the subject. The books by Harsanyi and Selten, Moulin, Roth, and van Damme are somewhat more specialized in scope. Harsanyi and Selten tackle the monumental problem of providing a selection theory (a means of choosing a unique solution in games having multiple equilibria) for noncooperative games. Moulin deals with cooperative game theory, applied mainly to welfare economics. Roth looks at axiomatic bargaining models. van Damme concentrates on recent developments, including equilibrium refinements, noncooperative bargaining, and evolutionary game theory.

Contents

7 *n*-person cooperative games with transferable utility 242

List of figures

List of tables

List of Tables

Game Theory with Applications to Economics

1
Introduction to games

Samuel Johnson (1755) defines the word *game* as "sport of any kind." Our modern notion would typically add that games usually have particular rules associated with them. Examples include athletic games such as soccer, golf, basketball, and tennis; card games such as bridge, poker, and cribbage; and board games such as chess, backgammon, and Go. Most of these games share an interactive and competitive element, that is, a player strives to outdo the other players in the game, and her success depends on both her own actions and the actions of the other players. For instance, in tennis, one does not try merely to hit the ball back to the other player, one tries to hit it back so that the other player cannot return it. Thus, where the ball is aimed will depend on where the other player is placed. An exception to the interactive and competitive element in the preceding list of games is golf, in which each player struggles against an absolute standard, and the actual performance of a player does not depend on the actions of the other players.

Several features typify most games. First, games have rules that govern the order in which actions are taken, describe the array of allowed actions, and define how the outcome of the game is related to the actions taken. Second, games involve two or more players, each of whom is struggling consciously to do the best he can for himself. Third, each player's outcome depends on the actions of the other players. The player knows this, and knows that choosing the best action requires an intelligent assessment of the actions likely to be taken by the other players.

These general characteristics typify many situations that are not *games* in the sense *sport*. For example, when the management of a company and the leadership of a union face each other over the bargaining table to negotiate a new contract, they are in a gamelike situation. The rules are less formal and less detailed than they are for chess, but there are nonetheless rules. Offers and counteroffers are put forward by each side with a view to making the final settlement as favorable as possible. What offer one should make best to further its interest depends on just how that offer will be received and responded to by the other side. In 1979, the new head of the

International Harvester Company bargained so hard with the United Auto Workers Union that union members had the impression that he wished to destroy their organization. After a long and bitter strike, a settlement was reached; however, the union became so intransigent at the subsequent contract negotiations 3 years later that they did not settle with the company until (coincidentally) that same head resigned from his job.

Most people face bargaining situations at one time or another. Buying or selling a house or a car, for example, represents a game in that each side wants to obtain as much as possible from the other.

The first important theorem in game theory, the *saddle point theorem* for two-person, zero-sum games, was published by von Neumann (1928). This was followed by the rich collaboration culminating in von Neumann and Morgentern (1944), which includes approaches to the treatment of many kinds of games along with much discussion of the potential applications of game theory. The early history of game theory, going back to von Neumann's precursors, is discussed in Rives (1975).

1 Examples of games

Several illustrative games are described in Section 1. The first is an oligopolistic market with three firms; the second and third are political science models of the election process. All of these are *noncooperative games,* which means that the players are unable to make contractual agreements with one another. The final example is a labor–management dispute whose outcome should be a contract signed by the two players.

1.1 A three-firm market for computers

Consider a market in which the number of active firms is small, but greater than 1. Imagine, for example, firms in the rapidly changing computer industry. Each firm must decide how to direct its efforts in developing and marketing new equipment, and the best plan for one to follow depends on the plans adopted by all of the others. A small- to middle-sized firm might be able to tackle seriously the market for just one size of computer, and, to survive, it must select a size that will not have many competing firms. Meanwhile, problems of product development require that it commit itself to a course of action years in advance, before it can possibly have a clear idea what the other firms have chosen. As a simple numerical illustration, imagine that there are three firms, each of which can choose to make large (L) or small (S) computers. The choice of firm 1 is denoted S_1 or L_1, and the choices of firms 2 and 3 are denoted S_i or L_i, where $i = 2$ or 3 indicates the firm. Table 1.1 shows the profit each firm would receive given the various choices the three firms could make. For example, the second entry in the first row is $(0, -10, 60)$ indicating profit of 0 for firm 1, -10 for firm 2, and 60 for firm 3. This results from the choices $(S_1 S_2 L_3)$. Firms 1 and 2 choose to produce small computers, while firm 3 produces large ones. Note

TABLE 1.1 Profits to three computer firms associated with various product decisions

	S_2S_3	S_2L_3	L_2S_3	L_2L_3
S_1	−10 −15 −20	0 −10 60	0 10 10	20 5 15
L_1	5 −5 0	−5 35 15	−5 0 15	−20 10 10

that this is the most profitable possible outcome for firm 3; however, if firm 2 had foreseen the choices of firms 1 and 3, then it would have done better by producing large computers.

Suppose that each of the firms has the information in Table 1.1. We can easily analyze the situation from the vantage point of a single firm, say firm 1. It is better off choosing S_1 if either or both of the others choose L, but it is best off with L_1 if both of the others decide on S. Whether firm 1 makes large or small computers thus depends on its assessment of what the other two firms do. For both of the other two firms, *large* is the better choice as long as at least one firm selects *small*. This immediately suggests a combination of choices that may be a plausible equilibrium outcome: S_1, L_2, and L_3. From Table 1.1, it is clear that this combination causes no ex post regret. That is, given the choices of the other two firms, no one firm could have done better by choosing something else. The same cannot be said for any other selection open to the firms. For any other outcome, at least one firm could have done better by making the other choice, given the choices of the rival firms. The outcome (S_1, L_S, L_3) is a *noncooperative equilibrium* for the game. Such equilibria are studied extensively in Chapters 2 to 5 in a variety of models. The defining feature of this equilibrium is that no single player would have obtained a larger payoff had she used an alternative strategy, given the strategies of the other players.

1.2 Two political election games

Politics provides ready examples of game theoretic situations. Imagine several candidates who are running for office, say, for mayor. Prior to the start of the campaign, each candidate must select a platform, which means that a stand must be taken on each of several issues. In one version of the example, the original platform is immutable, and, supposing each candidate merely seeks election and is equally willing to be elected on the basis of any platform, the best platforms for a candidate depends on the platforms adopted by the others. Figure 1.1 illustrates this for two issues: the annual budget for the police force, represented by the horizontal axis, and the jail sentence to be imposed for armed robbery, represented by the vertical axis.

Each voter has preferences regarding positions in this two-dimensional issue space. For simplicity, suppose each voter has a favorite position, that

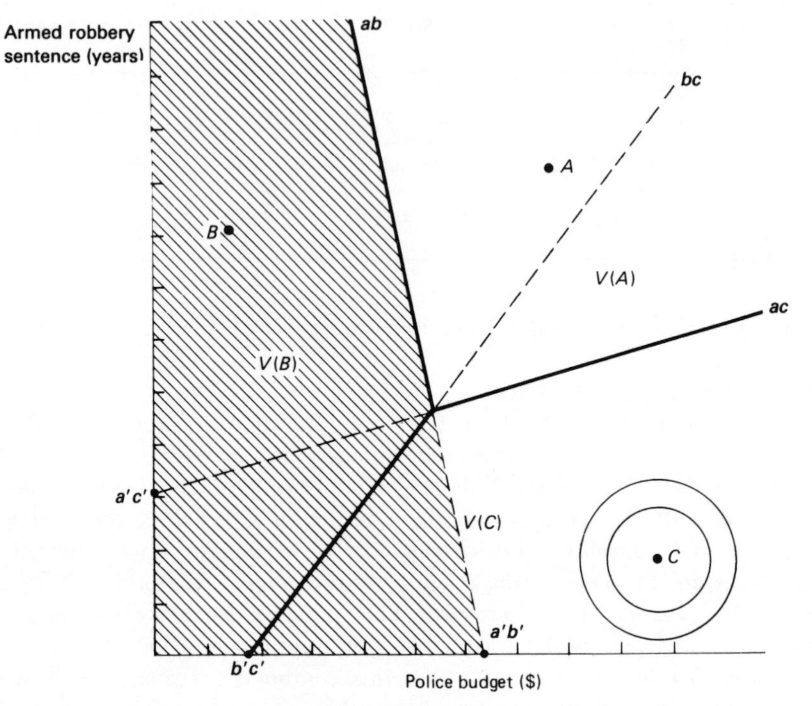

FIGURE 1.1 Voters and candidates in a policy space.

the farther a point is from this best point, the worse it is, and that all positions equidistant from this best point are equally good. A voter's favorite position can be thought of as her *location* in the issue space. For example, the voter located at C in Figure 1.1 has indifference contours that are concentric circles around C. Suppose a voter will vote for that candidate whose position stands highest on the voter's utility scale. Then the three straight lines in Figure 1.1 indicate voters who are indifferent with respect to the various pairs of candidates. For example, the voters on the line through ab and $a'b'$ are indifferent with respect to candidates A and B. The shaded region shows the voters who prefer A to B. The solid part of that line shows voters who find A and B superior to C. The broken part of that line shows voters who prefer C to A or B. The other two lines have similar meanings. Consequently, the space of voters can be divided into regions according to which voter is favored. $V(A)$ is the region of voters who prefer A to either of the other candidates. $V(B)$ and $V(C)$ have parallel meanings. Letting v_A, v_B, and v_C denote the number of voters in the three regions, the winner is the candidate having the largest number of votes. In this game, the players are the three candidates. Voters behave according to clearcut rules that imply they make no judgments. For each candidate, a location is a strategy; hence, choosing a strategy means choosing a location. A complete specification of this game would include a utility, or payoff, function for each candidate, based, ultimately, on the distribution of votes and on whether the candidate won or lost.

The example can be modified by supposing the voters are also players. As before, the candidates are players who choose platforms. Each voter chooses a candidate; however, a voter's utility depends on whether she has voted for one of the top two vote getters. The rationale for this is the view that voting for a third place candidate is throwing one's vote away. Thus a voter maximizes her payoff by voting for the candidate (i.e., platform) she prefers among the top two vote getters. In three-candidate presidential elections in the United States, it is usually clear during the campaign which candidate has the smallest following (i.e., which candidate would get fewest votes if each voter voted for the nearest candidate). If, for example, $v_A > v_B > v_C$, then those voters favoring C would not vote for him, on the supposition that he would run behind the other two. The actual vote would be split between A and B, with the winner being determined by the voting preferences of the voters in region $V(C)$.

The two voting models embody different views concerning the purpose of voting. The first model is consistent with the view that by voting a person identifies the candidate he believes to be best among all candidates. The second model supposes that voters wish to use their votes to bring about as good an outcome as they can. If a voter's favorite candidate is doomed to place a distant third, he will do more to bring about a favorable outcome by casting his vote for his second choice, thereby reducing the chance that his least favorite candidate will win. Abandoning the preferred candidate in this case does not hurt the voter because his preferred candidate cannot win anyway.

Whichever rule is used to determine who votes for whom, it is clear that the position chosen by a candidate will depend on more than the way voters are distributed; it will also depend on how the candidate presumes the other candidates will position themselves.

1.3 Labor–management: A cooperative game example

The computer firm example and the election examples are all instances of what are called *noncooperative games*. The essential feature distinguishing these games is that the decision makers (players) are unable to make legally binding contracts with one another. The computer firms cannot collude, nor can the voters or the candidates. This inability to collude is part of the *rules of the game*; and is hence assumed to be beyond the players' control. This may be contrasted with a labor–management situation in which the outcome of the game is a contract if the two sides (players) agree or a threat if they do not. The threat outcome could be a strike or could be a complete breakdown between the two sides, so that they go their separate ways and terminate relations altogether. In this case, the workers seek jobs elsewhere, refusing to go back to the company, and the company must acquire and train a whole new workforce. This is represented in Figure 1.2,

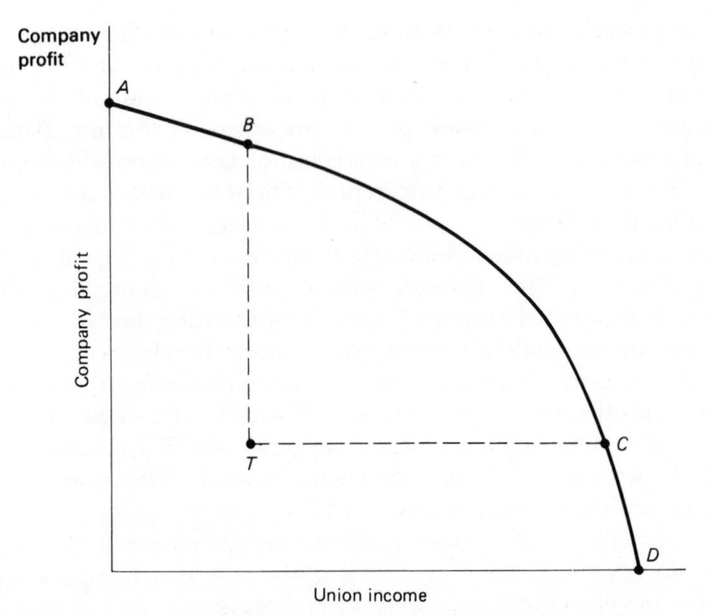

FIGURE 1.2 An example of labor–management bargaining.

where the horizontal axis is the total income of the union workers and the vertical axis is the company profit. Point T shows the threat payoffs that will obtain if the two sides cannot agree; the points on and beneath the curve $ABCD$ show the payoffs that are achievable through agreement.

Because the game is cooperative, that is, admitting of binding agreements, it is natural to focus attention on what the players ought, in some sense, to agree on. Two natural restrictions on their agreements are that the two players agree to an outcome that gives each player at least what he would get at T. For someone to accept less would be irrational. The reasoning behind this is that a player can guarantee that much without the help of the other player. Second players should agree on something that cannot be jointly improved upon. The first restriction, called *individual rationality,* limits attention to outcomes that lie above and to the right of T; the second, called *group rationality,* restricts outcomes to the payoff possibility frontier $ABCD$. In order to satisfy both restrictions, the outcomes must lie on the part of the frontier between B and C.

2 Forms in which games are represented

You may have noticed that the noncooperative game examples and the cooperative example were organized and discussed in very different ways. In the noncooperative example, attention is focused on the actions that each player is able to take, and on how these actions jointly determine each player's payoff. In the labor–management example, the actions of the players are largely suppressed. What matters is what the players are able to obtain, separately and together. How they obtain an outcome is not of central importance. That is, if they fail to agree, they are at T. It does not really matter what they do to achieve T. Similarly, if they agree, they are capable of attaining any point on or below the payoff frontier. Again, it is not important to know how they behave to achieve a particular outcome. What matters is that a particular set of outcomes is freely available to them if they choose to cooperate.

Noncooperative games are often expressed in a fashion that exposes each move a player can make; this is called the *extensive form*. Alternatively, such games are expressed in a way that suppresses individual moves but highlights the overall plans, or *strategies*, available to players. This form, illustrated by Table 1.1, has been called the *normal form*, but has recently been more aptly called the *strategic form*. Cooperative games are often shown in a way that underlines achievable outcomes; as in Figure 1.2. This form is called the *characteristic function form*, or the *coalitional form*. More is said on these matters in later chapters.

2.1 The extensive form and some basic concepts applying to all games

In this section, games are informally described in extensive, strategic, and coalitional forms. Although some elements are common to each form, there are substantial differences. The extensive form, represented by a *game tree*, is discussed first and is used as a vehicle with which to introduce concepts of wider interest. *Every game has a set of rational decision makers, called* **players**, *whose decisions are central to the study of games. The* **set of players** *is denoted* N *where* $N = \{1, 2, \ldots, n\}$. Throughout the book, n is generally finite, and each player knows how many players are in the game. Putting random actions aside for a moment, a game in extensive form starts with a particular player making a move. After the first player moves, some other player has a turn to move, and so on until the game terminates. When the game ends, the players receive their payoffs. Randomness can be added by having certain decision points at which *player* 0 moves. Player 0 is called *nature*, or *chance*, and chooses its move by drawing from a probability distribution that is known to all the other players (i.e., the players in N).

Simultaneous moves by two or more players are modeled using *information sets*. When it is the turn of a player to make a move, she is always located at a specific decision point, called a *node*. If she knows precise location of the node, then that node constitutes an information set. Suppose two players move simultaneously, that player 1 is treated by the extensive form representation as moving first, and that both players know which specific node they are moving from. If player 1 has m possible moves, then each move leads to a different mode. Player 2 will not know at which of these m nodes she is actually located, and, thus, these m nodes will constitute an information set.

2.1.1 *A game in extensive form*

As an example of a game in extensive form, imagine two people who are going to match pennies twice in succession for a stake of \$5. The procedure is that each will choose heads (H) or tails (T) without knowing what the other has chosen. They then reveal their choices to one another. If the two coins do not match (i.e., if there is one H and one T), then player 1 wins \$5 from player 2. If the coins match, player 2 wins \$5 from player 1. At the second stage, the player who lost at stage 1 has the right to decide whether they proceed to another matching or quit. Thus, the loser can now select quit (Q) or H or T. If he chooses H or T, then, simultaneously with that choice, the winner chooses H or T. This is illustrated in the game tree shown in Figure 1.3. To simplify the illustration, it is assumed that utility is measured by money. In the figure, an open circle denotes a point at which a player makes a decision, called a *decision node.* A filled circle, called a *terminal node,* is an endpoint of the game. Payoffs are written at each terminal node. The left-most node is reached by player 1 first choosing H, followed by player 2 choosing H, followed by player 1 electing to end the game (choosing Q). The first (top) payoff is that of player 1 (−5). A game tree represents the action of a game as if moves were always sequential. In the figure, player 1 is shown to move, after which player 2 moves. This sequential action is made equivalent to the simultaneous action described

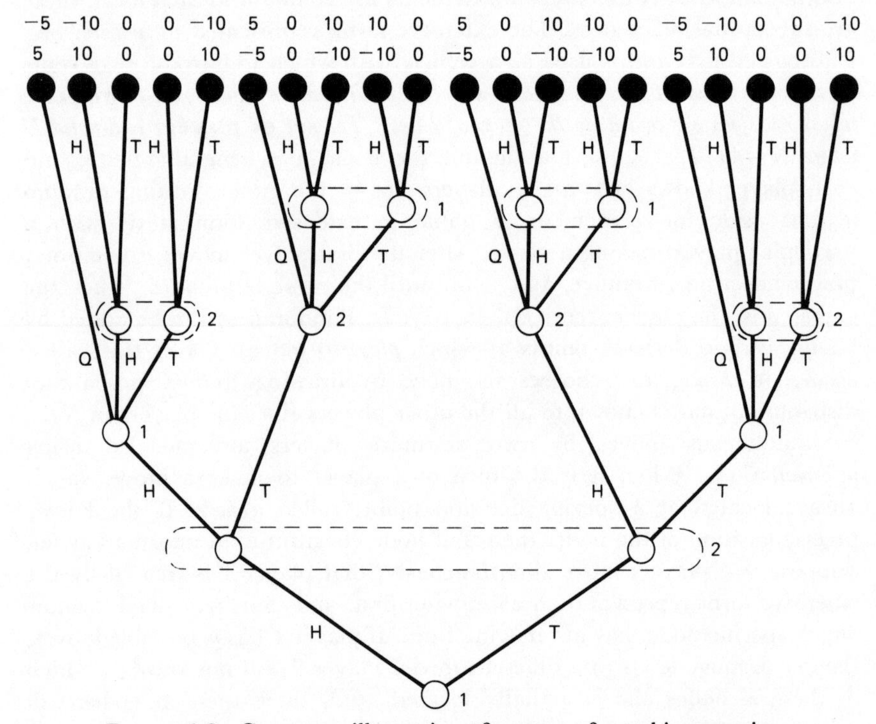

FIGURE 1.3 Game tree illustration of a game of matching pennies.

above by the *information sets*. The broken figures in Figure 1.3 that enclose two decision nodes are information sets. Next to each information set is the name of the player who moves there. If a node is not enclosed by a broken figure, then that node, by itself, comprises an information set. The player who moves from a particular information set knows only that she is in that particular set; therefore, if the set contains more than one node, the player cannot distinguish between them. In general, an information set can contain any number of nodes; however, each node of an information set must have the same number of branches leading from it. This is because the number of branches is the number of choices a player has at that point, and, if she does not have exactly the same number of choices at each node, she has a means of distinguishing between nodes. Also, if one node follows another in the game tree by one or more moves, the two nodes cannot be in the same information set.

2.1.2 Complete information, perfect information, and perfect recall

It is important to be clear concerning the information that a player possesses in a game, and several kinds of information must be distinguished. First, **complete information** rather than **incomplete information** *refers to whether or not each player knows (a) who the set of players is, (b) all actions available to all players, and (c) all potential outcomes to all players.* Complete information *obtains when each player knows (a), (b), and (c).* This is like saying each player knows the whole game tree, including the payoffs listed at each terminal node. In the game illustrated in Figure 1.3, complete information requires that each player have a copy of Figure 1.3 itself (or equivalent information). Suppose, by contrast, that player 1 knew only the payoffs in the top row (his own payoffs), while player 2 knew only the payoffs in the second row. This would be a case of incomplete information. If one or more players lacks knowledge of (a), (b), and/or (c), then the game is one of incomplete information. Most of the models in this book assume complete information, and that assumption should be understood to hold unless incomplete information is explicitly stated.

A second kind of information relates to the information sets. *If each information set in the game consists of just one node, then the game is one of* **perfect information**, *if that is not the case, the game is one of* **imperfect information**. Figure 1.3 depicts a game of complete and imperfect information.

Finally if the information sets are always consistent with a player never forgetting anything, then the game is one of **perfect recall**. *If this does not hold, the game is one of* **imperfect recall**. The game in Figure 1.3 is one of perfect recall, but the variant of it shown in Figure 1.4 is one of imperfect recall. Note that in Figure 1.4 player 1 has an information set that contains four nodes. Such an information set is possible only if player 1 has forgotten whether she chose H or T on her first move. If she recalled her first move, then the two

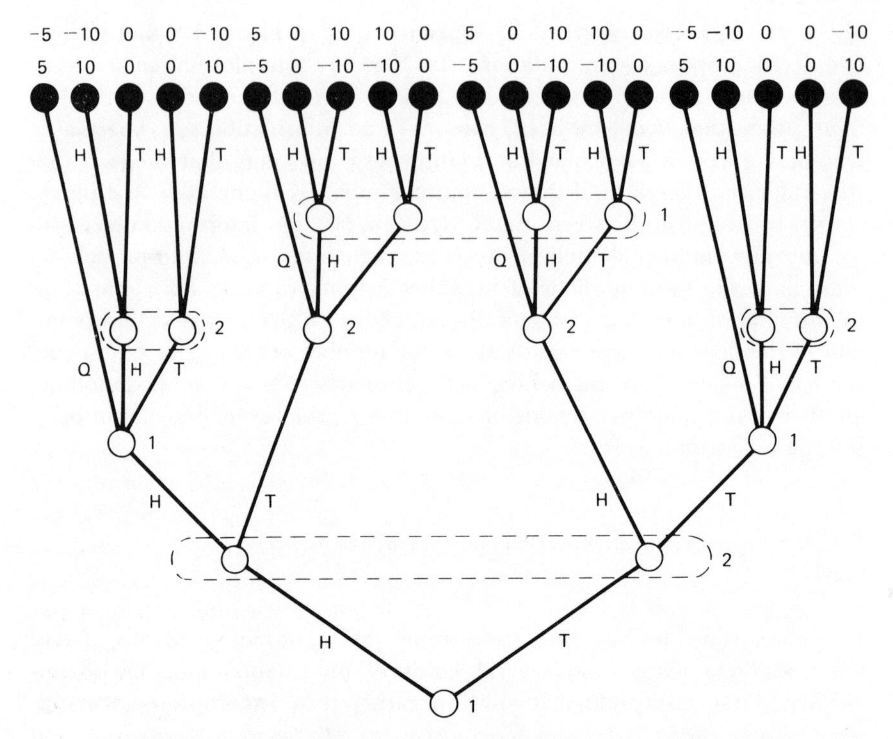

FIGURE 1.4 Matching pennies in a game without perfect recall.

right nodes and the two left nodes would be in different information sets, as in Figure 1.3.

Most of the games studied in this book assume imperfect information and perfect recall. This is because the selection of game theory models is biased towards models with applicability to the social sciences. Simultaneous-move games are commonly encountered in applications, and such games embody imperfect information. Perfect recall is also a natural choice for applications because players are generally either individuals or firms. If they are firms, then they are assumed to have centralized information and decision making. To see where imperfect recall might arise, consider the bidding process in a game of bridge. It is natural to regard bridge as a two-player game, with each player being a two-person team. Suppose Alice and Ben are partners. The early moves of the game involve each person picking up her or his cards and examining them. Alice will not recall what Ben saw, and vice versa. When it is Alice's turn to move (i.e., to bid), she does not know Ben's hand, but she knows her own. When that player (i.e., Alice and Ben) has its next move, it is Ben who bids, and he does not know the contents of Alice's hand.

2.1.3 The rules of the game, common knowledge, binding agreements, and commitments

The extensive form of the game shows the complete move, information set, and payoff structure. It does not quite give a total definition of the game. The missing elements, covered under the rubric *rules of the game,* include (a) whether information is complete or incomplete, (b) common knowledge, and (c) whether it is possible for the players to make binding agreements or commitments. Other aspects of information previously mentioned (perfect or imperfect information, perfect or imperfect recall) are captured by the extensive form. It would even be possible to represent incomplete information in the extensive form if a separate game tree were drawn up for each player showing, for each, the information she was to have. *Common knowledge* refers to those things that are known by all players, and known by each to be known to all of them, and so forth (see Aumann (1976) and Milgrom (1981)). In most of this book, games of complete information are characterized by each player knowing the entire structure and payoffs of the game, by each player knowing that all players possess this information, and by each player knowing that all players have this information. There is, for example, an important conceptual distinction to be made between (a) a complete information game in which complete information is common knowledge and (b) a complete information game in which each player does not actually know whether the other players also have complete information. In general, there is no reason to suppose that intelligent behavior and equilibrium will be the same in both cases.

To illustrate these informational considerations, look at the game in Figure 1.3. It is a game of complete information if all the information in the figure is known to both players; it is a game of incomplete information if everything in the figure, except some of the payoffs at some of the terminal nodes, is known to each player. The gaps of knowledge need not be the same for each player. The information situation is common knowledge if each player is correctly informed about what the other knows (and each knows that the other knows that he knows this, and so forth).

Concerning binding agreements, it is common knowledge among the players whether they may be made. In motivating the notion of a binding agreement, it is usual to note that the game requires an outside authority that enforces any such agreements. The situation is as if the parties to a binding agreement write and sign a contract that they register with this authority. The authority can monitor the agreement at no cost, can tell with certainty whether its terms are being carried out, and can, like an avenging angel, impose on violators sanctions so severe that cheating is absolutely out of the question. Related to binding agreement is commitement. *A* **commitment** *is an action taken by a single player that is binding on him* and that, to be of any use, must be known to the other players. A commitment is used to persuade others to take actions that will favor oneself by threatening to take a particular action which will be costly to onseelf if they do not do so. A rather fanciful example is that of the perfectly rational person who demands a payment from each of several others by threatening to kill himself if the payments are not made. The others would rather make the payments than see the person die, but are unlikely to believe he would carry out such a threat. The threat can be made credible if the person can make a contract with some outside agent who promises to kill him should the payments fail to be made. **Binding agreements** *and* **commitments** *are both instances of voluntary restrictions on the actions available, where those restrictions are enforced. With* **binding agreements,** *two or more players make jointly agreed upon restrictions, while* **commitments** *are unilateral restrictions.*

Another point should be noted about game trees: for all but very simple games they are unwieldy. Anyone doubting this can convince himself by making a complete game tree for the game of tic-tac-toe. At the first move, player 1 has nine choices. At the second move, player 2 has eight choices, but he can be at any one of nine nodes. Thus, at the third move, player 1 can be at any one of 72 nodes. The game must go at least five moves, and, at the fifth (the third move of player 1), player 1 could be at any one of $9 \cdot 8 \cdot 7 \cdot 6 = 3{,}024$ nodes. Game trees are handy for illustrating basic game theoretic concepts and for analyzing particularly simple games, but other forms are required to deal with many classes of games. Many games in later chapters have uncountably infinite strategy spaces. For these games, it is not clear that an extensive form could be fruitfully specified.

2.2 The strategic form

Imagine a game in extensive form such as the one described in Figure 1.3. Suppose a player was committed to play this game, but was unable to show up for the execution of it. He could find a person to stand in for him, but it would be necessary to give the stand-in a complete set of instructions on what choices to make. This amounts to giving the stand-in a decision rule for each information set in the game belonging to the player. Two information sets can have different decision rules associated with them. Suppose the player to be player 1. His instructions might then be to: choose H on the first move; then, if player 2 chose H on her first move, to choose Q; but if player 2 chose T on her first move, to choose H on the second move. Such a set of instructions is called a **strategy**. *What characterizes a strategy is that, at every point of decision for a player, the strategy dictates precisely what the player does.* In general, games are played by a sequence of moves, but individual moves are of interest insofar as they contribute to an overall plan of action—a strategy. Strategies can allow for players to randomize in choosing moves. For example, a valid strategy for player 1 is to choose H on his first move; then, if player 2 choses H, to choose Q with probability .4, H with probability .5, and T with probability .1; if player 2 chose T, player 1 chooses T. For a particular strategy for player 1 and a particular strategy for player 2, it is possible to calculate an expected payoff for each player. Given a pair of strategies, the probability of reaching any specific terminal node is determined. If a strategy has no randomly determined choices, it is called a *pure strategy*; otherwise it is called a *mixed strategy*.

Table 1.2 contains a complete enumeration of all the pure (i.e., nonrandomized) strategies of players 1 and 2 for the game in Figure 1.3. Table 1.3 is a double-entry payoff matrix giving the payoffs associated with any pair of pure strategies that the players might choose for this game. In Table 1.3, the strategies of player 1 are the row headings, while those of player 2 are the column headings. The numbers, 1, . . . , 12, correspond to the strategies as they are defined in Table 1.2. For any particular *strategy combination* (i.e., a pair of strategies, one for player 1 and one for player 2) the two corresponding numbers in the table are the payoffs of the two players; the figure that is higher and to the left is the payoff of player 1. For example, suppose player 1 chooses 8 and player 2 chooses 3. Then the payoff to player 1 is 10, and the payoff to player 2 is −10. Notice that the *strategic form* represented in Table 1.3 completely hides the underlying move structure of the game. In general, the strategic form of a game, although very handy to work with, suppresses information about the underlying move structure of the game that may be of interest. The treatment in

TABLE 1.2 Strategies for the strategic form of the game shown in Figure 1.3

	Strategies of player 1			Strategies of player 2		
	First move	Second move if the first move of player 2 is H	T	First move	Second move if the first move of player 1 is H	T
1	H	Q	H	H	H	Q
2	H	Q	T	H	T	Q
3	H	H	H	H	H	H
4	H	H	T	H	T	H
5	H	T	H	H	H	T
6	H	T	T	H	T	T
7	T	H	Q	T	Q	H
8	T	T	Q	T	Q	T
9	T	H	H	T	H	H
10	T	T	H	T	H	T
11	T	H	T	T	T	H
12	T	T	T	T	T	T

TABLE 1.3 Payoff matrix showing the strategic form of the game shown in Figure 1.3

Strategies of player 1 \ Strategies of player 2	1	2	3	4	5	6	7	8	9	10	11	12
1	−5 5	−5 5	−5 5	−5 5	−5 5	−5 5	5 −5	5 −5	0 0	0 0	10 −10	10 −10
2	−5 5	−5 5	−5 5	−5 5	−5 5	−5 5	5 −5	5 −5	10 −10	10 −10	0 0	0 0
3	−10 10	0 0	−10 10	0 0	−10 10	0 0	5 −5	5 −5	0 0	0 0	10 −10	10 −10
4	−10 10	0 0	−10 10	0 0	−10 10	0 0	5 −5	5 −5	10 −10	10 −10	0 0	0 0
5	0 0	−10 10	0 0	−10 10	0 0	−10 10	5 −5	5 −5	0 0	0 0	10 −10	10 −10
6	0 0	−10 10	0 0	−10 10	0 0	−10 10	5 −5	5 −5	10 −10	10 −10	0 0	0 0
7	5 −5	5 −5	0 0	0 0	10 −10	10 −10	−5 5	−5 5	−5 5	−5 5	−5 5	−5 5
8	5 −5	5 −5	10 −10	10 −10	0 0	0 0	−5 5	−5 5	−5 5	−5 5	−5 5	−5 5
9	5 −5	5 −5	0 0	0 0	10 −10	10 −10	−10 10	0 0	−10 10	0 0	−10 10	0 0
10	5 −5	5 −5	10 −10	10 −10	0 0	0 0	−10 10	0 0	−10 10	0 0	−10 10	0 0
11	5 −5	5 −5	0 0	0 0	10 −10	10 −10	0 0	−10 10	0 0	−10 10	0 0	−10 10
12	5 −5	5 −5	10 −10	10 −10	0 0	0 0	0 0	−10 10	0 0	−10 10	0 0	−10 10

Chapters 2 to 5 of noncooperative games and their equilibrium points brings this out.

Cournot (1838) duopoly provides a familiar example of a game in strategic form. To see this, suppose an inverse demand function $p = \max\{0, 100 - q_1 - q_2\}$ where p is market price and q_i is the output of firm i. In view of this inverse demand function, the firms may be thought to choose output levels from the interval $[0, 100]$. Supposing that there are no costs, the profit of firm i can be written $\pi_i = \max\{0, 100q_i - q_i^2 - q_1q_2\}$. Staying within the classic framework of *single-shot* duopoly, having no past and no future, only the current single period, Cournot equilibrium makes sense. It is a pair of output levels, (q_1^*, q_2^*), *such that the output level of each firm maximizes its profit given the output level of the other firm.* In the present case this occurs at $q_1^* = q_2^* = 33\frac{1}{3}$. Translating to a game, a strategic form game is typically denoted (N, S, P), where N is the set of players, $S = S_1 \times S_2$ is the strategy set, and $P = (P_1, P_2)$ denotes the payoff functions of the players. Relating this to the Cournot duopoly, $N = \{1, 2\}$ is the set of firms; $S_1 = S_2 = [0, 100]$ are the strategy sets of the firms with $S = [0, 100]^2$, and $P_i = \pi_i$ are payoff functions.

2.3 The coalitional, or characteristic function, form

The hallmark of cooperative games is that the players, or any subgroup of them, have the right to make contractual agreements that are 100 percent binding. A subset of players that has the right to make an agreement is called a *coalition,* and it is usually assumed that any subset of the players can form a coalition. In such a cooperative setting, the strategies are of less direct interest to someone wanting to analyze a game; rather, the payoff opportunities open to each player and to each coalition are of central concern. Indeed, some games are most naturally described in ways that make it unclear just how strategies ought to be described. Other games in which strategies are clearly present and easily described must have another description that easily captures the power and opportunities of the coalitions. Some examples should clarify these comments.

First, take a common example in which strategies make no explicit appearance. Suppose there are three people who are told that they can divide (up to) \$100 in any way they wish. The rules are that a division of the money, denoted $x = (x_1, x_2, x_3)$, (a) must satisfy the condition $x \geq 0$, (b) must allocate no more than \$100 (i.e., $x_1 + x_2 + x_3 \leq 100$), and (c) at least two of the three players must agree to an allocation. Suppose that utility is measured by money for each player. Then the payoff possibilities open to each coalition can be described by means of a *characteristic function,* denoted $v(K)$, that associates with each coalition K the total utility that the members of that coalition can achieve when they act in concert. Thus, for the game of dividing \$100, the characteristic function is (a) $v(\{i\}) = 0$ for $i = 1, 2, 3$, (b) $v(\{i, j\}) = 100$ for any two (distinct) players i and j, and (c) $v(\{1, 2, 3\}) = 100$.

One cooperative game solution concept, the *core,* is a generalization of Edgeworth's (1881) *contract curve,* and is based on the notion that an outcome agreeable to all players must give as much to each single player and to each coalition as it (the player or coalition) can achieve for itself. The reasoning behind this solution is that such requirements are needed to obtain the agreement of all players and coalitions. If, for example, a proposed solution gives to a player less than the player can guarantee himself on his own, then the player will not agree to the joint outcome. Similarly, if a coalition can, on its own, achieve an outcome that gives more to each player than each received under a proposed outcome, then the coalition will not agree to the proposal.

In the present game, this requires that each individual receive at least 0 and each pair of individuals receive at least \$100 between them. For the three-person game of splitting \$100, the core is empty. It is impossible to divide \$100 so that players 1 and 2 receive \$100, players 1 and 3 receive \$100, and players 2 and 3 receive \$100. These conditions could be met in a game in which any two players can obtain \$100, but a coalition of three players can obtain \$150 (or more). Then an outcome in which each player receives at least \$50 satisfies all the required conditions.

The situation is substantially the same if utility is not measured by money. Suppose that the utility of money for each player were $u_1(x_1) = x_1^{.5}$ for player 1, $u_2(x_2) = x_2$ for player 2, and $u_3(x_3) = \ln(1 + x_3)$ for player 3. It is no longer possible to describe with a single number the possibilities open to a coalition of two or more players. Instead, a set is used to describe them. Take, for example, the coalition $K = \{1, 2\}$, consisting of players 1 and 2. Each player alone can guarantee a payoff of 0; therefore, the only coalitional outcomes that are of interest are the ones giving each player at least these minima. What this coalition can achieve is illustrated in Figure 1.5, where the upper right frontier gives all the outcomes to players 1 and 2 corresponding to the division of the \$100 between the two of them. On and below this frontier, bounded underneath and at the left by the two axes, are all two-persons payoffs achievable by the two of them.

Duopoly can be used here as an example of how a characteristic function can be derived from a game in strategic form. Firms are assumed to be able to make binding agreements. In general, this is not typical of a duopoly market in the presence of antitrust law; however, we may assume the market is regulated by a government commission that has come under the control of the firms. First, consider the duopoly game examined earlier. The duopolists can achieve joint profit maximization if they choose their output levels so that $q_1 + q_2 = 50$. This affords them a joint profit of 2,500, which is split in proportion to their respective outputs. Thus, in a profit diagram, they have a *payoff possibility frontier* consisting of a straight line going from 2,500 on the π_1 axis to 2,500 on the π_2 axis. This is, in essence, a transferable utility game. It can be altered into a nontransferable utility game by adding cost functions for the firms. Suppose the firms' total cost functions are $C_1(q_1) = 5q_1$ and $C_2(q_2) = 10q_2 + .5q_2^2$, respectively. Letting $\pi_i = pq_i - C_i(q_i)$ denote the profit of firm i, the payoff possibility frontier for the two firms is calculated by maximizing $\lambda\pi_1 + (1 - \lambda)\pi_2$ with respect to q_1 and q_2 for all values of λ between 0 and 1 (subject to $q_1 \geqslant 0$ and $q_2 > 0$). This frontier is calculated assuming the firms cannot make direct money transfers among themselves. In game theoretic terms, *side payments* are ruled out. The strategy sets for each player can be taken as the output levels in the interval [0, 100], but the strategies themselves are not of any real interest. Clearly, either player can guarantee herself a payoff of at least 0 by producing nothing; therefore, the set of payoffs attainable by the coalition consists of the payoffs on and below the payoff possibility frontier that also give at least 0 to each player. For the present example, these are shown in Figure 1.6. In this two-player game, any outcome on the profit possibility frontier that gives at least 0 to each player is in the core.

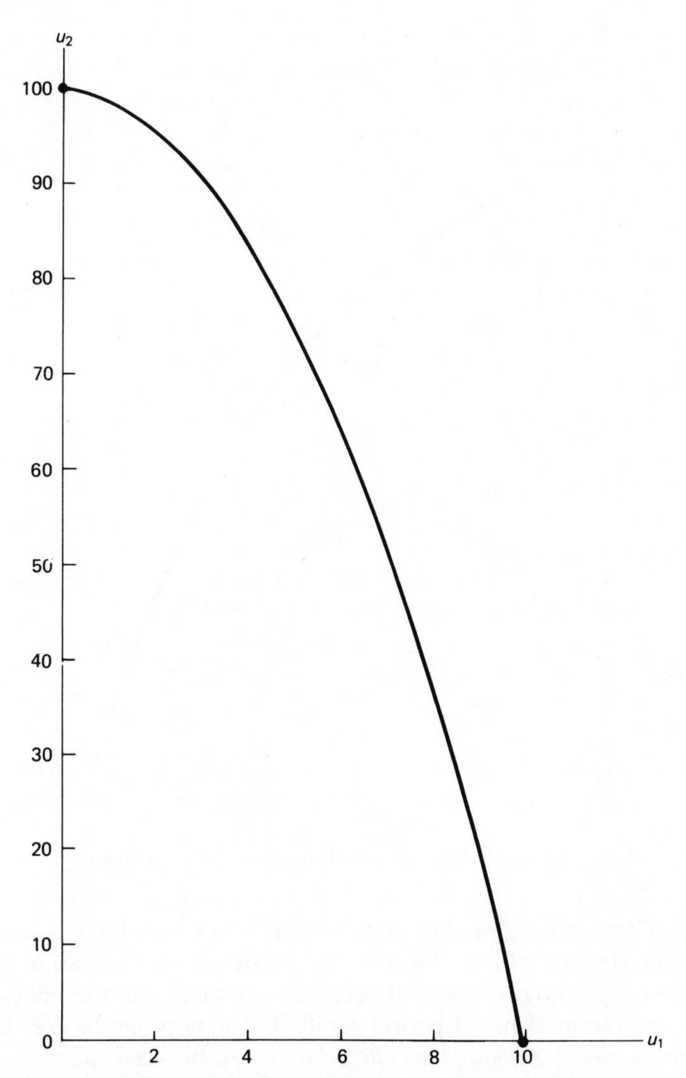

FIGURE 1.5 Attainable utility pairs for two players when utility is not transferable.

3 Outline of the book

The next three chapters are devoted to noncooperative games; the last four chapters deal with cooperative games. Many of these chapters contain a section that illustrates the models by applications to economics and, sometimes, politics.

3.1 Noncooperative games

Chapter 2 deals exclusively with finite games. These are games in which there are a finite number of players and in which each player has a finite

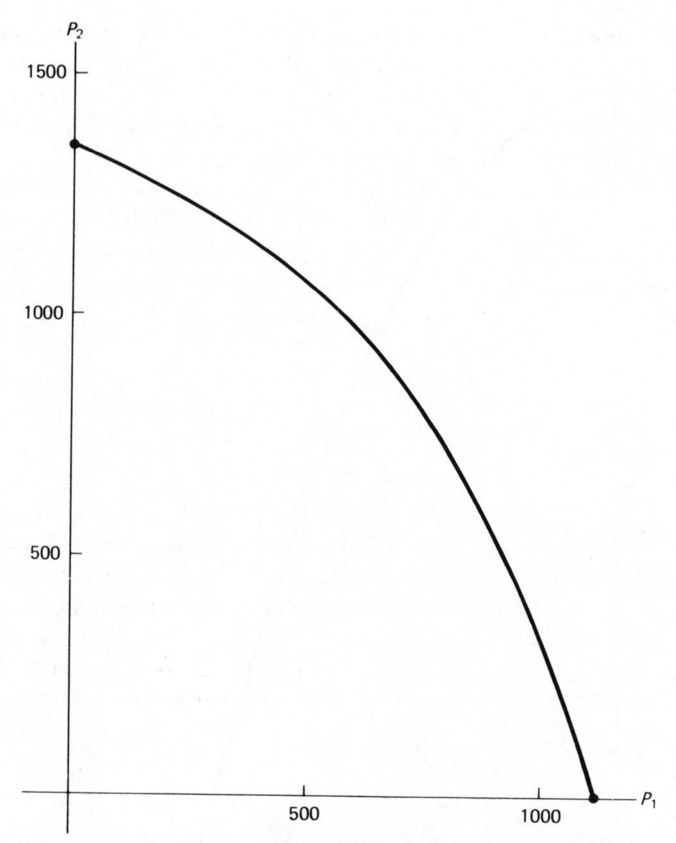

FIGURE 1.6 The payoff possibility frontier for a duopoly.

number of pure strategies. The *extensive form* is developed and some classic results are derived. These include the existence of pure strategy noncooperative equilibria in games of perfect information and the sufficiency of *behavior strategies* in games of perfect recall. The *strategic form* is derived from the extensive form, thus making the relationship between the two explicit. Then various *refinements of noncooperative equilibrium* are taken up. These include *subgame perfection, perfection, sequential equilibrium,* and *proper equilibrium.* The motivation of these refinements is that certain games will have multiple equilibria, some of which may seem behaviorally implausible.

Chapter 3 takes up noncooperative games in strategic form. Games are directly described in the strategic form instead of being formulated in the extensive form with the strategic form being derived from it. After some basics about such games, two-person, zero-sum games are discussed, along with von Neumann's (1928) saddle point (minimax) theorem, and two-person strictly competitive games. The latter are two-person games in which all outcomes are Pareto optimal; they form a generalization of two-person, zero-sum games that preserves the essential strategic feature of these games: in both two-person, zero-sum games and two-person strictly competitive games there is no room whatever for cooperation between players. In a two-person game that is not strictly competitive or in a three- or more-person zero-sum game, there is room for some cooperation between at least one pair of players. The remainder of the chapter deals with general n-person noncooperative games in strategic form, focusing on Nash's (1951) theorem on existence of equilibrium points for noncooperative games. The Nash theorem was stated in the context of finite games, that is, for games in which each player has only a finite number of pure strategies; however, generalizations are considered in Chapter 3 that go beyond Nash and that are particularly useful in economic and political applications.

Chapters 4 and 5 deal with games that can be thought of as either dynamic games or games in a form similar to the extensive form. For illustration, picture the foregoing duopoly as a noncooperative game in which each player will *once and only once* select an output level, with the two players choosing their output levels simultaneously. Such a game is more naturally described in strategic form than in extensive form. Now suppose that the two players will repeat this game in a countable sequence of time periods, $t = 1, 2, 3, \ldots$. This *repeated game*, or *supergame*, is dynamic in the sense that time explicitly enters. The game could be reduced to a strategic form, thus becoming subsumed in the material of Chapter 3, but this reduction would obscure elements of the game that may be of interest. In the duopoly example, one wants to know more than whether the game has an equilibrium. One wants to know something about the nature and characterization of the equilibrium and about the nature of the specific single-period choices that are part of an equilibrium strategy.

Clearly in the duopoly game, one choice of an output level by a firm constitutes a single move of that player. In the one-iteration version of the game, move and strategy coincide; however, in the repeated version, a strategy consists of an initial move and the rules by which each later move will be chosen as a function of the information that will come into the hands of the player. In this application, it would be very desirable to keep track of individual moves, rather than to have them buried in a strategic form. The extensive form is a little cumbersome because the players actually move simultaneously, while the extensive form requires that moves be treated formally as if they were sequential. The formulations in Chapters 4 and 5 are a compromise between the two forms in which the individual moves of the players remain visible.

The difference in material between the two chapters is based on *structural time dependence*. In Chapter 4, all models are either games in which each period repeats a structure identical to that of the preceding period, and, like the duopoly example, each period's activity could be thought of as an individual game, or games in which a different situation may be encountered each period, but each period's situation is independent of all other periods. This independence is *structural* in the sense that the payoffs in a single time period depend only on the moves of that period. The behavior of the players may depend on observations of past choices made by rival players. In other words, structural time dependence may be absent while behavioral time dependence is present. The models of Chapter 5 allow for structural time dependence by permitting the payoffs associated with each period t to depend on moves made in both period t and period $t-1$.

3.2 Cooperative games

Chapters 6, 7, and 8 deal with cooperative games. There is no single solution concept for cooperative games that has played a role comparable to that of the Nash equilibrium in noncooperative games. Consequently these chapters have less of a unifying theme than do Chapters 2 to 5. Because of the complications added by possible agreements among individual coalitions, there is a big difference between the analysis of two-person and $n>2$-person cooperative games. Therefore, Chapter 6 is devoted exclusively to two-person cooperative games, focusing mainly on the Nash (1950, 1953) models and related approaches.

Another difference between cooperative and noncooperative games is the transferable/nontransferable utility division. Transferable utility is a simplifying assumption that is undesirable in many applications but that is a great analytical convenience in many cooperative game models. Chapter 7 takes up transferable utility models, introducing the *core*, the *von Neumann–Morgenstern solution*, the *bargaining set*, and the *Shapley* (1953b) *value*. The value approach to cooperative games provides a strong contrast with the core concerning multiplicity of solutions. The core is often empty, as in the game of splitting $100; it is often very large, as in the game in which two players can achieve $100, but all three players can achieve $200. The Shapley value, in contrast, exists for a large class of games and is always unique when it exists. Chapter 8 is devoted to nontransferable utility models, taking up the core and a generalization of both the Nash bargaining model and the Shapley value.

There is link between the cooperative game chapters and some of the material on noncooperative games, because a central aspect of the *repeated games* literature (see Chapter 4) is that a repeated game allows a *cooperative outcome* to be supported by (i.e., be the result of) a *noncooperative equilibrium*. In a way, this causes a blurring of the distinction between cooperative and noncooperative games; however, this blurring need not be confusing if a distinction is kept in mind between cooperative versus noncooperative games (i.e., the structures) on the one hand, and cooperative versus noncooperative outcomes on the other. The presence or absence of binding agreements is the definitive element for cooperative versus noncooperative games. If binding agreements are possible, then the game (structure) is cooperative, otherwise it is noncooperative. A cooperative outcome can be defined as an outcome that is Pareto optimal in a game in which not all outcomes are Pareto optimal. A noncooperative outcome is merely an outcome supported by a noncooperative equilibrium. Thus the *cooperative* outcome of repeated games are *both cooperative and noncooperative outcomes*. Such outcomes are noncooperative because they are supported by noncooperative equilibrium strategies; they are cooperative because they are Pareto optimal. By contrast, a structure is either cooperative or noncooperative, but not both. With respect to structure, these properties are mutually exclusive and exhaustive; with regard to outcomes they are neither.

3.3 Application to economics and political science

Most, although not all, chapters conclude with a few pages devoted to applications in the social sciences. Most applications are to economics, a few are to political science. These examples are not intended to substitute for textbook presentations of the subjects with which they deal; however, they are serious applications in the sense that they are usually well-accepted models of the processes they describe. An attempt is made to keep these models relatively simple without making them so unreasonably simplistic that teachers of these subjects would never use them.

The collection of applications is somewhat idiosyncratic, and is not put forth as a representative sample of the ways that game theory has been used in economics and politics. Nevertheless, especially for economics, I think the examples are good individual representatives, even if the collection may leave out some important topics. Above all, these applications are intended to show the reader what some interesting uses of game theory look like and, in each instance, to link clearly a game theoretic model with an economic or political model.

4 A note on the exposition

If the ideas of game theory are presented in a largely nontechnical way, the reader can glean only a shadow of their meaning and power; therefore, a moderate level of technical sophistication must be assumed in order to present models with enough precision and generality. The bulk of the book should be within the grasp of readers with a moderate ability to follow logical argument and with enough familiarity with mathematics that they are comfortable with models formulated using symbols. The mathematical appendix at the end of the book contains some notation and definitions that are used in the book. The appendix is no substitute for a mathematics text, but it may serve as a handy resource for refreshing one's memory.

In addition to the sections containing applications, many examples are included in the purely game theoretic sections of the book. I hope that between these examples and considerable verbal explanation, the material presented in the remaining chapters is readable with nothing more than a solid grasp of basic calculus. Some of the proofs are not easy; however, the presentation and discussion of the model and results should be accessible and useful even where the reader encounters a proof too difficult to work through.

2

Finite noncooperative games

A game is *finite* if the number of players is finite, the number of times any player moves is finite, and the number of (nonrandomized) alternatives available to a player at any move is finite. A game is *noncooperative* if the players are not permitted to make *binding agreements*. A binding agreement is a contractual arrangement among some or all players that, once agreed upon, cannot be broken. It is perfectly and completely enforceable. A game is *cooperative* if binding agreements are possible. This chapter, along with Chapters 3 through 5, is concerned with noncooperative games.

Much of the early work in game theory was carried out for finite games. Prominent examples are von Neumann's (1928) saddle point equilibrium for two-person zero-sum games and Nash's (1951) proof of existence of noncooperative equilibrium for *n*-person noncooperative games. Many results for finite games have been generalized to go far beyond their original scope. Such generalizations are particularly important for economics, where the modeling traditions have tended toward assuming that economic actors have infinitely many choices. Common examples are models of the economy in which any consumer can choose a consumption bundle from a compact, convex (hence, uncountably infinite) set, and firms can choose production plans from closed, convex sets

Finite games have much more than historic interest. As game theory progresses, new results are often worked out first for finite games. The extensive literature on *refinements of noncooperative equilibrium* is a case in point. Furthermore, finite games in extensive form, which occupy much of this chapter, offer a superb setting in which to illustrate and understand the general nature of games, to see the fundamental distinction between *moves* and *strategies*, and to see the very comprehensive nature of a strategy.

In Section 1 the *extensive form* is developed in detail. The *strategic form* is then derived from the extensive form in Section 2. The *noncooperative equilibrium* is introduced in Section 3, where some basic results, such as the existence of pure strategy equilibrium for games of perfect information and von Neumann's saddle point theorem for two-person zero-sum games are also presented, Section 4 introduces *subgames*. A subgame is a portion of a game, running from some specific point in the game to the end, that has all the qualitative features of a game. Section 5 is devoted to *refinements of noncooperative equilibrium.* Many games have multiple equilibria, and certain equilibria in such games are intuitively unreasonable. Refinements of equilibrium are intended to weed out unreasonable equilibria.

1 The structure of finite extensive form games

A finite game in extensive form has a finite number, n, of players. The set of players is denoted $N = \{1, \ldots, n\}$. Each player moves (takes an action) a finite number of times, and the *move* is the basic, indivisible act taken by a player. Randomness may intervene at a finite number of points in the game. The random elements are treated as if there is an additional player, called *nature,* or *chance,* who is designated as player 0. Each point at which player 0 moves is a point at which a random draw is made from a discrete probability distribution; thus, nature does not exercise judgment. Each point at which nature moves may be associated with a different distribution.

The extensive form is developed formally in Sections 1.1 and 1.2. Section 1.1 develops the *game tree* and Section 1.2 completes the specification of the game by describing *information sets,* the *moves of nature,* and the *payoffs.* Section 1.3 describes important categories of games and Section 1.4 introduces the concept of *strategy* and specific types of strategies, such as *pure* and *mixed strategies.*

1.1 The game tree

A game tree is illustrated in Figure 2.1. It is characterized by a *set of decision nodes, D,* portrayed as unfilled circles; a *set of terminal nodes, T,* portrayed by filled circles; and lines showing connections between the nodes. Each decision node is associated with a player who moves at that node.[1] The one line to a node from below is the move that leads directly to the node and the several lines emanating upward from a node are the moves that can be taken by the player from that node. One node, denoted d_0, shown at the bottom of the diagram, is the *distinguished node,* indicating where the game begins, and any node in T is an end point of the game. The relationship between the nodes is described by the *immediate predecessor function, e_1,* which states for each node $d \neq d_0$ the unique node from which the game proceeds in a single move to d. For example, d'' in Figure 2.1 is the immediate predecessor of d' [i.e., $d'' = e_1(d')$]. Except for d_0, each node has exactly one immediate predecessor. Each node in D, including d_0, is the immediate predecessor of one or more nodes; however, nodes in T are never predecessors. They are the terminal nodes at which the game ends.

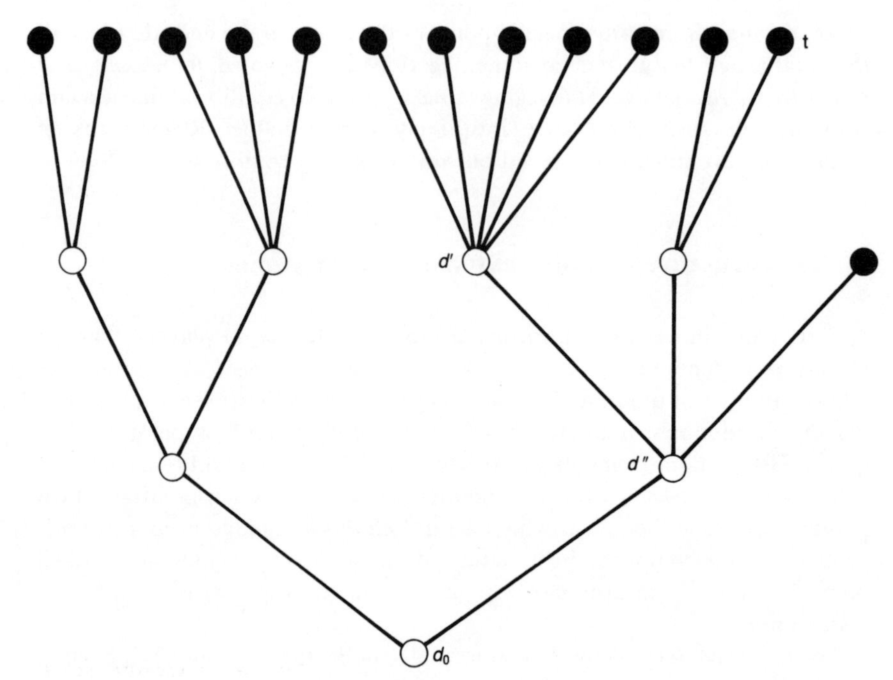

FIGURE 2.1 A game tree.

DEFINITION 2.1 *$Y = (D, T, e_1)$ is a* **finite game tree** *if D is a nonempty finite set, T is a nonempty finite set, $D \cap T = \varnothing$, and the predecessor function e_1 satisfies the following conditions*: (1) *There is a unique $d_0 \in D$ for which $e_1(d_0) = \varnothing$.* (2) *If $d \in D \cup T$ and $d \neq d_0$ then $e_1(d) \in D$ and $e_1(d)$ is unique.* (3) *If $d \in D \cup T$ and $d \neq d_0$ then either $e_1(d) = d_0$ or there exists a positive finite k and $\{d_1, d_2, \ldots, d_k\} \subset D$ such that $d_k = e_1(d)$, and $d_{j-1} = e_1(d_j), j = 1, \ldots, k$.*

Conditions on the function e_1 require (1) that the game tree has a unique starting point (d_0), (2) that each node other than d_0 has a unique prececessor, and (3) that there is a unique path from d_0 to any other node in the tree.

1.2 Extensive form games

For a complete definition of finite extensive form games, it remains to identify each node in D with the one player who owns it, the probability distributions governing the moves of chance (player 0), the information held by the players, and the n-vector of payoffs associated with each terminal node. At any node $t \in T$ the game is over and the payoffs $p(t) = (p_1(t), \ldots, p_n(t)) \in R^n$ are received by the players in N with $p_i(t)$ being the payoff to player i.[2] Thus $p: T \to R^n$ is the *payoff function*. The payoff to a player is the utility received by that player. It is assumed throughout that the von Neumann–Morgenstern axioms on decision making under risk apply, so that the *expected utility theorem* holds.[3]

D is partitioned into information sets, $I = \{I_{01}, \ldots, I_{0r_0}, I_{11}, \ldots, I_{ir_1}, \ldots, I_{n1}, \ldots, I_{nr_n}\}$, where the information sets $I^i = \{I_{i1}, \ldots, I_{ir_i}\}$ $\{i \in N \cup \{0\})$ are those at which player i moves. Thus, each information set I_{ij} is a set of nodes that all belong to player i, and player i cannot distinguish among them. That is, should player i reach a node in I_{ij}, he will not be able to tell which of the nodes in I_{ij} he has attained; but will know only that he is at some node in I_{ij}. Each set I_{ij} $(j = 1, \ldots, r_i)$ is nonempty and at each node in I_{ij} player i has precisely the same number, m_{ij}, of moves as at any other node in I_{ij}. The partition I is subject to further restrictions: (1) $\{d_0\} \in I$; that is, the game begins at an information set containing precisely one node, so that the player who moves first knows which node he starts from, and (2) each information set at which nature moves consists of just one node.

For each $I_{0j} \in I^0$, $q_j = \{q_{j1}, \ldots, q_j, q_{j,m_{0j}}\}$ is a probability distribution over the choices at I_{0j}, so that $q_j \geqslant 0$ and $\sum_{k=1}^{m_{0j}} q_{jk} = 1$. $q = \{q_1, \ldots, q_{r_0}\}$ is the *set of probability distributions for nature.* Thus, the description of a finite extensive form game can be summarized by N, Y, p, I, and q. The following notation will prove useful: for two sets A and B, $A \backslash B \equiv \{x \in A \mid x \notin B\}$. So $A \backslash B$ consists of members of A other than those that are also in B.[4]

Definition 2.2 $\Gamma = (N, Y, p, I, q)$ is a **finite game in extensive form** *where N is a finite set of players, Y is a game tree, p is a payoff function, I is an information set partition, and q is the set of probability distributions for nature.*

From any node in D there is a (strictly positive) finite number of choices or branches, with each branch leading to another node. The node $d_0 \in D$ is unique and is the node at which the game begins. $e_1(d)$ is that unique node from which d is reached in a single move. Let $e_2(d) = e_1[e_1(d)]$ be the node that precedes d by two moves. Then $e_k(d)$, the node that precedes d by k moves, can be defined recursively in the obvious way until some k is reached for which $e_k(d) = d_0$. For any $d \in D \cup T \backslash \{d_0\}$, there is a unique *path from d_0 to d.* The path from d_0 to d is denoted $e(d) = \{d_0, e_{k-1}(d), \ldots, e_1(d), d\}$ (where $e_k(d) = d_0$).

In a literal sense a game in extensive form does not allow two or more players to make simultaneous moves. The structure of the game presumes that one player moves at a time, although the move taken by a player may not become immediately known to any other player. To represent simultaneous moves, the players are arbitrarily ordered and each player ignorant of what was chosen by his predecessors.

Figure 2.2 provides an example of the game of matching pennies. As this game is played, two players simultaneously select heads or tails. When both have chosen, they simultaneously reveal their choices. They will have decided in advance that, say, when the two coins match, player 1 wins. There are two information sets, $I_{11} = \{d_0\}$ and $I_{21} = \{d_1, d_2\}$. Note that player 1 moves first, but player 2 does not know the move selected by player 1 when it is her turn to move.

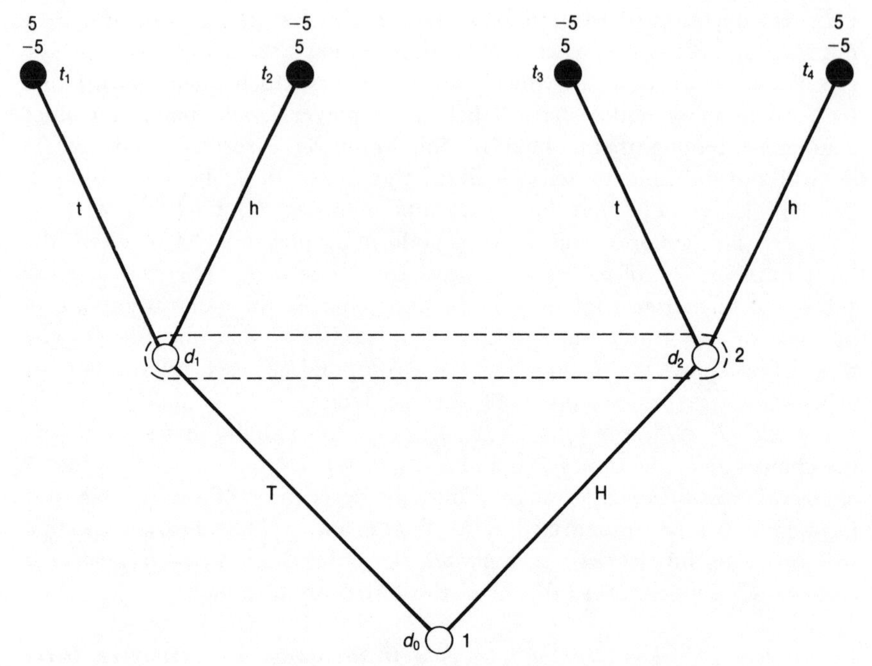

FIGURE 2.2. One round of matching pennies.

1.3 Classifying games in terms of information

The information distinctions defined here relate to two different matters: what the players know about the payoff structure and what form the information set partition takes. It is assumed throughout that the set of players, N, the game tree, Y, and the probability distributions, q, governing the moves of nature are known to all players. The concepts *complete information* and *incomplete information* pertain to whether the payoff function, p, is completely known to all players. In a game of *complete information* each player knows p. In a game of *incomplete information* at least one player lacks some information about p. A typical form of incomplete information exists when each player knows his own payoff at each terminal node but is not informed of the payoffs of other players.

DEFINITION 2.3 $\Gamma = (N, Y, p, I, q)$ is a **game of complete information** *if each player is informed of the complete specifications of the game. That is, each player knows N, Y, p, I, and q. Otherwise it is a* **game of incomplete information.**

The game of matching pennies in Figure 2.2 is one of complete information if each player has a copy of Figure 2.2 (or equivalent information) and both players know they both have such a copy.[5] Imagine that before the game is played, the two players are sitting near each other in a room and someone hands each player a copy of Figure 2.2 in clear view of both of them.

The information distinctions pertaining to the partition of D into information sets are *perfect information* and *perfect recall.* In a game of perfect information, each information set consists of exactly one node. This means that a player always observes all moves made by all players and no one ever forgets anything he ever observed about a move he or anyone else ever made.

DEFINITION 2.4 *The game* $\Gamma = (N, Y, p, I, q)$ *is a* **game of perfect information** *if each information set* $I_{ij} \in I$ *contains only one node. Otherwise it is a* **game of imperfect information.**

If one or more information sets contain at least two nodes, the game is one of *imperfect information.* The game in Figure 2.2 is a game of imperfect information.

In a *game of perfect recall* a player never forgets any move that he, himself made nor any other information that has once been acquired. For any $d \in D$ let $F_1(d, k)$ denote the node that is achieved when the kth move is taken from node d. $F_1(d, k)$ is called the kth *successor node to d.* For example, in Figure 2.2 $F_1(d_0, 2) = d_2$ and $F_1(d_1, 1) = t_1$. Thus, $e_1[F_1(d, k)] = d$ and, for $d \in I_{ij}$, $e_1^{-1}(d) = \{F_1(d, 1), \ldots, F_1(d, m_{ij})\} = \bar{F}_1(d)$ is the set of successor nodes to d. Let

$$F_1(I_{ij}, k) = \bigcup_{d \in I_{ij}} F_1(d, k) \tag{2.1}$$

be the set of nodes that can be achieved from the information set I_{ij} when the kth move is taken. If player i moves at I_{ij} and can remember her moves, then she knows that, following her kth move, some node in $F_1(I_{ij}, k)$ must have been the next node.

DEFINITION 2.5 *The game* $\Gamma = (N, Y, p, I, q)$ *is a* **game of perfect recall** *if, for any* $i \in N$, *any* I_{ij} *and* $I_{il} \in I^i$, *any move k available from* I_{il}, *and any* d', $d'' \in I_{ij}$, (1) $e(d') \cap I_{il} = \varnothing$ *if and only if* $e(d'') \cap I_{il} = \varnothing$ *and* (2) $e(d') \cap F_1(I_{il}, k) = \varnothing$ *if and only if* $e(d'') \cap F_1(I_{il}, k) = \varnothing$.

Condition (1) requires that, if two nodes are in the same information set I_{ij} of player i, then all paths to either of those nodes must have visited precisely the same collection of information sets of player i prior to reaching I_{ij}. This means that player i knows which information sets *of his own* have previously been visited. Condition (2) looks at any information set of player i (say, I_{il}) that has been visited on the way to I_{ij} (the present information set) and requires that each node in I_{ij} have on its path a node reached by the same move from I_{il}. This means that the player never forgets a move that he made in the past.

The matching pennies game in Figure 2.2 is a game of perfect recall; the game in Figure 2.3 is one of imperfect recall. This can be seen by noting the information set I_{32} at which player 3 moves, where player 3 cannot recall whether or not he visited his information set $\{d_1\}$. One path to I_{32} passes through $\{d_1\}$ and the other path does not. This violates condition (1) in Definition 2.5.

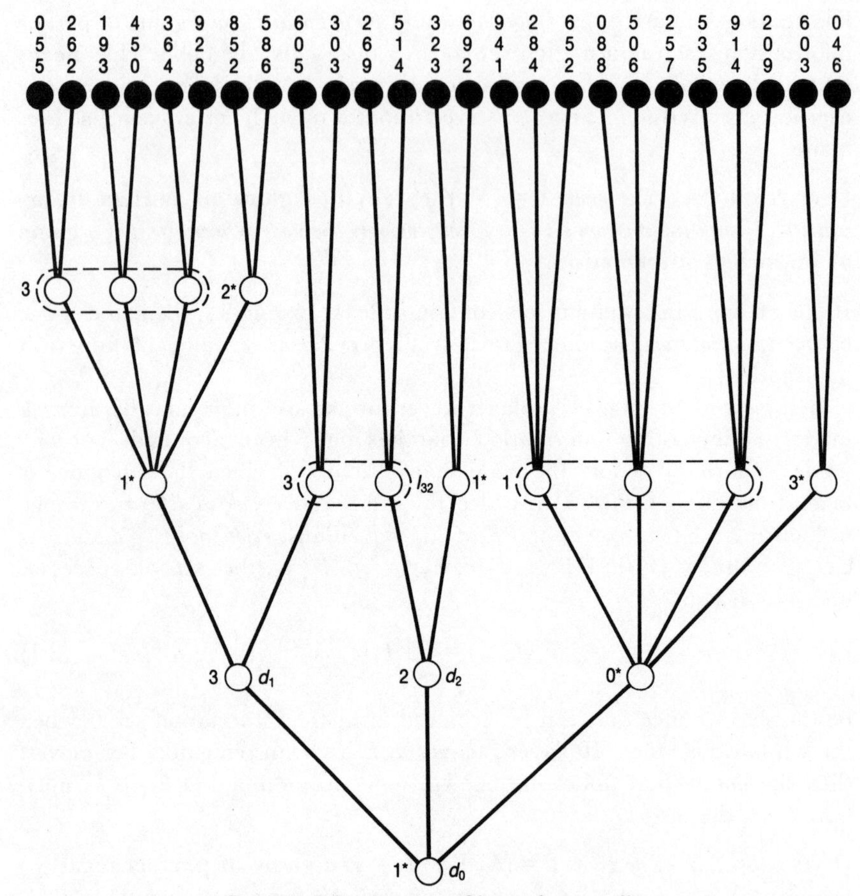

FIGURE 2.3 A game of imperfect recall.

The game in Figure 2.4 is a game of perfect recall. Suppose, however, that I_{22} and I_{23} were combined into one information set. This would imply that player 2 forgot whether his own first move was t or h. This violates condition (2) of Definition 2.5, which requires that, if two nodes are in the information set $I_{22} \cup I_{23}$, then all paths to nodes in this set must have used the same move in I_{21} (i.e., h or t, but not h for one node and t for the other).

The great majority of games studied in this book are games of complete information, perfect recall, and imperfect information. This means that each player knows the payoffs of all players, each player recalls all of his own moves, or which of his own information sets he has visited, and some information sets contain more than one node.

1.4 Pure, mixed, and behavior strategies

A players strategy is that player's plan for choosing moves at each of her information sets $I_{ij} \in I^i$ in the game. At I_{ij} she will know she has m_{ij}

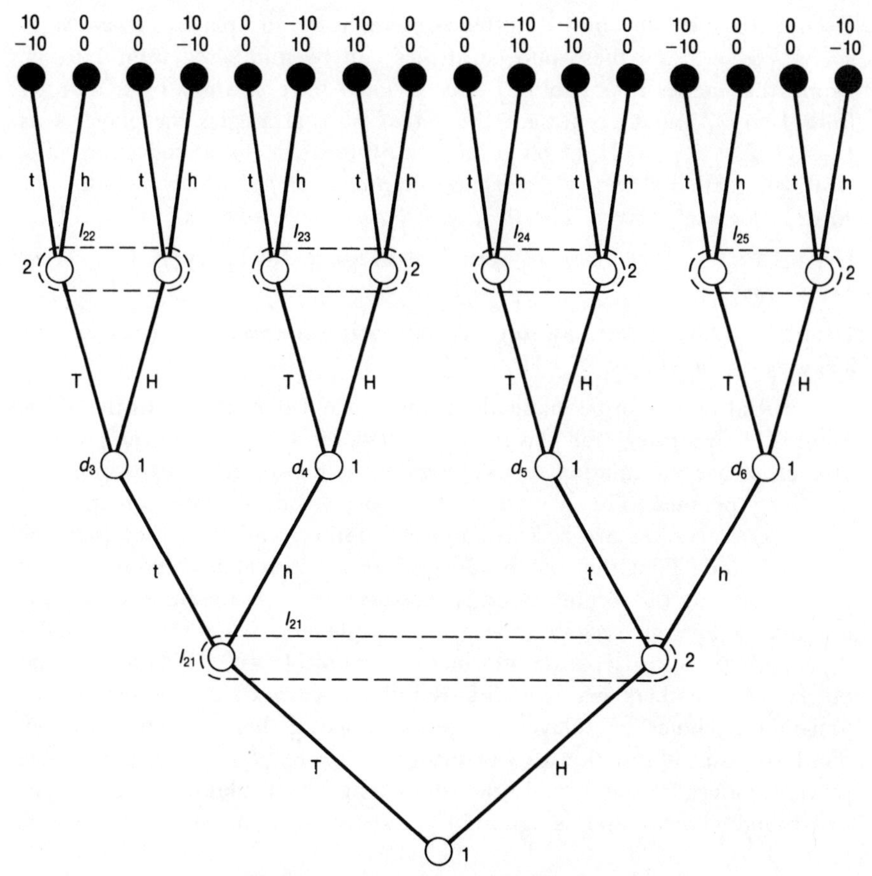

FIGURE 2.4 Two rounds of matching pennies.

available choices and, associated with each choice will be one specific successor node for each node in I_{ij}. Three kinds of strategies–*pure strategies, mixed strategies,* and *behavior strategies*–are developed. A pure strategy is a nonstochastic set of choices. That is, in a pure strategy, player *i* associates a specific alternative with each information set at which she moves. A mixed strategy is a randomization over the set of pure strategies. A behavior strategy is a method of designing a randomized strategy, in which the randomization is done separately over each individual move. These concepts are developed at length later.

1.4.1 Pure strategies

Under a pure strategy, a player selects a specific move for each information set. In general, a different move can be selected at each information set.

DEFINITION 2.6 *A* **pure strategy for player *i*** *is $\pi_i = (\pi_{i1}, \ldots, \pi_{ir_i})$, where each π_{ij} is an integer indicating the move to be made in the information set I_{ij} $(j = 1, \ldots, r_i)$. Thus, $\pi_{ij} \in \{1, 2, \ldots, m_{ij}\}$ and $\pi_i \in \times_{j=1}^{r_i} \{1, 2, \ldots, m_{ij}\}$.*

The number of pure strategies available to player i is $m_i = m_{i1} \cdot m_{i2} \cdots m_{ir_i}$ and these pure strategies can be numbered from 1 to m_i. Doing this makes it possible to refer to each pure strategy by a number from 1 to m_i and to represent the set of pure strategies for player i as $\Pi_i = \{1, 2, \ldots, m_i\}$. It is often necessary to refer to a collection of n strategies, called a *strategy combination,* or *strategy profile,* which includes one strategy for each player. The *set of pure strategy combinations* is $\Pi = \times_{i=1}^{n} \Pi_i$.

DEFINITION 2.7 *The* **set of pure strategies for player** i *is* $\Pi_i = \{1, \ldots, m_i\}$.

DEFINITION 2.8 *A* **pure strategy combination** *is denoted* $\pi = (\pi_1, \ldots, \pi_n)$ *and is an element of* $\Pi = \times_{i \in N} \Pi_i$.

A crucial point can be made about the concept of *strategy* with the aid of Figure 2.4. A strategy for player i describes precisely what player i will do at each of her information sets. The game in Figure 2.4 is two rounds of matching pennies. The players each choose heads or tails for the first round, their choices are revealed to one another, and then they play the second round. That the first round choices are revealed to both of them prior to playing the second round is shown by the way the information sets are drawn for the second moves of the players. Note that the nodes $d_3, \ldots, d_6$ are all in separate information sets of player 1. This means that player 1 remembers her own first-round choice and has discovered the first-round choice of player 2 before making her second selection. Similarly, the information sets of player 2, labeled $I_{22}, \ldots, I_{25}$, at which player 2 moves for the second time, show that player 2 knows both players' first-round choices and is ignorant only of the second-round choice of player 1.

One might think, naively, that a pure strategy for player 1 would be merely one of the following: (a) Choose H in the first round and H in the second. (b) choose H in the first round and T in the second, (c) choose T in the first round and H in the second, and (d) choose T in the first round and T in the second. However, (a)–(d) is not a complete list of the pure strategies open to player 1. A complete list is given in Table 2.1.

Note, again, that a strategy covers every contingency that can conceivably arise for the player in the game. Saying that a strategy specifies a player's move at *each of her information sets* implies that a choice of move is made even on information sets that she cannot possibly reach. For example, look at strategy 6 in Table 2.1. With this strategy player 1 cannot possibly reach the information sets $\{d_3\}$ and $\{d_4\}$. Nonetheless, a strategy must specify moves on these information sets. In addition, the strategy specifies a move for a player as a function of the information that will have been accumulated by the time action is demanded. For example, the second move of player 1 can be made conditional on the revealed history of play that will be known to her when her second choice must be made. Each of the four information sets (nodes) at which player 1 might make her second move corresponds to different information about the way the game progressed up to the time the second decision had to be made.

TABLE 2.1 Complete list of pure strategies for player 1 in the two-round penny matching game

Strategy	First round	Second round choice if first round pair is			
		H h	H t	T h	T t
1	H	H	H	H	H
2	H	H	H	H	T
3	H	H	H	T	H
4	H	H	T	H	H
5	H	T	H	H	H
6	H	H	H	T	T
7	H	H	T	H	T
8	H	H	T	T	H
9	H	T	H	H	T
10	H	T	H	T	H
11	H	T	T	H	H
12	H	H	T	T	T
13	H	T	H	T	T
14	H	T	T	H	T
15	H	T	T	T	H
16	H	T	T	T	T
17	T	H	H	H	H
18	T	H	H	H	T
19	T	H	H	T	H
20	T	H	T	H	H
21	T	T	H	H	H
22	T	H	H	T	T
23	T	H	T	H	T
24	T	H	T	T	H
25	T	T	H	H	T
26	T	T	H	T	H
27	T	T	T	H	H
28	T	H	T	T	T
29	T	T	H	T	T
30	T	T	T	H	T
31	T	T	T	T	H
32	T	T	T	T	T

1.4.2 *Mixed strategies*

A *mixed strategy* for a player is merely a probability distribution over that player's pure strategies. Thus, a mixed strategy for player i is some m_i-coordinate vector, $s_i \geqslant 0$, whose coordinates sum to 1.

DEFINITION 2.9 *The* **set of mixed strategies for player *i*** is

$$S_i = \left\{ s_i \in R_+^{m_i} \mid \sum_{j=1}^{m_i} s_{ij} = 1 \right\} \tag{2.2}$$

The **set of mixed strategy combinations** *is* $S = \times_{i \in N} S_i$. *If a mixed strategy places strictly positive probability on each pure strategy (i.e., if* $s_i \gg 0$*), then it is called* **completely mixed.**

There is a subtle point to be made about strategy choice and the nature of strategies, players' behavior, and their information. It is often said that a player can be thought to choose his strategy at the start of the game, before any actions have been taken by anyone. Does this mean that a player is forbidden from changing his strategy part way through the game? Would it not be in a player's interest sometimes to change his strategy part way through the game? In general, a player may alter his strategy part way through the game, but if he has analyzed the game thoughtfully and completely will not wish to make any change. A strategy includes the specification of the player's behavior in each and every conceivable circumstance that can arise in the game. If a strategy were to be changed midcourse, then the point in time at which it would be changed could be anticipated at the start of the game and the player could specify appropriate behavior at the outset when he selected his strategy.

To take an example, suppose you are player 1 in a two-person game of 250 successive rounds of matching pennies. You win whenever the coins match. Suppose you contemplate a strategy of always randomizing at each of your information sets, placing a probability of .5 on heads and .5 on tails. At your 101st move you could be in an information set corresponding to player 2 having always selected heads in her first 100 moves. If you were there, you might suppose that player 2 were not randomizing, but was choosing heads with certainty each time, and you might wish to counter this by simply choosing heads and continuing to choose heads as long as you see player 2 doing the same. If this is what you think is optimal, you can forsee it at the start. You know before anyone chooses anything that this information set exists, and you know what player 2 will have done in her first 100 moves if you find yourself there. Thus, if choosing heads at that point makes sense to you, then *always randomizing no matter what* does not make sense to you, and this can be forseen at the start.

1.4.3 Behavior strategies

A *behavior strategy* is a special sort of randomized strategy that is made up of a collection of independent probability distributions, one for each of the player's information sets.

DEFINITION 2.10 *Let* $\beta^i_j = (\beta^i_{j1}, \beta^i_{j2}, \ldots, \beta^i_{jm_{ij}})$ *be a probability distribution over the moves in* I_{ij}. *Thus* $\beta^i_j \geq 0$ *and the elements of* β^i_j *sum to* 1. *Then* $\beta^i = (\beta^i_1, \beta^i_2, \ldots, \beta^i_{r_i})$ *is a* **behavior strategy for player *i*** *and* $\beta = (\beta^1, \beta^2, \ldots, \beta^n)$ is a behavior strategy combination. *The* **set of behavior strategies for player *i*** *is denoted* B_i and the **set of behavior strategy combinations** *is denoted B. The behavior strategy* β^i is **completely mixed** *if* $\beta^i \gg 0$.

Interest in behavior strategies stems from two sources. First, behavior strategies are an intuitively appealing form of randomized strategy and are easier to work with. Second, in a game of perfect recall, any mixed strategy is essentially equivalent to a unique behavior strategy. This is proved in Section 3.4.

2 The strategic form derived from the extensive form

The *strategic form* of a game, $\Gamma = (N, S, P)$, represents the game in terms of the set of players N, the strategy spaces S, and the players *payoff functions* P. The set of players has, of course, been introduced very early and the appropriate strategy sets, the sets of mixed strategies, were developed in Section 1.4.2. In order to complete the picture, we must express the payoff of each player as a function of the strategies each chooses. This is done in two steps. First, payoffs are expressed as a function of the pure strategies of the members of N. This expression is then extended to mixed strategies.

First consider a game in which nature has no moves (i.e., there is no randomness in the structure of the game). Then a given pure strategy combination π dictates a unique path from the initial node d_0 to some specific terminal node $t_\pi \in T$. That terminal node specifies a payoff vector $p(t_\pi)$ and consequently the payoff to each player i, $p_i(t_\pi)$. Let $a^i_\pi \equiv p_i(t_\pi)$ and let the matrix $A^i = \|a^i_\pi\|$, having order $m_1 \times m_2 \times \cdots \times m_n$, represent the payoff to player i for any pure strategy combination $\pi \in \Pi$. The vector of pure strategy payoffs associated with π in the game is denoted $a_\pi = (a^1_\pi, \ldots, a^n_\pi) \equiv p(t_\pi)$ and the set of pure strategy payoffs is represented by the matrix $A = \|a_\pi\|$. Each a_π entry of A is a payoff vector in R^n, thus the order of A is $n \times m_1 \times m_2 \times \cdots \times m_n$. Table 1.1 shows a *payoff matrix A* for a three-person game in which each player has two pure strategies.

For games in which nature is active, the payoff a^i_π is the *expected payoff to player i when the pure strategy combination π has been chosen*. The expectation is, of course, over the choices of player 0. Nature, player 0, has r_0 information sets, at the information set I_{0j} there are m_{0j} choices, and the choice over the nodes in I_{0j} is governed by the probability distribution $q_j = (q_{j1}, \ldots, q_{j,m_{0j}})$. Let $\Theta_j = \{1, \ldots, m_{0j}\}$, let $\Theta = \times_{j=1}\Theta_j$, and denote an element of Θ by θ. Then $\theta = (\theta_1, \theta_2, \ldots, \theta_{r_0})$ is a vector of r_0 positive integers, with the jth coordinate being a nonrandom move from the information set I_{0j}. The probability that nature makes the specific sequence of moves indicated by a particular θ is readily calculated from q. Therefore, $(\theta, \pi) \in \Theta \times \Pi$ is a specific path from d_0 to some unique $t_{\theta,\pi} \in T$. The (expected) payoff to a player i associated with an arbitrary pure strategy combination π is

$$a^i_\pi = \sum_{\theta_1=1}^{m_{01}} \cdots \sum_{\theta_{r_0}=1}^{m_{0r_0}} q_{1,\theta_1} \cdots q_{r_0,\theta_{r_0}} p_i(t_{\theta,\pi}) \tag{2.3}$$

for all $\pi \in \Pi$ and all $i \in N$. From equation (2.3) are derived the payoff matrices $A^i = \|a^i_\pi\|$, the a_π, and the set of pure strategy payoffs $A = \|a_\pi\|$, as in the case above where player 0 is inactive. The game can be represented as $\Gamma = (N, A)$, which is the *payoff matrix representation of the game*. As will be seen immediately, there is yet another useful representation.

Writing payoffs in terms of mixed strategy combinations is now straightforward. The payoff to player i associated with the mixed strategy combination s is denoted $P_i(s)$. Wherever probabilities enter the scene it is assumed that a player's payoff is her *expected payoff*. Thus, von Neumann–Morgenstern utility is assumed throughout this book. This topic is such standard fare in economics that I assume an understanding of it. Readers wishing to refresh their acquaintance with the subject can look at Luce and Raiffa (1957, Chapter 2), Varian (1984, Sections 3.18 and 3.19), or Arrow (1971).)

$$P_i(s) = \sum_{\pi_1=1}^{m_1} \cdots \sum_{\pi_n=1}^{m_n} s_{1\pi_1} \cdots s_{n\pi_n} a^i_\pi \tag{2.4}$$

for all $\pi \in \Pi$, all $s \in S$, and all $i \in N$. The game can now be expressed in the *strategic form*, given by $\Gamma = (N, S, P)$.

DEFINITION 2.11 *The* **strategic form** *of a game is given by* $\Gamma = (N, S, P)$, *where* N *is the set of players,* $S = \times_{i \in N} S_i$ *is the strategy space of the game, and* $P = (P_1, \ldots, P_n)$ *is the vector of payoff functions of the players.*

The strategic form is used in later chapters, where it will be seen that games other than finite games can be represented in much the same way. The strategic form presented here is specifically derived from finite games and explicitly admits mixed strategies. One could, for example, specify a game as $\Gamma = (N, B, P)$, in which case strategies would be the set of behavior strategies B.

3 Some basic topics and results on finite games

This section examines various facts relating to noncooperative equilibrium behavior. In Section 3.1 the *noncooperative equilibrium*, a concept dating back to Cournot (1838) and including von Neumann's (1928) saddle point equilibrium, is defined and briefly discussed. Section 3.2 deals with *dominant strategies* and with *dominant strategy equilibria*. A strategy of player i is dominant when it is as good as or better than any other strategy available to the player, no matter what strategies might be selected by other players. Next are results due to Kuhn (1953) that prove in Section 3.3 that any finite game of complete and perfect information has a noncooperative equilibrium in pure strategies and in Section 3.4 that attention may be restricted to behavior strategies in games of perfect recall. Section 3.5 contains an exposition of the saddle point equilibrium theorem for two-person zero-sum games.

3.1 The definition of noncooperative equilibrium

The *noncooperative equilibrium* obtains when each player in a game independently chooses a strategy in a way that maximizes his payoff. All players do this *independently* in the sense that they do not act in concert, and one player cannot affect the strategy choice of any other player. At the same time, each and every player is thought to take the probable choices of other players into account in choosing his own strategy.

As an example, consider the game in Table 2.2, where player 1 chooses the row and receives the first payoff listed in a box and player 2 chooses the column. Allowing for mixed strategies means that player 1 can randomize over the two rows and player 2 can randomize over the three columns. The pure strategy combination (a_2, b_3) has the feature that player 1 maximizes his payoff by choosing a_2, given that player 2 selects b_3, and player 2 maximizes her payoff by choosing b_3, given that player 1 chooses a_2. This gives (a_2, b_3) a self-enforcing character and a rational expectations character. If, say, player 1 proposes (a_2, b_3) to himself as a potential outcome, he thinks that he wants to choose a_2 if he believes that player 2 will choose b_3, and he notes that player 2 would want to choose b_3 if she thought player 1 would choose a_2. If each expects (a_2, b_3) and then each proceeds to select that strategy that maximizes his or her own payoff given what is expected of the other, then their expectations will be realized.

Compare (a_2, b_3) with any other possible outcomes—with (a_1, b_2), for example. If (a_1, b_2) is contemplated as a possible outcome, player 1 reasons that he would rather select a_2 (receiving a payoff of 15 instead of 8) if he believes player 2 will select b_2. But player 2 can see that player 1 would not have an incentive to choose a_1 if he thought player 2 were going to select b_2. Therefore, player 2, contemplating (a_1, b_2) as a possible outcome, sees that it is not plausible. Thus each player's choice is based on (1) selecting a strategy that maximizes the player's own payoff, given the strategies selected by the others, and (2) the belief that the other players will also attempt to select strategies that maximize their payoffs given the strategies selected by their rivals.

The *noncooperative equilibrium* is also known as the *equilibrium point,* the *Nash noncooperative equilibrium,* and the *Nash–Cournot equilibrium.* It is the most important equilibrium concept in noncooperative game theory. In discussing behavior it is often necessary to look at choices by one player while

TABLE 2.2 A two-person finite game

		Player 2		
		b_1	b_2	b_3
Player 1	a_1	10, 5	8, 8	−6, 4
	a_2	−2, −1	15, 4	3, 5

holding the choices of all others fixed. To facilitate this with convenient notation $s\backslash u_i \equiv (s_1, \ldots, s_{i-1}, u_i, s_{i+1}, \ldots, s_n)$ will be used. Thus, $s\backslash u_i$ is s with its ith coordinate replaced by u_i.

DEFINITION 2.12 *An* **equilibrium point** *is a strategy combination $s^* \in S$ such that $P_i(s^*) \geq P_i(s^*\backslash s_i)$ for all $s_i \in S_i$ and all $i \in N$.*

Complete information is central to the rationale of noncooperative equilibrium in the sense that the equilibrium is justified by reasoning processes along the lines of those used earlier in the example. These reasoning processes suppose that each player assumes rationality on the part of the other players and uses that assumption to analyze the game from the standpoint of the other players. Viewing the game from the perspective of one's rival players in the same manner as one views it from one's own perspective is most naturally carried out with complete information. The technical conditions in Definition 2.12 do not depend upon complete information, but the justification of those conditions as characterizing an interesting concept of equilibrium behavior does.[6]

Under some circumstances complete information is not needed. In the *dominant strategy equilibria* described in Section 3.2, for example, the obviously correct strategy for a player to choose does not depend on the strategy choices of the other players. Conversely, even complete information and the existence of a noncooperative equilibrium are not enough to ensure an equilibrium outcome will occur or to dictate or predict which outcome will occur. The game in Table 2.3 is a case in point. It has two equilibria in pure strategies, (a_1, b_1) and (a_2, b_3). Player 1 prefers the latter equilibrium and player 2 prefers the former. A player cannot independently select a strategy based on one equilibrium and feel certain that the other player will choose a strategy based on the same equilibrium.

Where does this leave us? First, if the equilibrium is unique, it is intuitively plausible that rational players in a complete information game will regard it as the obvious outcome and play accordingly. Second, the same would hold if there were several equilibria, but one of them Pareto dominated (i.e., gave at least as high a payoff to each player as) all the other equilibria. Finally, for situations such as the game in Table 2.3, where there are multiple equilibria and no one of them dominates the others, the equilibrium conditions could be regarded as necessary condi-

TABLE 2.3 A two-person finite game with multiple equilibria

		Player 2		
		b_1	b_2	b_3
Player 1	a_1	0, 10	8, 8	−6, 7
	a_2	−2, 3	15, 4	3, 5

tions for a *solution* (i.e., a unique outcome) to the game. Conditions guaranteeing the existence of equilibrium points in noncooperative games are given in Chapter 3.

3.2 Dominant strategies, dominated strategies, and dominant strategy equilibria

Dominant strategies and *dominated strategies* are, in a sense, opposite extremes. A strategy s_i' is dominated for player i if there is some alternate strategy s_i'' that gives player i at least as great a payoff as he receives using s_i', no matter what is chosen by the other players. Domination can be *weak* or *strong*.

DEFINITION 2.13 *The strategy $s_i' \in S_i$ is a* **weakly dominated strategy** *if there is some $s_i'' \in S_i$ such that $P_i(s \backslash s_i'') \geqslant P_i(s \backslash s_i')$ for all $s \in S$ and there is at least one $s \in S$ for which $P_i(s \backslash s_i'') > P_i(s \backslash s_i')$. If $P_i(s \backslash s_i'') > P_i(s \backslash s_i')$ for all $s \in S$, then s_i' is strongly (or strictly) dominated.*

A weakly dominant strategy for a player i ensures her at least as great a payoff as does any alternate strategy, no matter what strategies are selected by the other players.

DEFINITION 2.14 *The strategy $s_i' \in S_i$ is a* **weakly dominant strategy** *if, (a) for all $s \in S$, and all $s_i' \in S_i$, $P_i(s \backslash s_i') \geqslant P_i(s \backslash s_i'')$, and (b) for each $s \in S$ there is one $s_i'' \in S_i$ for which $P_i(s \backslash s_i') > P_i(s \backslash s_i'')$. If for all $s \in S$ and all $s_i'' \in S_i$, $P_i(s \backslash s_i') > P_i(s \backslash s_i'')$ then s_i' is strongly (or strictly) dominant.*

If a strategy is undominated, it need not be dominant. For example, in the game of Table 2.2 neither a_1 nor a_2 is dominated or dominant for player 1. The same is true of any mixed strategies of player 1. For player 2, however, b_1 is dominated by b_2. In the game of Table 2.3, no pure strategy of either player is dominated by either players pure strategies; however, b_2 is dominated by a mixed strategy consisting of (.5, 0, .5) where a probability of .5 is placed on b_1 and b_3.

A *dominant strategy equilibrium* is a strategy combination consisting of dominant strategies. In general, games do not have any dominant strategy equilibria, although they may occur in some applications of interest. See, for example, the discussion of the *free rider problem* in Chapter 3.

DEFINITION 2.15 *In the game $\Gamma = (N, S, P)$ s^* is a* **dominant strategy equilibrium** *if $s^* \in S$ and for each player $i \in N$, s_i^* is a dominant strategy.*

The prisoners' dilemma is a famous example of a game with a dominant strategy equilibrium. As Table 4.1 (page 109) reveals, *confess* dominates *not confess* for both players, and the equilibrium payoff is not Pareto optimal; both players have higher payoffs if both choose *not confess*. Thus, a dominant strategy equilibrium is a compelling outcome at which inefficient payoffs can be received by the players. A final point to note on this topic is that, if a player has two or more dominant strategies in a game, then these dominant strategies must be equivalent.

LEMMA 2.1 *Let $\Gamma = (N, S, P)$ be a finite noncooperative game of complete information and let player i have two dominant strategies, s_i' and s_i''. Then, for each $s \in S$, $P_i(s \backslash s_i') = P_i(s \backslash s_i'')$.*

Proof. If both strategies are dominant, then $P_i(s \backslash s_i') \geqslant P_i(s \backslash s_i'')$ for all $s \in S$ and $P_i(s \backslash s_i') \leqslant P_i(s \backslash s_i'')$ for all $s \in S$, which, together, imply the lemma. QED

3.3 Pure strategy equilibria in games of perfect information

A fundamental result due to Kuhn (1953) is that a finite game of complete and perfect information has a pure strategy equilibrium point. This result depends critically upon perfect information, which implies that the result of any random move of nature is observed by all players; that each player observes the moves of all players; that two or more players never move simultaneously; and that no player ever forgets anything. All these conditions must be fulfilled in order for perfect information to obtain—that is, each information set consists of just one node.

THEOREM 2.1 *A finite game of perfect information has an equilibrium point in pure strategies.*

Proof. Recall

$$\bar{F}_i(d) = \bigcup_{k=1}^{m_{ij}} F_1(d, k) \tag{2.5}$$

and

$$E_1 = \{d \in D \mid \bar{F}_1(d) \in T\} \tag{2.6}$$

and $T_1 = T \cup E_1$. E_1 consists of all those decision nodes from which the next move must lead to a terminal node. At each node in E_1 select the move that maximizes the payoff of the player to whom the node belongs. If the maximizing move is not unique, take the one with the lowest index.[7] Now each node in $d \in E_1$ has a unique terminal node, t_d, associated with it, and, via t_d, a unique payoff vector $p(t_d)$. The payoff function, p, can be extended from T to $T_1 = T \cup E_1$ by assigning $p(d) \equiv p(t_d)$ for $d \in E_1$.

Let $T \equiv T_0$ and suppose that T_{k-1} and E_{k-1} are defined and that p is extended to $T_{k-1} = T_{k-2} \cup E_{k-1}$. Let

$$E_k = \{d \in D \mid \bar{F}_1(d) \in T_{k-1}\} \tag{2.7}$$

and $T_k = T_{k-1} \cup E_k$. At each node in E_k all moves lead to a node to which a payoff vector is assigned. At each node in E_k select the move that maximizes the payoff of the player whose turn it is to move. As before, if there are several maximizers, select the one with the smallest index. This inductive procedure must stop in a finite number of steps, at which point each decision node has associated with it a unique move to be taken by the player who owns the node.

The collection of nodes at which player i moves and the moves to be taken there form a complete description of a pure strategy π_i^* of player i.

Denote the associated strategy combination by π^*. There is no node at which any player, by introducing randomization, could increase his payoff, nor could a player increase his payoff by any randomization over his set of pure strategies; therefore, π^* is an equilibrium point. QED

Games of perfect information are not common in economic applications of game theory. The existence of equilibrium points in finite complete information games, without restriction to pure strategies, was proved by Nash (1951). Nash's theorem is not proved here because it is a special case of a theorem that is proved in Chapter 3.

3.4 Behavior strategies in games of perfect recall

Another result due to Kuhn (1953) is that a mixed strategy is equivalent to a behavior strategy in a game of perfect recall in the sense that, in a game of perfect recall, the conditional probability with which any move k at information set I_{ij} is chosen, given that one is at I_{ij}, is independent of the path by which play proceeded to I_{ij}.

THEOREM 2.2 *In a game of perfect recall, an arbitrary mixed strategy $s_i \in S_i$ induces a unique behavior strategy.*

Proof. Let $\{E_0^i, \ldots, E_{l*}^i\}$ be a partition of I^i that is defined later. $E_0^i \subset I^i$ is the collection of information sets of player i that is preceded by no other of his information sets. That is, if $I_{ij} \in E_0^i$ and $d \in I_{ij}$, then $e(d) \cap I_{ik} = \varnothing$ for all $I_{ik} \notin I_{ij}$. For $l \geqslant 1$ let $E_l^i \subset I^i$ be the collection of information sets of player i that is preceded by exactly l of his information sets. That is, if $I_{ij} \in E_l^i$ and $d \in I_{ij}$, then $e(d) \cap I_{ik} \neq \varnothing$ for precisely l of the I_{ik} other than I_{ij} itself. For each pure strategy $\pi_i = (\pi_{i1}, \ldots, \pi_{ir_i})$ let $\delta(\pi_i) = \{\pi_{i1}, \ldots, \pi_{ir_i}\}$ denote the set of moves prescribed by the pure strategy π_i. Select an arbitrary mixed strategy, s_i, for player i and let $c(s_i) = \{\pi_i \in \Pi_i \mid s_{i\pi_i} > 0\}$, called the *carrier of* s_i, be the set of pure strategies to which s_i assigns positive probability.

Set $A_l(s_i)$, which form another partition of I^i, are now defined. $A_0(s_i) \subset I^i$ consists of every information set of player i that is either (a) in E_0^i or (b) that has a strictly positive probability of being reached under s_i starting from an information set in E_0^i. The remaining $A_l(s_i)$ are defined recursively. $A_1(s_i)$ is the next level of information sets of player i. These sets are not in $A_0(s_i)$, but they have a positive probability of being reached under s_i starting from a member of E_0^i. Then $A_l(s_i)$ is the next level after $A_{l-1}(s_i)$ and consists of information sets that are not already assigned to some $A_k(s_i)$ for $k < l$. In addition, to be in $A_l(s_i)$, I_{ij} must either (a) be in E_l^i or (b) have a strictly positive probability of being reached under s_i starting from an information set in E_l^i.

Now let $\gamma_{lj}(s_i)$ be the pure strategies in $c(s_i)$ that reach I_{ij}. Therefore, for $I_{ij} \in A_l(s_i) \cap E_l^i$, $\gamma_{lj}(s_i) = c(s_i)$ and for any other $I_{ij} \in A_l(s_i)$, $\gamma_{lj}(s_i)$ consists of those members of $c(s_i)$ that contain a path from at least one information set in E_l^i to I_{ij}. The behavior strategy induced by s_i is defined in equations (2.8) and (2.9). Equation (2.9) says that a move in an information set has zero probability attached to it, given the set is reached, if no pure strategy receiving positive probability under s_i selects this move in this information set starting from the appropriate level. In equation (2.8) l is the lowest level from which at least one node in I_{ij} can be reached with positive probability. For a node k in I_{ij} that can be reached with positive probability conditional upon starting from level l, the numerator is the probability of reaching this node starting from level l, and the denominator is the sum of such probabilities over all nodes reachable with positive probability from level l.

$$\beta^i_{jk} = \frac{\sum_{\substack{\pi_i \in \gamma_{lj}(s_i) \\ k \in \delta(\pi_i)}} s_{i\pi_i}}{\sum_{\nu=1}^{m_{ij}} \sum_{\substack{\pi_i \in \gamma_{lj}(s_i) \\ \nu \in \delta(\pi_i)}} s_{i\pi_i}} \tag{2.8}$$

if $k \in \delta(\pi_i)$ for at least one $\pi_i \in \gamma_{lj}(s_i)$ and

$$\beta^i_{jk} = 0 \tag{2.9}$$

otherwise. Equations (2.8) and (2.9) define uniquely the behavior strategy that is implied by the mixed strategy s_i. QED

This theorem is very useful because the requirement of perfect recall is met in a wide and interesting variety of games, and behavior strategies are a very appealing and natural way to think about mixed strategies. It will be seen in Section 5 that sequential equilibrium is framed in terms of behavior strategies. Behavior strategies have a decentralized character to them in the sense that they are made up of r_i independent components. Each component is a probability distribution over the moves in a single information set, and thus each piece can be totally described by itself. A strategy is a collection of these components. Of course, the optimality of one component by itself is contingent on the specification of the other components (as well as on the strategies of other players), but it is an appealing simplification to be able to describe a mixed strategy merely by describing separately the behavior at each of the player's information sets.

3.5 Two-person zero-sum games

Two-person zero-sum games have some special structural characteristics that determine additional characteristics of equilibrium points and that affect the meaning of equilibrium. Two-person zero-sum games can easily be defined in terms of the representations developed above. Obviously such games have player sets $\{1, 2\}$ and, at any terminal node $t \in T$, $p_1(t) = -p_2(t)$.

DEFINITION 2.16 $\Gamma = (\{1, 2\}, Y, p, I, q) = (\{1, 2), A) = (\{1, 2\}, S, P\}$ *is a* **two-person zero-sum game** (1) **in extensive form** *if, for each terminal node* $t \in T$, $p_1(t) = -p_2(t)$, (2) **in matrix game form** *if, for each* $\pi \in \Pi$, $a^1_\pi = -a^2_\pi$, *and* (3) **in strategic form** *if, for each* $s \in S$, $P_1(s) = -P_2(s)$.

At the heart of the analysis of these games is the perfect opposition of the players' interests. If s' gives a higher payoff to player 1 than does s'', the reverse holds for player 2. Consequently, trying to maximize one's own payoff is precisely the same as trying to minimize the payoff of the other player.

For each player, a *security level* can be defined that is the largest payoff that the player can guarantee herself no matter what strategy choice the other player may make. Let $V_1(s_1) = \min_{s_2 \in S_2} P_1(s_1, s_2)$ and $V_2(s_2) = \min_{s_1 \in S_1} P_2(s_1, s_2)$. The security level of player 1 is $v_1 = \max_{s_1 \in S_1} V_1(s_1)$ and the security level of player 2 is $v_2 = \max_{s_2 \in S_2} V_2(s_2)$. When $v_1 = v_2$ the common value is denoted v and is called the *value of the game*. The famous *saddle point*, or *minimax, theorem*, due to von Neumann (1928), states that any two-person zero-sum finite game has an equilibrium and a value. The following theorem, which states these results, is not proved here because it is a special case of results shown in Chapter 3. An elegant, brief, proof that relies on elementary mathematics can be found in Owen (1967).

THEOREM 2.3 *Any finite two-person zero-sum game* $\Gamma = (N, Y, p, I, q) = (N, S, P)$ *has an equilibrium point. If* s^* *is an equilibrium point of* Γ, *then* $v_1 = V_1(s_1^*) = v_2 = V_2(s_2^*) = v$.

The special antagonism of interests does not carry over to three- or more person games, because, even with the zero-sum property, in three-person games it is possible that two players can make common cause against the third. The zero-sum property is not actually important; what really matters is that $P_1(s') \geq P_1(s'') \leftrightarrow P_2(s') \leq P_2(s'')$. This theme is explored more fully in Chapter 3.

4 Subgames

A *subgame* of Γ is a game that is contained within Γ and that proceeds from one node in Γ to a subset of the terminal nodes of Γ. Consider first a *subtree of the game tree* Y, denoted Y_d. For $d \in D$ let $F(d)$ be the set of nodes that follow d.

$$F(d) = \{d' \in D \cup T \mid d \in e(d') \text{ and } d' \notin e(d)\} \tag{2.10}$$

For any $d \in D$, $F(d)$ is nonempty and the subtree Y_d is that part of the tree Y that includes d plus all nodes that follow d.

DEFINITION 2.17 *The* **subtree** $Y_d = (D_d, T_d, e_1)$ *where* $D_d = [\{d\} \cup F(d)] \cap D$ *and* $T_d = F(d) \cap T$.

If certain conditions are met, a subgame Γ_d consists of a subtree Y_d, the information sets occurring in Y_d, the payoff function restricted to the terminal nodes in Y_d, and the chance moves in Y_d at nodes assigned to chance. These conditions are, first, that d is the only node in its information set in Γ and, second, that any information set in Γ is either totally included in Γ_d or is totally excluded. For $d \in D$ let (1) $p_d: T_d \rightarrow R^n$ with $p_d(t) = p(t)$ for

$t \in T_d$, (2) $I_d^i = \{I_{ij} \in I^i \mid I_{ij} \cap D_d \neq \varnothing\}$, (3) $I_d = \bigcup_{i \in N} I_d^i$, and (4) let q_d be the set of probability distributions that is followed by player 0 (chance) at the various nodes in D_d where it is player 0's turn to move. Then p_d is the payoff function restricted to T_d, I_d^i is the collection of information sets in I^i that have nodes in Y_d, and q_d is the set of probability distributions of chance from Γ for chance moves at nodes in Y_d.

DEFINITION 2.18 $\Gamma_d = (N, Y_d, p_d, I_d, q_d)$ *is a* **subgame** *of* Γ *if* (*a*) $d \in D$, (*b*) $\{d\} = I_{ij}$ *for one* $i_{ij} \in I$, *and* (*c*) *for any* $I_{ij} \in I_d$, $I_{ij} \cap D_d = I_{ij}$.

The number of information sets in I_d^i is r_{di}, which cannot exceed r_i.

DEFINITION 2.19 *Let* Γ_d *be a subgame of* Γ. Γ_d *is a* **proper subgame** *of* Γ *if* $d \neq d_0$.

In Definition 2.18, condition (a) requires that the initial node of a subgame be a decision node of the original game; condition (b) requires that the information set containing d contain no other nodes; and condition (c) requires that, if some decsion node $d' \in D_d$, then D_d contains all other nodes in the information set containing d'.

The matching pennies game (Figure 2.2) has no subgames other than the original game itself. Indeed, the initial node is the only decision node that is alone in its information set. The game in Figure 2.3 has several subgames in addition to the game itself. The initial nodes of these subgames are marked by an asterisk (*) placed next to the player who moves at that node. Of course, no subgame can start from the information sets containing two or more nodes. Of the single-node information sets, all but two are the initial nodes of subgames. The exceptions are the node of player 3 (d_1) following the choice by player 1 of left from d_0 and the node of player 2 (d_2) following the choice by player 1 of center from d_0. Both of these nodes can lead to the same information set (I_{32}), and I_{32} fails condition (c) in Definition 2.18. Some, but not all, of its nodes can be reached from d_1, and the same is true from d_2.

5 Refinements of noncooperative equilibrium

It is easy to invent examples of games having multiple equilibria certain of which are intuitively unreasonable. The basic thrust of refinements is the attempt to delineate criteria that can be used to separate the *reasonable* from the *unreasonable* equilibria. As will be seen, these attempts have met with considerable, although by no means complete, success. A related goal is to predict or prescribe a specific outcome of a game. Multiple equilibria pose a difficulty here, consequently some effort at equilibrium selection is natural. A complete equilibrium selection theory that would be generally acceptable is an important goal for which to strive; however, outside a narrow setting it will be difficult to attain. (see Harsanyi and Selten, 1988).

5.1 Overview of refinements

The refinements considered below are (a) subgame perfect equilibrium, due to Selten (1965); (b) perfect equilibrium, due to Selten (1975); (c) proper equilibrium, due to Myerson (1978); and (d) sequential equilibrium, due to Kreps and Wilson (1982). All are related to one another in varying degrees, although there is variety in the way they are couched and in the ease with which they can be applied. They do not represent a complete list of all the refinements that have been invented, but rather represent some of the earliest and most influential. The reader interested in delving further into the literature on refinements is directed to van Damme (1987) and Kohlberg and Mertens (1986) for an exposition of many refinements and additional references.

To distinguish briefly among the refinements covered later a strategy combination s^* is a *subgame perfect equilibrium point* of a game Γ if the use of s^* by the players results in equilibrium play in each subgame of Γ. In considering a particular equilibrium point s^* the behavior of the players under s^* can be traced through the game tree. Certain information sets will be reached with positive probability and others may be reached with zero probability (i.e., will not be reached). The strategies, by their very nature, will dictate how players will behave if they are at any information set including those that have zero probability of being reached. Thus, some subgames will be reached and others may not. Any equilibrium point s^* will prescribe equilibrium play on subgames that are reached by s^* with positive probability, but need not prescribe equilibrium play on unreached subgames. It is possible that the behavior called for by a strategy combination s on unreached subgames will influence the optimality of s on parts of the tree that are reached. Such behavior on unreached subgames may act as a threat by one player on others; however, being nonequilibrium play within the unreached subgame, such threats lack credibility. Subgame perfection requires equilibrium play on all subgames, whether reached or not. Subgame perfection may thus be thought to reduce the possibility of noncredible threats.

Perfect equilibrium plugs a loophole in subgame perfection. There may be parts of the game tree that are unreached but that are not part of a subgame. Nonequilibrium behavior in these parts of a game tree is defineable and can have the same influence as nonequilibrium behavior on an unreached subgame.

Perfect equilibrium requires optimal behavior by each player i from each of her information sets I_{ij}. It makes great use of the fact that a completely mixed strategy combination reaches every node of a game with positive probability. An equilibrium point in completely mixed strategies is a perfect equilibrium because there are no unreached information sets and any equilibrium, perfect or not, must specify equilibrium behavior from each information set that is encountered with positive probability. If one adds small *mistakes* to the chosen behavior of the players, then the game must have a positive probability of reaching any node. An equilibrium point in a game with mistakes is approximately an equilibrium in the original game if the mistakes are small. By letting the mistake probabilities converge to 0 a sequence of equilibrium points for games with mistakes (*perturbed games*) is obtained whose limit is a perfect equilibrium point for the original game.

Proper equilibrium modifies perfect equilibrium by requiring that the mistake probabilities attaching to the use of a strategy be related in size to the extent to which the strategy is desirable to use. Thus, s^* is a ***proper equilibrium point*** if it is perfect and larger mistakes are much less probable than smaller mistakes. The justification for proper equilibrium is similar to the justification for other refinements: is it possible to cite examples in which some perfect equilibria are intuitively unreasonable in ways that proper equilibrium can eliminate.

Finally, s^* is a *sequential equilibrium* if, for each player i and each information set $I_{ij} \in I^i$, s_i^* is a best reply to s^* relative to *consistent beliefs* on the part of the players. Sequential equilibrium is almost a restatement of perfect equilibrium in the sense that any perfect equilibrium is a sequential equilibrium and almost all sequential equilibria are perfect. The description of sequential equilibrium is quite different from that of perfect equilibrium and beliefs of the players have an explicit role that they lack in the other refinements covered here.

These refinements are nested. Any proper equilibrium is a perfect equilibrium, any perfect equilibrium is a sequential equilibrium, any perfect equilibrium is a sequential equilibrium, and any sequential equilibrium is a subgame perfect equilibrium. The converse of each of these statements is not true in general.

5.2 Subgame perfection

Imagine a game Γ and an arbitrary strategy combination, s. Γ has some subgames and, for each subgame, s prescribes the behavior that will be followed by the players if that subgame is reached. In general, following the moves dictated by s from the start of the game will cause certain subgames to be reached; however, not all subgames need be reached. For an equilibrium point s^* (1) if a given subgame is reached using s^*, then s^* prescribes equilibrium behavior on that subgame, and (2) if a given subgame is not reached by s^*, then s^* need not prescribe equilibrium behavior on that subgame. This is restated formally in Lemma 2.2. If a strategy combination violates (1), then there is a player who is not using a best reply to the given strategy combination. On the other hand, payoffs are totally insensitive to the behavior specified by strategies on the unreached parts of the game.

If, however, s^* is subgame perfect, then it will (by definition) prescribe equilibrium behavior on all subgames, whether or not they are actually reached when s^* is played from the start. Subgames that are not reached by an equilibrium strategy combination are said to be *off the equilibrium path.* Behavior that is prescribed for parts of the game that are off the equilibrium path can affect whether or not a player's behavior is optimal on the equilibrium path. Sustaining an equilibrium in part by nonequilibrium behavior off the equilibrium path is comparable to sustaining the equilibrium by threats lacking in credibility.

Examine the game in Figure 2.5 and note that there are two pure strategy equilibria: (i) Strategy combination s', under which player 1 chooses L; player 2 chooses r if R was selected by player 1 and r if L was selected by player 1. (ii) strategy combination s'', under which player 1 chooses R; player 2 chooses r if R was selected by player 1 and l if L was selected by player 1. The associated payoffs are $P(s') = (14, 8)$ and $P(s'') = (10, 15)$. This game has three subgames, and it is easy to analyze the game by looking at them. The two proper subgames begin at the two decision nodes of player 2. If player 2 were at the left node, the obvious choice for her would be r. If she were at the right node, the obvious choice would also be r. If player 1 figured this out, he would expect player 2 to choose r no matter what he did. The game would reduce, for him, to that shown in Figure 2.6 and he would consequently select L because doing so would lead to his receiving 14, rather than 10. The preceding describes equilibrium (i), which is subgame perfect because it specifies equilibrium play in each of the three subgames. Note that one of the three subgames will never actually be encountered while playing equilibrium (i). As player 1 chooses L, the subgame beginning at the right hand node of player 2 will never be reached. Nonetheless, the presumptive behavior of player 2 at this

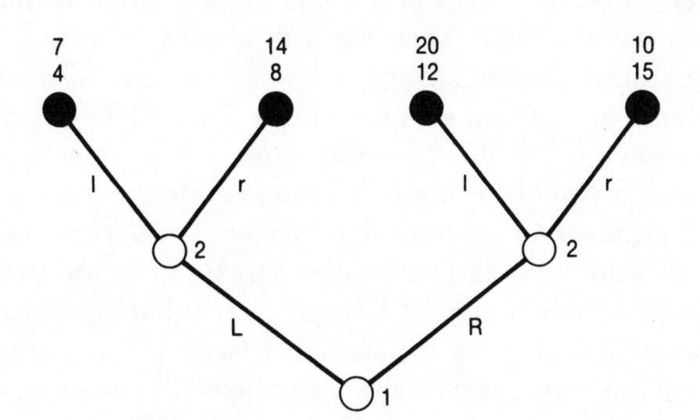

FIGURE 2.5 A game illustrating subgame perfection.

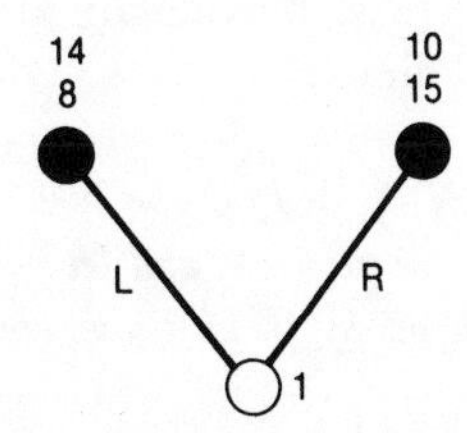

FIGURE 2.6 An illustration of backward induction.

node is taken into account in determining the behavior of player 1 earlier in the game.

Now look at equilibrium (ii). First, assure yourself that it is an equilibrium point. Suppose player 1 chooses L instead of R. Then, given the strategy of player 2, he gets 7 instead of 10. Given R for player 1, if player 2 uses a different strategy, she gets the same payoff if she changes her move at the left node only, or she gets 12 instead of 15 if she changes her move at the right node. Thus, neither player can unilaterally alter his or her strategy and receive a larger payoff. So what is wrong? This strategy combination is an equilibrium, in part, because it calls upon player 2 to select l if player 1 chooses L. On the one hand, player 1 will not actually select L, so the eventuality will not arise, but, on the other hand, it would be irrational for player 2 to choose l if the eventuality did arise. Consideration of this possibility, even though it will not happen, has contributed to making s'' an equilibrium point. This equilibrium presumes that, should player 2 find herself at the left node, she will not do the optimal thing. Consequently, player 1, who obtains a larger payoff at s' than at s'', may believe that if he chooses L, player 2 will choose optimally by selecting r.

It is not intuitively plausible that player 1 would take s'' seriously as a reasonable equilibrium for the reasons cited above, and it is not believable that a rational, thoughtful player 2 would take it seriously either, because player 2 can carry out the sort of analysis that player 1 would do. How, then, might one imagine s'' actually occurring? Suppose the two players are permitted some conversation before they choose their moves and player 2 says to player 1 that she insists on s''. That is, she threatens player 1 that if player 1 does not choose R, then she will choose l. Such a threat is not convincing, because, should player 1 actually choose L, it would not benefit player 2 to carry out the threat, since player 2 would do better to choose r. From interpretations of the preceding sort, it is sometimes said that (subgame) perfect equilibria eliminate noncredible threats.

It is not legitimate to propose that player 2 will carry out her threat because it will lend credibility to similar threats in future games. Such an argument supposes that Figure 2.5 is not a complete representation of the game, but rather is embedded in something larger. Were Figure 2.5 merely part of a larger game, the correct procedure would be to analyze that larger game.

A precise specification of the subgame strategy that a given strategy prescribes for (*induces upon*) a particular subgame follows. Subgame perfection is then formally defined.

DEFINITION 2.20 *Let $\Gamma = (N, Y, p, I, q)$ be a finite game of perfect recall, let $\beta^i \in B_i$ be any behavior strategy for player i, and let Γ_d be an arbitrary subgame of Γ. Then β^i_d is the* **strategy induced** *by β^i on Γ_d where*

$$\beta^i_d = (\beta^i_{k_1}, \beta^i_{k_2}, \ldots, \beta^i_{r_{di}}) \quad \text{and} \quad I^i_d = (I_{ik_1}, \ldots, I_{ir_{di}}).$$

The set of behavior strategies for player i in the subgame Γ_d, defined earlier, is denoted B_{di}; the associated set of behavior strategy combinations is denoted B_d; and the element of B_d induced by $\beta \in B$ is denoted β_d.

DEFINITION 2.21 *Let $\Gamma = (N, Y, p, I, q)$ be a finite game of perfect recall and let $\beta \in B$. If β_d is an equilibrium point of Γ_d, then β* **induces an equilibrium point on the subgame Γ_d**.

Let $D_S = \{d \in D \mid \Gamma_d$ is a subgame of $\Gamma\}$. D_S is the set of decision nodes at which the subgames of Γ begin.

DEFINITION 2.22 *Let $\Gamma = (N, Y, p, I, q)$ be a finite game of perfect recall. Then $\beta^* \in B$ is a* **subgame perfect equilibrium point** *of Γ if it induces an equilibrium point on Γ_d, for each $d \in D_S$.*

LEMMA 2.2 *Let $\Gamma = (N, S, P)$ be a finite noncooperative game and let s^* be an equilibrium point of Γ. Then* (1) *if a given subgame is reached using s^*, then s^* prescribes equilibrium behavior on that subgame*; (2) *if a given subgame is not reached, then s^* need not prescribe equilibrium behavior on that subgame.*

Proof. The proof is left to the reader.

An important feature of subgame perfection is that it need not make any explicit or implicit use of probabilities; it can therefore be applied to games having an uncountably infinite number of subgames. This point will be made more forcefully in Chapter 4; it should be noted, however, that this distinguishes subgame perfection from the other refinements discussed here.

5.3 Perfect equilibrium points

The direct intuitive justification for perfect equilibrium that is used here differs from somewhat from the justification used for subgame perfection, in that it is based on the robustness of each player's equilibrium strategy with respect to small *mistakes*, or *trembles*, on the part of the players. The idea of the tremble is that a player cannot perfectly control his choices. At each information set from which he moves, there is a small probability of selecting a move different from that intended. This is analogous to dialing an incorrect telephone number occasionally even when one knows the correct number.

Although perfection is formally stated in terms of mistakes, the improvement that perfection makes over subgame perfection is that, at a perfect equilibrium point, the strategy of each player i prescribes optimal behavior for that player from each *information set* I_{ij} from which the player moves. By contrast, a subgame perfect equilibrium point prescribes optimal behavior for player i at I_{ij} only if I_{ij} contains just a single node that is the distinguished node of a subgame. This aspect of perfect equilibrium comes through to some extent in the examples; however, as will be seen in Section 5.4, it is entirely explicit in the formulation of sequential equilibrium.

5.3.1 *The basics of perfect equilibrium*

Suppose that $s^* = (s_1^*, \ldots, s_n^*)$ are strategies that the players in a game Γ wish to choose and that $u = (u_1, \ldots, u_n)$ are completely mixed strategies that the players might select accidentally. So player i, who intends to choose s_i^*, will actually do so with probability $1 - \varepsilon$, and he will, with probability ε, make a mistake and choose u_i. Thus, u and ε are part of the game structure; only the s_i^* are *active choices* of the players. Consider the family of games, $\Gamma_{u\varepsilon}$, that varies according to the error probability ε (for a given u). The strategy combination s^* is a perfect equilibrium point of the original game Γ ($\equiv \Gamma_{d_0}$) if there exists a completely mixed $u \in S$ and an $\varepsilon' > 0$ such that s^* is an equilibrium point for all games $\Gamma_{u\varepsilon}$ with $\varepsilon \in [0, \varepsilon']$. Because perfect equilibrium is commonly described, as here, by means of error probabilities, it is sometimes called *trembling hand perfect equilibrium* to distinguish it from *subgame perfect equilibrium.*

Of course, a general finite game of perfect recall can have many information sets containing more than one node and the same sort of implausible behavior illustrated by the game in Figure 2.5 can arise at multiple node information sets, as in Figures 2.7 and 2.8. Figure 2.7 is from Selten (1975). It has no proper subgames, but has several equilibria. Let α_1, α_2, and α_3 denote the probabilities placed on right (i.e., on R, r, and ρ) by players 1, 2 and 3, respectively. Then $(\alpha_1, \alpha_2, \alpha_3)$ is an equilibrium if (a) $\alpha_1 = 0$, $\alpha_2 \in [\frac{1}{3}, 1]$, and $\alpha_3 = 1$, or (b) $\alpha_1 = 1$, $\alpha_2 = 1$, and $\alpha_3 \in [0, \frac{1}{4}]$. The category (a) equilibria are implausible. Consider, for example, (0, 1, 1) from this category. Because player 1 chooses L, the choice of player 2 is not actually operative; however, if there is any chance that player 1 will select R by mistake, then player 2 would do better to choose l than r. The choice of ρ by player 3 means that player 2 would receive a payoff of 4 instead of 1. That (0, 1, 1) is an equilibrium requires that the information set from which player 2 moves will not be visited when (0, 1, 1) is played. The same holds for all of the category (a) equilibria.

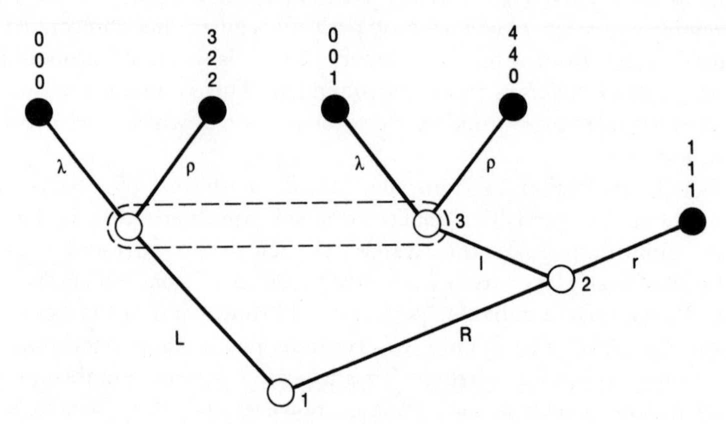

FIGURE 2.7 A three-person game with an imperfect equilibrium.

Now consider the category (b) equilibrium (1, 1, 0). Suppose that each player has a certain error probability, which may be different from player to player. Letting the error probabilities be ε_i for $i = 1, 2, 3$ and stating the payoff of each player as a function of his own strategy, given the equilibrium strategies of the others, the payoffs are

$$P_1 = \alpha_1(1 - \varepsilon_2 - 3\varepsilon_3 + 4\varepsilon_2\varepsilon_3) + 3\varepsilon_3 \tag{2.11}$$

$$P_2 = 2\varepsilon_3(2 - \varepsilon_1) + \alpha_2(1 - \varepsilon_1 - 4\varepsilon_3 + 4\varepsilon_1\varepsilon_3) \tag{2.12}$$

$$P_3 = 1 - \varepsilon_1 + \alpha_3(2\varepsilon_1 - \varepsilon_2 + \varepsilon_1\varepsilon_2) \tag{2.13}$$

From these payoffs it is possible to judge whether, as the error probabilities go to 0, the players' optimal strategies converge to (1, 1, 0). For player 1, clearly the payoff is rising in α_1 when the errors are very small, so choosing R is optimal in the limit. A similar conclusion holds for player 2. For player 3 the situation is a little more complicated, as it depends upon the relative sizes of ε_1 and ε_2. If $\varepsilon_1/\varepsilon_2$ is less than 1/2, then, as both errors go to 0, player 3 is best off choosing left. Therefore, the equilibrium point (1, 1, 0) is perfect.

Figure 2.8 presents a more straightforward situation. Two equilibria are (L, l) and (R, r). The former is implausible and suggests that player 2 is threatening player 1 in a noncredible manner in order to receive a payoff of 2. If player 2 should reach his information set, r would strictly dominate l: if player 1 chooses C, player 2 receives 1 rather than 0; if player 1 chooses R, player 2 receives 0 rather than -4.

In the following sections, perfect equilibrium is developed in three settings. In Section 5.3.2 attention is restricted to extensive form games of perfect recall in which only behavior strategies are considered. Section 5.3.3 takes up finite games generally and Section 5.3.4 examines further games in strategic form. Unlike subgame perfection; perfection has something to contribute to the analysis of strategic form games. Subgame perfection cannot usefully be addressed to such games because subgames cannot necessarily be discerned from the strategic form.

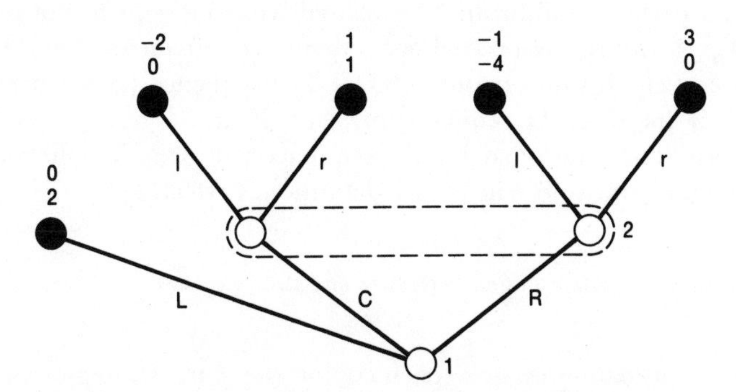

FIGURE 2.8 A two-person game with an imperfect equilibrium.

5.3.2 Perfect equilibrium in finite extensive form games with perfect recall

Attention is limited in this section to behavior strategies and to games of perfect recall. A game with error probabilities, called a *perturbed game,* will be defined. The errors associated with the perturbed game will be denoted η. These errors spell out the chance of error at each information set of each player; therefore, η is a vector with one coordinate for each move in each information set. A convergent sequence of perturbed games is then considered for which the errors go to 0 in the limit. Letting $\eta(k)$ denote the kth member of the sequence and denoting the kth perturbed game by $\Gamma_{\eta(k)}$, a sequence of equilibrium points is chosen, $\{\beta(k)\}$, where $\beta(k)$ is an equilibrium point of $\Gamma_{\eta(k)}$. If the sequence $\{\beta(k)\}$ has a limit, then that limit is a perfect equilibrium point of the original game.

To characterize the error probabilities, let $\eta = (\eta^1, \eta^2, \ldots, \eta^n)$ and let $\eta^i = (\eta^i_1, \eta^i_2, \ldots, \eta^i_{r_i})$, where each $\eta^i_j = (\eta^i_{j1}, \eta^i_{j2}, \ldots, \eta^i_{j,m_{ij}})$ is an error distribution over the moves in I_{ij}. It is required that

$$\eta^i_j \gg 0 \quad \text{and} \quad \sum_{l=1}^{m_{ij}} \eta^i_{jl} < 1 \tag{2.14}$$

These error probabilities are the minimum probabilities that can be associated with each move in each information set. Because of these minimum probabilities, the players' choices are restricted. Player i can select any behavior strategy subject to the restriction that $\beta^i_{jl} \geqslant \eta^i_{jl}$. The restricted set from which player i chooses is denoted $B_{i\eta}$ and the corresponding set of behavior strategy combinations is denoted B_η. Comparing the error specification η with (u, ε) error specification used in Section 5.3.1, in a game of perfect recall any (u, ε) implies a unique η. The natural way to associate a unique η with a given (u, ε) is explored in Exercise 7 at the end of the chapter.

Note that the usual assumptions, put forth in Chapter 3, hold in a perturbed game (compact, convex strategy sets $B_{i\eta}$, a player's payoff is concave in his own strategy); therefore, a perturbed game has an equilibrium point. This immediately implies that any finite game of perfect recall has a perfect equilibrium. Associated with the sequence of perturbed games $\{\Gamma_{\eta(k)}\}$ and the associated convergent error sequence $\{\eta(k)\}$ there is a sequence of equilibrium points $\{\beta(k)\}$. The sequence $\{\beta(k)\}$ need not be unique, nor need it be convergent, but it must have a convergent subsequence. The limit of such a subsequence will be an equilibrium point of the original game and will be, by definition, perfect.

5.3.3 Perfect equilibrium in finite extensive and strategic form games

If the error probabilities are specified for the pure strategies of a finite game, then perfection is easily described in terms that apply to both the extensive and strategic forms. In this section the exposition will be in terms of strategic form games; in Section 5.3.4 it will be in terms of extensive form games. The error probabilities for player i are denoted ξ_i, where $\xi_i = (\xi_{i1}, \ldots, \xi_{im_i}) \gg 0$ and $\sum_j \xi_{ij} < 1$.

DEFINITION 2.23 *A* **perturbation,** *or a* **tremble,** *is* $\xi \gg 0$ *where* $\xi = (\xi_1, \ldots, \xi_n)$, $\xi_i = (\xi_{i1}, \ldots, \xi_{im_i}) \in R_+^{m_i}$, $\sum_{j=1}^{m_i} \xi_{ij} < 1$, *and* $i \in N$.

A perturbation gives the minimum probabilities with which the various pure strategies of the players will be chosen. Let $\Gamma = (N, Y, p, I, q)$ be a finite extensive form game and let $\Gamma = (N, S, P)$ be its strategic form representation.

DEFINITION 2.24 *A* **perturbed version** *of* $\Gamma = (N, S, P)$ *or a* **perturbed game** *is* $\Gamma_\xi = (N, S_\xi, P)$, *where* ξ *is a perturbation,* $S_\xi = \times_{i \in N} S_{\xi i}$ *and* $S_{\xi i} = \{s_i \in S_i \mid s_i \geqslant \xi_i\}$.

A strategy combination $s^* \in S$ is a perfect equilibrium point of $\Gamma = (N, S, P)$ if there exists a sequence of perturbations $\{\xi(k)\}$ converging to 0 and a sequence of associated equilibrium points of the perturbed game $\Gamma_{\xi(k)}$, $\{s^*(k)\}$, that converges to s^*.

DEFINITION 2.25 *Let* $\Gamma = (N, S, P)$ *be a finite game. Then* $s^* \in S$ *is a* **perfect equilibrium point** *of* Γ *if there exists* $\{\xi(k)\}$ *converging to* 0 *such that* $s^*(k)$ *is an equilibrium point of* $\Gamma_{\xi(k)}$, $k = 1, 2, \ldots$, *and* $\{s^*(k)\}$ *converges to* s^*.

THEOREM 2.4 *Let* $\Gamma = (N, S, P)$ *be a finite game. Then* Γ *has a perfect equilibrium point.*

Proof. Choose an arbitrary sequence of trembles, $\{\xi'(k)\}$, that converges to 0. Let $s'(k)$ be an equilibrium point of the perturbed game $\Gamma_{\xi'(k)}$. Then the sequence $\{s'(k)\}$ has a point of accumulation, s^* because $\{s'(k)\}$ is contained in a compact set. Let $\{s(k)\}$ be a convergent subsequence of $\{s'(k)\}$ that goes to s^* and let $\{\xi(k)\}$ be the associated sequence of trembles. Then s^* satisfies the conditions of Definition 2.25 with the sequence of trembles $\{\xi(k)\}$ and the associated sequence of perturbed games, which means that s^* is a perfect equilibrium point of Γ. QED

A key to understanding perfect equilibrium is that each strategy in each member of the sequence $\{s(k)\}$ is completely mixed; therefore, every information set is visited with positive probability in each game $\Gamma_{\xi(k)}$. This implies, in turn, that a player can never be using a weakly dominated strategy at a perfect equilibrium point. To see this, look at the (L, l) equilibrium for the game in Figure 2.8. Suppose a perturbed version of this game in which there is a probability δ that player 1 chooses C and a similar probability that he chooses R. As long as either of these probabilities is greater than 0, the best reply of player 2 is r. For player 2, l is a dominated strategy and it cannot be optimal if the information set of player 2 has any chance of being visited.

Some would argue that any equilibrium should be perfect if it is to be acceptable, because games are, in fact, played by fallible agents. This argument has more scope than the one based on ruling out noncredible threats, for there are imperfect equilibria that do not depend on threats. An obvious class of examples is games in strategic form in which *move* and *pure strategy* coincide. Perfection can be applied, but threat plays no role, because each player has exactly one move and the players move simultaneously.

5.3.4 *ε-perfect equilibrium and perfect equilibrium*

Perfect equilibrium points can be characterized in yet another way as the limit of a sequence of *ε-perfect equilibrium points*. This approach is interesting for two reasons. First, it might occasionally appear a more natural way to determine whether a given equilibrium is perfect. Second, it parallels the simplest, most direct way of describing proper equilibria, which are discussed in Section 5.5. A strategy combination is ε-perfect if it is completely mixed and a probability of no more than ε is attached to any pure strategy that is not a best reply. That is,

DEFINITION 2.26 *Let* $\Gamma = (N, Y, p, I, q)$ *be a finite game and let* s^* *be a completely mixed strategy combination. Then* s^* *is an* **ε-perfect equilibrium point** *if, for all* $\pi_i \in \Pi_i$ *and all* $i \in N$, $s^*_{i\pi_i} \leq \varepsilon$ *whenever* $P_i(s^*\backslash\pi_i) < P_i(s^*\backslash\pi'_i)$ *for some* $\pi'_i \in \Pi_i$.

LEMMA 2.3 *Let* (a) $\Gamma = (N, Y, p, I, q)$ *be a finite extensive form game,* (b) $\{\varepsilon(k)\}$ *be a sequence of positive scalars that converges to* 0, (c) $s(k)$ *be an* $\varepsilon(k)$*-perfect equilibrium point of* Γ, *and* (d) $\{s(k)\}$ *have* s^* *as a limit. Then* (1) s^* *is a perfect equilibrium point and* (2) *if* s^* *is any perfect equilibrium, there is a sequence of* $\varepsilon(k)$*-perfect equilibria,* $\{s(k)\}$, *with* $\{\varepsilon(k)\}$ *converging to zero, whose limit is* s^*.

Proof. To prove (1), for each value of k define the vector of error probabilities $\xi(k)$ as follows: first, assign error probabilities to the pure strategies that are not best replies. Thus, if $P_i(s(k)\backslash\pi_i) < P_i(s(k)\backslash\pi'_i)$ for $\pi_i = j$ and for some π'_i, then $\xi_{ij}(k) = s_{ij}(k)$. Next assign error probabilities to the pure strategies that are best replies. To that end, let δ be the sum of the $\xi_{ij}(k)$ defined above (i.e., the sum over the pure strategies that are not best replies). Then let $\xi_{ij}(k) = (1 - \delta)s_{ij}(k)/k$.

The strategy combination $s(k)$ is an equilibrium point of the perturbed game $\Gamma_{\xi(k)}$. This is true because in the strategy combination $s(k)$, for any value of k, a pure strategy of a player that is not a best reply receives a weight equal to its error probability. Thus, no player i could receive a larger payoff by using some other strategy in $S_{\xi(k)i}$. Meanwhile $\{\xi(k)\}$ converges to 0 and $\{s(k)\}$ converges to s^*; therefore, the conditions of Definition 2.25 are met and s^* is a perfect equilibrium.

In order to prove (2), a construction is made that is the reverse of the construction above. If s^* is perfect, then there is an error sequence $\{\varepsilon(k)\}$

that converges to 0 and an associated sequence of perturbed games, $\{\Gamma_{\varepsilon(k)}\}$, with an associated sequence of equilibria, $\{s(k)\}$, that converges to s^*. Let the scalar $\varepsilon(k)$ be equal to the probability $s_{ij}(k)$ that is the largest probability in $s(k)$ assigned to a pure strategy that is not a best reply. Then $s(k)$ is an $\varepsilon(k)$-perfect equilibrium of Γ. QED

Lemma 2.3 will prove useful in showing that a perfect equilibrium cannot utilize any weakly dominated strategies.

LEMMA 2.4 *Let $\Gamma = (N, Y, p, I, q)$ be a finite game and let s^* be a perfect equilibrium of Γ. Then s^* is not weakly dominated.*

Proof. Suppose that s^* is weakly dominated. This means that there is a player i, a strategy $s_i' \in S_i$, and a strategy combination $s'' \in S$ such that $P(s\backslash s_i') \geq P_i(s\backslash s_i^*)$ for all $s \in S$ and $P_i(s''\backslash s_i') > P_i(s''\backslash s_i^*)$. Therefore there is some pure strategy, j, of player i that is not a best reply to s'' and to which positive probability, $s_{ij}^* > 0$, is attached. Consider a sequence $\{\varepsilon(k)\}$ that converges to 0 and for which each $\varepsilon(k) < s_{ij}^*$. Let $s(k)$ be an $\varepsilon(k)$-perfect equilibrium of Γ. Any $s \in S$ induces a probability distribution, $\mu^i(s)$, over the pure strategy combinations of the players $N\backslash(i)$. Thus, for each k, $\mu^i(s(k))$ can be written $\mu^i(s(k)) = \delta(k)\mu^i(s'') + (1 - \delta(k))\mu^i(u(k))$ for some $\delta(k) \in (0, 1]$ and $u(k) \in S$, because $s(k)$ is completely mixed. Domination of s_i^* by s_i' implies $P_i(u(k)\backslash s_i') \geq P_i(u(k)\backslash \pi_i)$ for $\pi_i = j$, so that $P_i(s(k)\backslash s_i') = \delta(k)P_i(s''\backslash s_i') + (1 - \delta(k))P_i(u(k)\backslash s_i') > \delta(k)P_i(s''\backslash \pi_i) + (1 - \delta(k))P_i(u(k)\backslash \pi_i) = P_i(s(k)\backslash \pi_i)$ for $\pi_i = j$. Then $s_{ij}(k) \leq \varepsilon(k) < s_{ij}^*$ and $\{s(k)\}$ cannot converge to s^* because $s_{ij}(k)$ will necessarily converge to zero. QED

5.4 Sequential equilibrium

We return to finite games in extensive form with perfect recall. Sequential equilibrium, proposed by Kreps and Wilson (1982b), is very close to perfect equilibrium in that any perfect equiibrium is a sequential equilibrium and almost any sequential equilibrium is a perfect equilibrium.

As a prelude to examining sequential equilibrium, imagine an equilibrium point, β, in completely mixed behavior strategies. Being completely mixed, the equilibrium is perfect. Any information set, I_{ij} can be arbitrarily selected and one can ask whether the strategy of player i is optimal for him from that point in the game onward. Even though I_{ij} does not begin a subgame, the question is meaningful; for it merely asks whether β^i is a best reply to β from that point on *based upon all players having used the strategy combination β from the start of the game.* Knowing β and knowing the point at which the game begins, it is possible to calculate the probability of being at each node in I_{ij} conditional upon having reached I_{ij}. If player i thought the other players would use the β^j $(j \neq i)$ and were, himself, considering β^i, these conditional probabilities on the nodes of I_{ij} would constitute reasonable beliefs for player i. Consequently, given the $\beta^j(j \neq i)$, it is possible to calculate the payoff associated with β^i conditional upon having these beliefs and upon having reached I_{ij} and to do the same for any other (behavior) strategy available to player i. The preceding optimality calculation has an *as if* character associated with it. Optimality is calculated *as if* each player believes the other players will be using their associated equilibrium strategies.

Sequential equilibrium explicitly takes into account the players' beliefs. Suppose for the moment that a player's beliefs could be anything. The equilibrium is stated in terms of an *assessment,* which is (β, λ), where β is a behavior strategy combination and λ is a *system of beliefs.* These beliefs consist of a family of probability distributions of which a typical member is λ_j^i, a probability distribution over the nodes in the information set I_{ij}. Denote by v_{ij} the number of nodes in I_{ij}. Then $\lambda_j^i = (\lambda_{j1}^i, \lambda_{j2}^i, \ldots, \lambda_{jv_{ij}}^i)$, $\lambda^i = (\lambda_1^i, \lambda_2^i, \ldots, \lambda_{r_i}^i)$ are the beliefs of player i, and $\lambda = (\lambda^1, \lambda^2, \ldots, \lambda^n)$ is the system of beliefs.

DEFINITION 2.27 *Let $\Gamma = (N, Y, p, I, q)$ be a finite extensive form game of perfect recall. A* **system of beliefs** *for the players in Γ is a family of probability distributions, denoted λ, where $\lambda = (\lambda^1, \lambda^2, \ldots, \lambda^n)$, $\lambda^i = (\lambda_1^i, \lambda_2^i, \ldots, \lambda_{r_i}^i)$, and each $\lambda_j^i = (\lambda_{j1}^i, \lambda_{j2}^i, \ldots, \lambda_{jv_{ij}}^i)$ is a probability distribution over the nodes in I_{ij}. Consequently*

$$\lambda_j^i \geq 0 \quad \textit{and} \quad \sum_{l=1}^{v_{ij}} \lambda_{jl}^i = 1 \tag{2.15}$$

Thus, λ_j^i is the probability distribution that player i believes to be the true probability that he is at each of the various nodes in I_{ij}, given that the information set I_{ij} has been reached.

DEFINITION 2.28 *An* **assessment** *is a pair (β, λ), where β is a behavior strategy combination and λ is a system of beliefs.*

An assessment is a sequential equilibrium if β is a *sequential best reply* to (β, λ) and λ is consistent with β. For β to be a sequential best reply means that, from each information set of the game, a player's strategy is a best reply to the strategies of the others, given his beliefs.

DEFINITION 2.29 *Let $\Gamma = (N, Y, p, I, q)$ be a finite extensive form game of perfect recall and let (β, λ) be an assessment. Then β^i* is a **sequential best reply to (β, λ) for player i** *if if β^i maximizes the payoff of player i form each information $I_{ij} \in I^i$, given the beliefs, λ^i of player i.*

DEFINITION 2.30 *Let $\Gamma = (N, Y, p, I, q)$ be a finite extensive form game of perfect recall and let (β, λ) be an assessment. Then β is a* **sequential best reply to (β, λ)** *if it is a sequential best reply for each player $i \in N$.*

The consistency of beliefs, λ, is defined in two steps. Let β be a completely mixed strategy. Then there is a unique system of beliefs that is consistent with it. Such a completely mixed β implies a unique probability of reaching each node in D that follows the initial node. From these probabilities, a probability distribution over nodes in each information set, conditional on reaching that set, can be calculated. These probabilities describe the unique beliefs that are consistent with β. If a strategy combination is not completely mixed, then certain information sets have a probability 0 of being reached. With such information sets consistent beliefs are defined as the limit of beliefs associated with a sequence of completely mixed strategy combinations that converge to the given strategy combination.

DEFINITION 2.31 *Let $\Gamma = (N, Y, p, I, q)$ be a finite extensive form game of perfect recall and let (β, λ) be an assessment. Then λ is* **is consistent with β** *if β is completely mixed and λ is the unique system of beliefs whose probabilities equal to those implied by β. If β is not completely mixed, then λ is consistent with β if there exists a sequence $\{\beta(k)\}$ of completely mixed behavior strategy combinations that converge to β, with $\{\lambda(k)\}$ being the companion sequence of unique consistent beliefs, and with λ being the limit of $\{\lambda(k)\}$.*

DEFINITION 2.32 *Let $\Gamma = (N, Y, p, I, q)$ be a finite extensive form game of perfect recall and let (β, λ) be an assessment. Then (β, λ) is a* **sequential equilibrium** *if λ is consistent with β and β is a sequential best reply to (β, λ).*

Next it is proved that any perfect equilibrium is a sequential equilibrium. Recall that a perfect equilibrium is a strategy combination that meets certain conditions, while a sequential equilibrium is an assessment that meets certain conditions. Therefore, to say that a perfect equilibrium is sequential must mean that a perfect equilibrium, combined with appropriate beliefs to form an assessment, is a sequential equilibrium.

THEOREM 2.5 *let $\Gamma = (N, Y, p, I, q)$ be a finite game of perfect recall, let s^* be a perfect equilibrium of Γ, and let β^* be the behavior strategy induced by s^*. Then there exists a λ such that (β^*, λ) is a sequential equilibrium.*

Proof. There is a unique behavior strategy induced by s^* that may be denoted β. There must be a sequence of completely mixed strategy combinations, $\{s(k)\}$, that converge to s^* and that are equilibria of perturbed games, because s^* is perfect. Let $\beta(k)$ be the unique behavior strategy combination that is induced by $s(k)$ and let $\lambda(k)$ be the unique system of beliefs that is consistent with $\beta(k)$. $\{\beta(k)\}$ must converge to β and the sequence $\{\lambda(k)\}$ has either a limit or cluster points. Let λ be that limit, or, if there is no limit, be a cluster point of $\{\lambda(k)\}$. Then (β, λ) is a sequential equilibrium. QED

Comparing sequential equilibrium with subgame perfect equilibrium and perfect equilibrium, several things are apparent. First, a sequential equilibrium is always subgame perfect. Second, sequential equilibrium is very close to perfect equilibrium, but does not coincide with it. That is, suppose a game Γ in which (β, λ) is a sequential equilibrium. It is not always true that β is a perfect equilibrium. Look at the game in Figure 2.9, taken from van Damme (1987). In this game only (R, l) is a perfect equilibrium, but any assessment (β, λ) is a sequential equilibrium if it

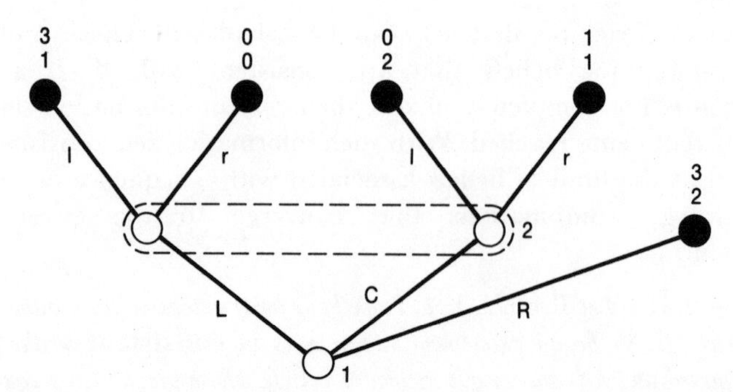

FIGURE 2.9 A game with sequential equilibria that are not perfect.

directs player 1 to put zero probability on C and player 2 to choose l, and has beliefs requiring player 2 to put probability 1 on the left node of his information set. That is, $\beta^1 = (\alpha, 0, 1-\alpha)$ for $\alpha \in [0, 1]$, $\beta^2 = (1, 0)$, and $\lambda^1 = (1, 0)$. Finally, in comparing sequential with perfect equilibria, the former provides an explicit role for beliefs of the players, while, in the tradition of noncooperative equilibrium, the latter does not discuss beliefs; however, it is clear that beliefs have an implicit role in perfect equilibrium. This is true of noncooperative equilibria in general.

5.5 Proper equilibrium

The motivation for Myerson's (1978) introduction of proper equilibrium parallels the motivation for other refinements. Proper equilibrium is, in particular, a refinement of perfect equilibrium just as perfect equilibrium is a refinement of subgame perfect equilibrium. It is possible to come up with examples in which some perfect equilibria are implausible outcomes, just as some noncooperative equilibria may be implausible outcomes. Proper equilibrium is a refinement that excludes some of these implausible outcomes. Myerson (1978) provides the examples reproduced in Tables 2.4 and 2.5.

Look first at the game in Table 2.4. There are two noncooperative equilibria, (a_1, b_1) and (a_2, b_2), but only one perfect equilibrium, (a_1, b_1). As long as player 2 places the slightest probability on b_1, player 1 wants to put the maximum possible probability on a_1. A parallel argument holds for

TABLE 2.4 A game with an obvious perfect equilibrium

		Player 2	
		b_1	b_2
Player 1	a_1	1, 1	0, 0
	a_2	0, 0	0, 0

TABLE 2.5 A game with an unreasonable perfect equilibrium

		Player 2		
		b_1	b_2	b_3
	a_1	1, 1	0, 0	−9, −9
Player 1	a_2	0, 0	0, 0	−7, −7
	a_3	−9, −9	−7, 7	−7, −7

player 2 choosing to place maximum probability on b_1. Now examine the game in Table 2.5. It differs from the first game by the addition of a new strategy for each player that is, for each of them, strictly dominated. Indeed, intuition suggests that the third row for player 1 should be treated as if it were not even there. Similarly for third column for player 2: a_3 is so disastrously bad for player 1 that he surely would never choose it under any circumstances, suggesting that it should be irrelevant to his behavior. The same applies to b_3 for player 2. Yet in the game of Table 2.5 the strategy combination (a_2, b_2) is a perfect equilibrium because of the new strategies. To see this, let all error probabilities have the common value ε and let ε go to 0. For player 1 the best reply to $s_2 = (\varepsilon, 1 - 2\varepsilon, \varepsilon)$ is $s_1 = (\varepsilon, 1 - 2\varepsilon, \varepsilon)$. In the limit this converges to (0, 1, 0) for each player, or (a_2, b_2).

Proper equilibrium eliminates (a_2, b_2) by requiring that if one pure strategy of player i has a smaller payoff associated with it than another, given the strategies of the other players, then the probability placed on the worse strategy must be no greater than ε times the probability placed on the better strategy. That is to say, suppose neither pure strategy j nor pure strategy l are best replies to s, and that the payoff associated with j is greater than that associated with l. Then in a perturbed game it is required that $s_{il} \leqslant \varepsilon s_{ij}$. Proper equilibrium places more restrictions on the error probabilities in the sequences $\{\varepsilon(k)\}$, restricting them so that the *worse* a strategy is, the lower the error probability associated with it.

Proper equilibrium is formally defined as the limit of a convergent sequence of ε-proper equilibria.

DEFINITION 2.33 *Let* $\Gamma = (N, Y, p, I, q)$ *be a finite extensive form game and let* s^* *be completely mixed strategy combination. Then* s^* *is an* **ε-proper equilibrium point** *if* $s^*_{i\pi_i} \leqslant \varepsilon s^*_{i\pi_i'}$ *whenever* $P_i(s^* \backslash \pi_i) < P_i(s^* \backslash \pi_i')$ *for all* $\pi_i, \pi_i' \in \Pi_i$ *and all* $i \in N$.

DEFINITION 2.34 *Let* (a) $\Gamma = (N, Y, p, I, q)$ *be a finite extensive form game,* (b) $\{\varepsilon(k)\}$ *be a sequence of positive scalars that converges to* 0. (c) $s(k)$ *be an* $\varepsilon(k)$*-proper equilibrium point of* Γ, *and* (d) $\{s(k)\}$ *have* s^* *as a limit. Then* s^* *is a* **proper equilibrium point** *of* Γ.

Myerson (1978) proves the following existence theorem for proper equilibrium.

THEOREM 2.6 *Let $\Gamma = (N, Y, p, I, q)$ be a finite extensive form game. Then Γ has a proper equilibrium.*

Proof. Let $\varepsilon \in (0, 1)$, $m = \max_{i \in N}\{m_i\}$, let $\delta = \varepsilon^m/m$, $S_{\delta i} = \{s_i \in S_i \mid s_{ij} \geq \delta, j = 1, \ldots, m_i\}$, and $S_\delta = \times_{i \in N} S_{\delta i}$.[8] Now define a correspondence ϕ from S_δ to subsets of S_δ as follows:[8] $\phi = (\phi_1, \ldots, \phi_n)$ and, for each $s \in S$ and each $i \in N$

$$\phi_i(s) = \{s_i' \in S_{\delta i} \mid \text{if } P_i(s\backslash\pi_i) < P_i(s\backslash\pi_i') \text{ then } s_{i\pi_i}' \leq \varepsilon s_{i\pi_i'}' \text{ for all } \pi_i, \pi_i' \in \Pi_i\} \quad (2.16)$$

For each s, $\phi(s)$ is a compact, convex set in the domain of ϕ. The domain S_δ is closed and convex, so the Kakutani fixed-point theorem applies and ϕ has a fixed point. Any fixed point of ϕ is a proper equilibrium. Denote by s^ε a fixed point of ϕ and generate a sequence of such fixed points by taking a sequence of values of ε that converges to 0. The sequence $\{s^\varepsilon\}$ has a convergent subsequence and the limit of such a sequence is a proper equilibrium of Γ. QED

Any finite game thus has a proper equilibrium, any proper equilibrium is perfect, any perfect equilibrium is sequential, and any sequential equilibrium is subgame perfect. In general, the converse statements are not true, that is, subgame perfect equilibria need not be sequential equilibria and so forth.

6 Concluding comments

There are many other refinements in addition to those discussed in this chapter. A new refinement generally emerges as a solution to a fault inherent in the old stock of refinements. These faults generally show up as equilibria that seem unreasonable as outcomes. Thus, subgame perfect equilibrium, perfect equilibrium, sequential equilibrium, and proper equilibrium have been introduced to eliminate equilibria that appear implausible or undesirable. The coverage of refinements in this chapter is very brief and the reader seeking fuller coverage, including both additional results on the characteristics of the refinements covered here and an exposition of many more refinements, is urged to consult van Damme (1987).

Except for subgame perfect equilibrium, the refinements discussed here are defined for finite games. To my mind this is a serious, albeit possibly surmountable, limitation. The limitation stems from the role played by probabilities in the definition of the refinement. In moving to infinite games (i.e., games having an infinite number of pure strategies for each player) it is unclear whether one wants to use discrete or continuous probability distributions to characterize the required errors and it is not clear what difference this issue makes. Fortunately, for many economic applications, subgame perfection seems to be all that is needed and it can be used in conjunction with infinite strategy spaces.

In the next several chapters, attention will be focused mainly on infinite games in which players would not, in fact, want to use mixed strategies, because such strategies will be dominated by pure strategies. The models employed are nearer to neoclassical economics in form and analysis using them is therefore more easily integrated into the main body of economic theory.

Exercises

1. Among tennis, blackjack, football, chess, and Monopoly, which are games of complete information? Perfect information? Perfect recall?
2. Draw a game tree for a three-move situation in which the first move is a choice of R and L, the second move is a choice of r and l, and the third move is a choice of ρ and l.
3. Suppose a game is based upon the game tree from problem 2, with the following structure of information sets: player 1 moves first, then player 2 moves, and then player 1 moves a second time. At his move, player 2 has observed the move made by player 1. At his second move, player 1 remembers what he did at his first move, but does not observe the move of player 2. How many pure strategies does each player have? What are they? Given a mixed strategy s_1 for player 1 and a mixed strategy s_2 for player 2, what is the probability that player 1 chooses λ on his second move, given that he is in the information set corresponding to choosing L at his first move?
4. Using the game tree from problem 2, suppose that as before, player 2 observes the first move of player 1. At the second move of player 1, he has only the following information about the previous two moves in the game: (1) the moves of players 1 and 2 matched (i.e., the pair of moves was (L, l) or (R, ρ)), but he cannot distinguish between them, or (2) the moves did not match (i.e., the pair of moves was (R, l) or (L, ρ)), but he cannot distinguish between them. Given a mixed strategy s_1 for player 1 and a mixed strategy s_2 for player 2, what is the probability that player 1 chooses λ on his second move, given that he is in the information set corresponding to (1)?
5. Does an arbitrary mixed strategy for player 1 in the game described in problem 3 induce a unique behavior strategy in that game? Does it do so for the game described in problem 4?
6. In a game of imperfect recall, an arbitrary behavior strategy cannot be expressed as a mixed strategy (i.e., as a probability distribution over the set of pure strategies). Does this mean that the set of mixed strategies is somehow inadequate?
7. For the game in Figure 2.4 (two rounds of matching pennies) compare the three error specifications that are given for perfect equilibrium in Section 5.3. Start with a completely mixed error strategy combination, u, and a scalar error probability, ε. Suppose some arbitrary intended strategy combination s. The triplet (u, ε, s) specifies what mixed strategy combination, s', is actually played in the game. What error probability vector ξ is implied by (u, ε, s)? Is this ξ a function only of (u, ε) or does it depend on s as well? The strategy combination

s' induces a unique behavior strategy, β'. Find this behavior strategy. For fixed (u, ε), let s vary over S and determine what are the minimum probabilities attaching to each node in each information set. Use these minima to construct η.

Notes

1. As a matter of convenient terminology, a player may be said to "own" an information set or a node (or an information set or node may "belong" to a player) when it is that player's turn to move there.
2. The notational convention followed for vector inequalities is, for $x, y \in R^n$, $x \geqslant y$ means that $x_i \geqslant y_i$ for all i, $x > y$ means $x \geqslant y$ and $x \neq y$, and $x \gg y$ means that $x_i > y_i$ for all i.
3. The expected utility theorem is standard fare in economics and I thus do not cover it here. Readers seeking an exposition can find one in many sources, including Luce and Raiffa (1957).
4. Similar notation with a different meaning is used for vectors. In particular for a vector x, $x \backslash y_i \equiv (x_1, \ldots, x_{i-1}, y_i, x_{x+1}, \ldots, x_n)$.
5. In this stipulation I am combining into *complete information* two elements that could be separated: knowledge of p and common knowledge that p (and other aspects of the game) are known to everyone and everyone knows that they are known to everyone, and so forth. In general I am assuming that players are always correctly informed about the sort of knowledge possessed by other players and that this fact is common knowledge. For example, if a game is one of incomplete information in which each player knows only his own payoffs, then each player is aware that all players know only their own payoffs and all players know that each knows this, and so forth.
6. I overstate the case. A Bayesian analysis might be done when complete information is lacking. See Section 6 of Chapter 3 on games of incomplete information.
7. When several optimal choices are available any nonrandom way of choosing among them is acceptable.
8. For some discussion of correspondences and fixed point theorems, see Section 2.2 of Chapter 3.

3

Noncooperative games in strategic form

The games studied in this chapter are often characterized as *one-shot,* or *single-period,* games. The real significance of this characterization is that the temporal structure of the game—that is, the extensive form—is considered to be uninteresting. Consequently, the usual practice, followed here, is to ignore the extensive form and define games in strategic form from the outset. To see why the extensive form might lack interest, imagine an n-person game in which each player has only one move and all players must move simultaneously. The extensive form of this game would impose on it a temporal structure that was artificial, because the players would have to be treated as if they moved sequentially. The games covered in Chapters 4 and 5 are explicitly intertemporal, with a set of players in a *supergame* in which a sequence of ordinary games, each like a game from this chapter, is played. In these supergames, the nature of individual moves is interesting; therefore, they are studied in a *semiextensive form* that lies between the strategic form and a true extensive form; however, the semiextensive form retains all the essential parts of the move structure.

In most complete-information noncooperative games examined in Chapters 3 to 5, only one concept of equilibrium is studied: the noncooperative equilibrium due to Nash (1951), referred to variously as *equilibrium point, noncooperative equilibrium, Nash-Cournot equilibrium,* and the context ensures there will be no confusion with Nash's (1950, 1953) cooperative solution, the *Nash equilibrium.* Nash is the great pioneer in noncooperative game theory, although the Cournot (1838) equilibrium is a precursor of the Nash (noncooperative) equilibrium. The fundamental idea behind the equilibrium point is that each player in a complete information game has chosen a strategy that maximizes his own payoff given the strategies of the other players. Of course, it is commonplace in economic theory to postulate that each decision maker chooses, in equilibrium, behavior that maximizes his objective function; however, it is also standard in much of economic theory to assume that the behavior of a single decision maker has no effect on the circumstances facing any other decision maker. In game theory, this latter simplification is abandoned; the behavior of any one player may well affect the payoff functions of all others.

The games studied in this chapter reveal many of the basic insights to be gained from the study of noncooperative games. They have many valuable direct applications and, additionally, provide a basis for and introduction to the models in Chapters 4 and 5. Indeed, as the chapters on cooperative games reveal, noncooperative game theory can aid in the analysis of cooperative games.

Section 1 of this chapter contains basic definitions, assumptions, and rules used throughout the chapter, as well as in Chapters 4 and 5. In Section 2, a general n-person noncooperative game is formulated and it is shown that such games have equilibrium points. In Section 3, *strictly competitive two-person games* are examined. These are games in which the interests of the two players are strictly opposed in the sense that anything that benefits one player is necessarily detrimental to the other. The *saddle point equilibrium* for two-person zero-sum games has some special properties that carry over to strictly competitive games but that do not carry over to two- or more-person noncooperative games in general. In general, equilibrium is not unique; however, conditions ensuring uniqueness are given in Section 4. Section 5 deals with noncooperative games under incomplete information; that is, games in which a player does not know the payoff functions of the other players. Harsanyi's (1967, 1968a, 1968b) work in this very difficult area is followed. In Section 6, several examples applying noncooperative games are examined. Section 8 contains concluding comments.

1 Basic concepts for noncooperative games

This section contains definitions of the basic concepts needed in the chapter. In addition to defining notation and various terms, assumptions usually made in noncooperative game theory are also stated. The material of this section provides a basic framework from which all other sections draw heavily. The following concepts and notation recur throughout this chapter:

$N = \{1, \ldots, n\}$ *is the* **set of players.**

S_i *is the* **strategy space of player** i *and, for this chapter, is a subset of the Euclidean space* R^m.

$S = S_1 \times \cdots \times S_n$ *is the Cartesian product of the individual strategy spaces and is the* **strategy space of the game.**

$s_i \in S_i$ *denotes a* **strategy of player** i, *and is therefore an element of* S_i.

$s = (s_1, \ldots, s_n) \in S$ *is called a* **combination**, *or more formally, a* **strategy combination**, *and consists of n strategies, one for each player.*

P_i *is the* **payoff function of player** i *and is scalar valued.*

$P(s) = (P_1(s), \ldots, P_n(s)) \in R^n$ *is the* **payoff vector.**

Recall that $s \backslash t_i$ denotes $(s_1, \ldots, s_{i-1}, t_i, s_{i+1}, \ldots, s_n)$, the combination s with t_i substituted in place of s_i.

An **equilibrium point** is a combination, s^*, that is feasible (i.e., is contained in S) and for which each player maximizes his own payoff with respect to his own strategy choice, given the strategy choices of the other players. More formally:

DEFINITION 3.1 *An* **equilibrium point** *is a combination* $s^* \in S$ *that satisfies* $P_i(s^*) \geq P_i(s^*\backslash s_i)$ *for all* $s_i \in S_i$ *and for all* $i \in N$.

The equilibrium point was introduced by Nash (1951). The saddle point equilibrium of von Neumann (1928) is a special instance of it, as is the Cournot (1838) equilibrium in oligopoly theory.

Many models considered in the next three sections adopt the following set of common assumptions: the strategy sets are both compact and convex, the payoff functions are defined, continuous, and bounded on S, and each payoff function, P_i, is either concave or quasiconcave with respect to s_i. Concave functions of one variable are illustrated in Figure 3.1.

A function $y = f(x)$ *is* **concave** *if, for any* x^1 *and* x^2 *in the domain of the function, and any scalar* $\lambda \in [0, 1]$

$$f[\lambda x^1 + (1 - \lambda)x^2] \geq \lambda f(x^1) + (1 - \lambda)f(x^2) \tag{3.1}$$

Quasiconcave functions of one variable are illustrated in Figure 3.2. A common example of a quasiconcave function is the utility function under standard assumptions, including in particular *convex preferences*. Convex preferences imply that the set of commodity bundles preferred or indifferent to a given bundle is a convex set. Translated into a restriction on utility functions, this is precisely quasiconcavity. (See standard references, such as Debreu (1959) or Varian (1984).) Of course, any concave function is quasiconcave, but the converse does not hold.

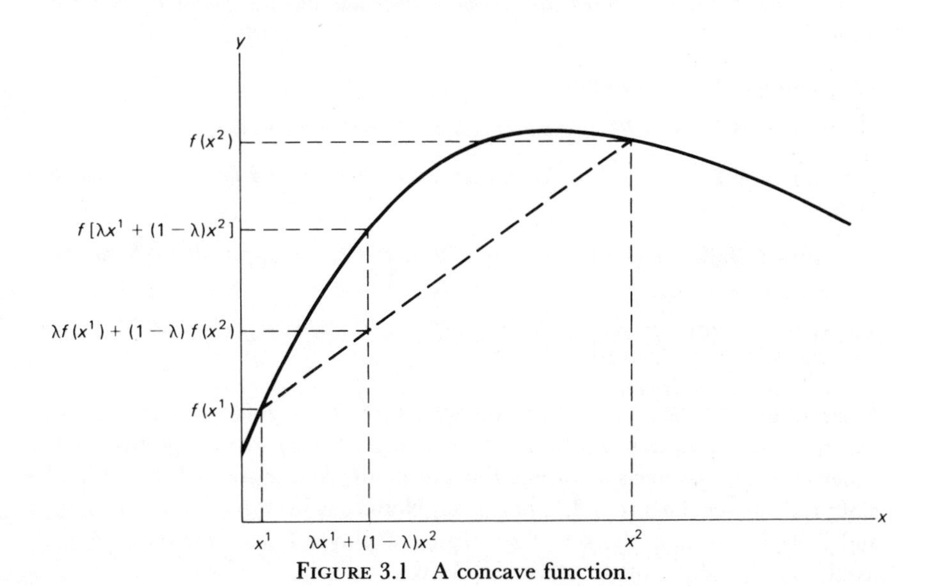

FIGURE 3.1 A concave function.

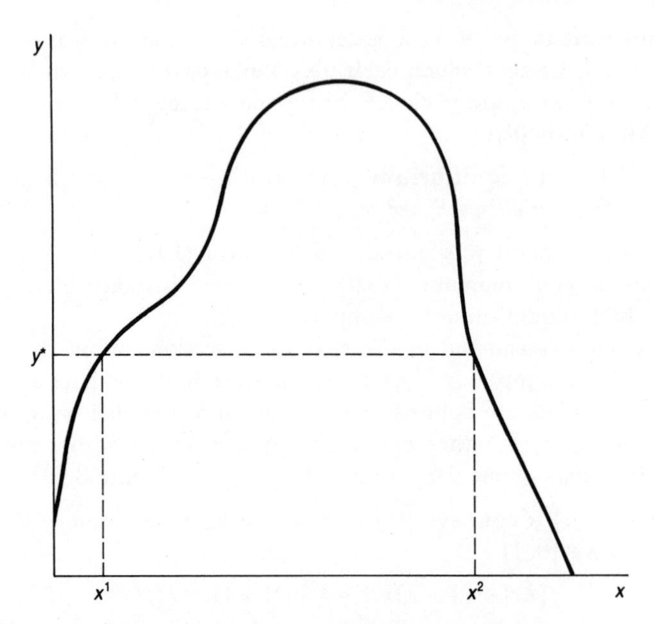

FIGURE 3.2 A quasiconcave function.

A function $y = f(x)$, defined on a domain D, is **quasiconcave** *if, for any x, $x' \in D$ such that $f(x) = f(x')$ and any $\lambda \in [0, 1]$,*

$$f[\lambda x + (1 - \lambda)x'] \geq f(x) \tag{3.2}$$

A **compact set** *in R^n is a set that is both* **closed** *(i.e., contains its own boundary) and* **bounded** *(i.e., can be contained within a ball of finite radius). A* **convex set** *has the property that the straight line segment connecting any two points in the set is also in the set.*

The assumptions are as follows:

ASSUMPTION 3.1 *$S_i \subset R^m$ is compact and convex for each $i \in N$.*

ASSUMPTION 3.2 *$P_i(s) \in R$ is defined, continuous, and bounded for all $s \in S$ and all $i \in N$.*

ASSUMPTION 3.3C *$P_i(s \backslash t_i)$ is concave with respect to $t_i \in S_i$ for all $s \in S$ and all $i \in N$.*

ASSUMPTION 3.3Q *$P_i(s \backslash t_i)$ is quasiconcave with respect to $t_i \in S_i$ for all $s \in S$ and all $i \in N$.*

Assumptions 3.1–3.3Q pertain to the structure of the game. There are two versions of Assumption 3.3 because each is needed at various points later. To ensure the existence of an equilibrium point, Assumptions 3.1, 3.2, and 3.3Q suffice (see Theorem 3.1, page 72). Note that both Assumptions 3.3Q and 3.3C imply that the set of strategies of player i that maximize P_i for fixed s_j $(j \neq i)$ of the other players is convex.

Under Assumption 3.3Q the strategy sets S_i should be regarded as sets of pure strategies. Under Assumption 3.3C, the S_i could consist of either pure or mixed strategies. Two salient points should be noted here. First, if (N, S, P) is a finite game, then Assumption 3.3C is automatically satisfied. This is saying that P_i is necessarily a concave function of s_i when P_i is given by equation (2.4). The reader might prove this as an exercise. The second point is that a player i always has an optimal strategy in his pure strategy set when S_i is the set of pure strategies and either Assumption 3.3C or Assumption 3.3Q holds. There is no need to bother with mixed strategies in such circumstances, and, if strict concavity or strict quasiconcavity holds, a player's best choice is never a mixed strategy. Additional conditions relate to the rules of the game and to the information conditions.

RULE 3.1 *The players are not able to make binding agreements.*

RULE 3.2 *The strategy choice made by each player can be regarded as having been made prior to the beginning of the play of the game, and without prior knowledge of the strategy choices made by other players.*

DEFINITION 3.2 *A* **game of complete information** *is a game in which each player i knows all the strategy sets S_j and each knows all payoff functions P_j, $j \in N$.*

Rule 3.1 defines a noncooperative game. In a game $\Gamma = (N, S, P)$, the conditions given in Definition 3.1 define an equilibrium point whether or not the rules of the game permit binding agreements. That is, a game may be cooperative and, at the same time, have noncooperative equilibria. Indeed, the strategy sets in a cooperative game in strategic form would incorporate the possibilities for agreement (see Rubinstein (1982) and Chapter 6, Section 4). Rule 3.2 states that players may be thought of as choosing their strategies simultaneously; however, this places no restriction on the structure of the game but only underscores the definition of *strategy*. For example, it does not mean that a particular move of a player takes no account of the unknown past moves of other players. Quite the contrary, even at the start of a game, a player can anticipate the whole array of situations in which he might find himself at, say, his fifth move, and his choice of a fifth move can be different according to which of these situations actually takes place. All of this is easily spelled out prior to the start of play.

In a game satisfying Assumptions 3.1 and 3.2, the *set of attainable payoffs*, $H = \{P(s) \mid s \in S\}$, is necessarily compact. Related to this set is its upper frontier, the *Pareto optimal set*, also called the *payoff possibility frontier*. Formally, it is defined by $H^* = \{y \in H \mid z \in H$ implies $y_i \geqslant z_i$ for at least one $i \in N\}$. That is, the payoff vector y is Pareto optimal if there is no other attainable payoff vector that gives a higher payoff to each player.

Definition 3.2 provides an opportunity to raise an important issue that largely escaped notice until relatively recently; namely, what is *common knowledge* in a game. My purpose here is to give a heuristic sense of its meaning and then to rely on that heuristic sense to make a specific common knowledge assumption that will be maintained, except where specific notice to the contrary is given. Suppose (N, S, P) is a two-person game of complete information. This game could occur under many information situations, a few of which are partially sketched here. Situation 1: Player 1 does not know whether player 2 has complete information. Situation 2: Player 1 knows that player 2 has complete information, but player 1 does not know whether player 2 knows that player 1 knows that player 2 has complete information. Situation 3: Player 1 knows that player 2 has complete information. Player 1 knows that player 2 knows that player 1 possesses this knowledge, but player 1 does not know whether player 2 knows that player 1 knows that player 2 knows that player 1 knows that player 2 has complete information. These situations differ materially and Definition 3.2 does not address them. Suppose that the players of a game were in a room together and an official entered carrying complete copies of (N, S, P), one copy for each player. In the presence of all other players, each player sees that the n copies are identical. Each player then receives a copy in full view of all others. We conclude that each player knows that all players have complete information, that all players know that all players have complete information, that all players know that all players know that all players have complete information, and so forth. This *knowing* proceeds to any arbitrary number of layers deep. Such infinitely deep knowing constitutes common knowledge of complete information.

If Definition 3.2 does not hold, then the game is one of *incomplete information.* Suppose, for example, that each player knows all the S_j, but only player i knows P_i. This information would be common knowledge if each player knew that each player i knew all the S_j, but only her own P_i, that each player knew that each player knew, and so forth.

2 Existence of equilibrium points for n-person noncooperative games

The principal purpose of this section is to prove that all n-person noncooperative games of complete information that satisfy Assumptions 3.1, 3.2, and 3.3Q have equilibrium points. Throughout this section, any game $\Gamma = (N, S, P)$ is understood to satisfy all the preceding conditions and is referred to simply as a noncooperative game.[1] The first existence proof for equilibrium points in n-person noncooperative games is due to Nash (1951). The concept of equilibrium point is a natural, although not obvious, generalization of von Neumann's (1928) saddle point equilibrium for zero-sum games. In his generalization, Nash had the insight to abandon the minimax approach of von Neumann. In von Neumann's context, a player who strives to increase her own payoff is necessarily striving to decrease the payoff of the other player. Outside two-person strictly competitive games, this is not true in general; hence, for player i to act as if other players in an n-person noncooperative game wish to minimize her payoff is usually incorrect and excessively pessimistic. As the

Cournot oligopoly example in Section 6 makes clear, Nash was generalizing the Cournot (1838) oligopoly equilibrium as well as generalizing von Neumann's (1928) saddle point equilibrium. Nash dealt with finite games; important generalizations of the Nash model appear in Nikaido and Isoda (1955) and Berge (1957).

Even the reader who is totally uninterested in existence proofs is urged to read enough of this section to become acquainted with *best-reply mappings* and their relationship to equilibrium points. The best-reply concept lends a great intuitive appeal to the method of proof, and, more importantly, is very useful and widely used. It is introduced and discussed next, after which existence of equilibrium points is proved. The existence argument is that any noncooperative game has at least one equilibrium point; however, nothing is said concerning either the number of equilibrium points or how such equilibria might be related. Without making further restrictions on the model, nothing more can be said.

2.1 Best-reply mappings and their relationship to equilibrium points

The best-reply mapping might better be called the optimal strategy mapping; however, the former usage is firmly entrenched and is followed here. Suppose player i performs the following thought experiment. He contemplates a particular strategy assignment to the other players—$(s'_1, s'_2, \ldots, s'_{i-1}, s'_{i+1}, \ldots, s'_n)$—and wonders what strategy choice on his own part would maximize P_i given the $s'_j\,(j \neq i)$. It is this thought experiment that defines the best-reply mapping for player i. In Definition 3.3, the best-reply mapping is defined as a mapping from S (rather than from $S_1 \times \cdots \times S_{i-1} \times S_{i+1} \times \cdots \times S_n$) to subsets of S_i. This is merely a technical convenience. The best replies to s are the same as the best replies to any $s \backslash t_i$ for player i.

DEFINITION 3.3 *The* **best-reply mapping for player i** *is a set-valued relationship associating each strategy combination $s \in S$ with a subset of S_i according to the following rule*: $r_i(s) = \{t_i \in S \mid P_i(s \backslash t_i) = \max_{s'_i \in S_i} P_i(s \backslash s'_i)\}$.

The strategy t_i is a best reply for player i to the strategy combination s if t_i maximizes the payoff of player i, given the strategy choices of the others. In general, t_i need not be unique. If P_i were *strictly* quasiconcave in s_i, then $r_i(s)$ would be just one point. The strategy combination $t \in S$ is a (joint) best reply to $s \in S$ if each component, t_i, of t is a best reply for player i.

DEFINITION 3.4 *The* **best-reply mapping** *is a set-valued relationship associating each strategy combination $s \in S$ with a subset of S according to the rule $t \in r(s)$ if and only if $t_i \in r_i(s)$, $i \in N$. Thus, $r(s)$ is the Cartesian product $r_1(s) \times r_2(s) \times \cdots \times r_n(s)$.*

The best-reply mapping provides a natural way to think about equilibrium points because all equilibrium points satisfy the condition that

s^* is an equilibrium point if and only if $s^* \in r(s^*)$. That is, an equilibrium point is a best reply to itself, and any strategy combination that is a best reply to itself is an equilibrium point. Thus,

LEMMA 3.1 *Let $\Gamma = (N, S, P)$ be a noncooperative game. $s \in S$ is an equilibrium point of Γ if and only if $s \in r(s)$.*

Proof. First it is shown that if s^* is an equilibrium point, then $s^* \in r(s^*)$. Recall the definition of equilibrium point. s^* is an equilibrium point if $s^* \in S$ and $P_i(s^*) = \max_{s_i \in S_i} P_i(s^* \backslash s_i)$, $i \in N$. The latter condition is precisely the condition for $s_i^* \in r_i(s^*)$.

Now suppose that $s^* \in r(s^*)$. To see that s^* must be an equilibrium point, refer to the definition of the best-reply mapping: $s_i^* \in r_i(s^*)$ if $P_i(s^*) = \max_{s_i \in S_i} P_i(s^* \backslash s_i)$, $i \in N$. But the latter condition states that no player could achieve a greater payoff by using a different strategy, given the strategies of the other players. This, of course, is what defines an equilibrium point. QED

2.2 Fixed points of functions and correspondences

When $s \in r(s)$, s is called a **fixed point** of r, and Lemma 3.1 can be restated by saying that the set of fixed points of r coincides with the set of equilibrium points. This allows the question of existence of equilibrium to be stated in terms of the existence of fixed points of the best-reply mapping r. Before this is directly addressed, some facts about set-valued mappings are stated, and a theorem relating to the existence of fixed points is introduced. *For mappings from R^n to R^m it is customary in the economics literature to call a set-valued mapping, such as the best-reply mapping, a* **correspondence** *and to save the word* **function** *for mappings that associate a point in R^m with a point in the domain.* For example, the payoff mappings, P_i, are functions. This custom is followed here.

DEFINITION 3.5 *Let A be a subset of R^n and B be a compact subset of R^m. Suppose the correspondence ϕ is defined for any $x \in A$ and that $\phi(x) \subset B$ and is closed. Then ϕ is* **upper semicontinuous** *if $y^0 \in \phi(x^0)$ whenever* (a) $x^0 \in A$, (b) $x^k \in A$, $k = 1, 2, \ldots$, (c) $\lim_{k\to\infty} x^k = x^0$, (d) $y^k \in \phi(x^k)$, $k = 1, 2, \ldots$, *and* (e) $\lim_{k\to\infty} y^k = y^0$.

Thus, a correspondence ϕ is upper semicontinuous if, for any convergent sequence in the domain of ϕ whose limit is also in the domain of ϕ, and any convergent companion sequence of points, y^k in the image sets $\phi(x^k)$, the limit of the companion sequence, y^0, is in the set $\phi(x^0)$. Examples are shown in Figure 3.3. Note that in Figure 3.3*c* the sets $\phi(x)$ are always convex, whereas, in Figure 3.3*a* and *b*, they are not. Convexity of the image sets, $\phi(x)$, is required for the following remarkable theorem that gives conditions guaranteeing existence of a fixed point.

KAKUTANI (1941) FIXED-POINT THEOREM *Let $\phi(x)$ be an upper semicontinuous correspondence, defined for $x \in A \subset R^n$, with $\phi(x) \subset A$ for all $x \in A$. If A is compact and convex and, for all $x \in A$, $\phi(x)$ is nonempty and convex, then ϕ has a fixed point. That is, there is $x^* \in \phi(x^*)$ for some $x^* \in A$.*

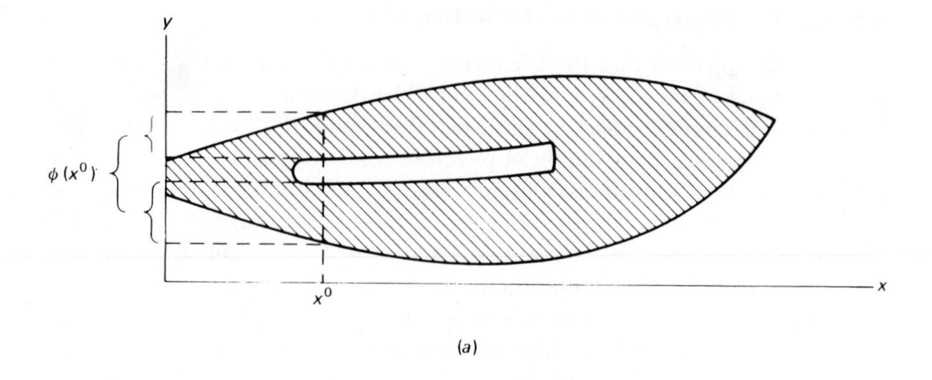

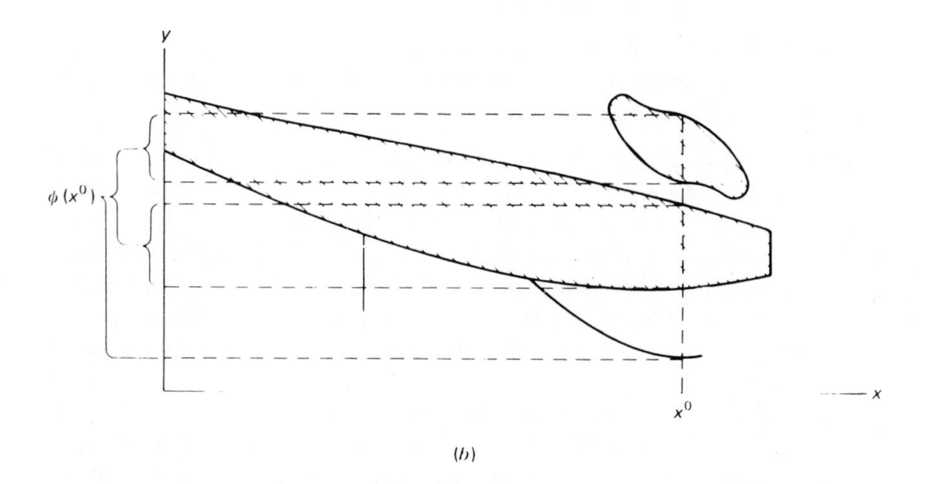

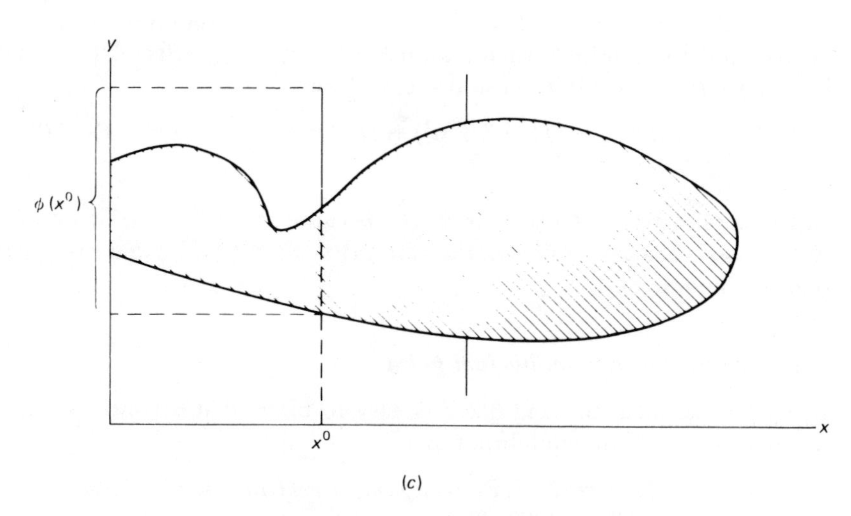

FIGURE 3.3 Upper semicontinuous correspondences.

2.3 Continuity properties of the best-reply mapping

It is readily apparent that the best-reply correspondence r satisfies many of the conditions of the Kakutani theorem. The domain of r is S, which is a compact and convex subset of the Euclidean space R^{nm}. $r(s)$ is nonempty for each s because $r_i(s)$ is the set of payoff maximizing values of s_i (given the s_j, $j \neq i$), and a continuous function (the payoff function) defined on a compact set (S_i) must achieve a maximum on that set. Furthermore, $r(s)$ is convex and contained in S for all $s \in S$. Convexity of $r(s)$ follows from two facts. First, each $r_i(s)$ is convex due to the quasiconcavity of P_i in s_i. Second, the Cartesian product of convex sets is convex. That $r(s) \subset S$ follows from the very definition of $r(s)$. The one requirement that is not obviously satisfied is upper semicontinuity of r; however, that is proved in Lemma 3.2. The other properties are established in Theorem 3.1.

LEMMA 3.2 *Let* $\Gamma = (N, S, P)$ *be a noncooperative game of complete information that satisfies Assumptions* 3.1, 3.2 *and* 3.3Q. *The best-reply correspondence of* Γ, $r: S \rightarrow S$, *is upper semicontinuous.*

Proof. Let (a) $\{s^k\}_{k=1}^{\infty} \subset S$ be any convergent sequence of points in S, (b) s^0 be the limit of the sequence, (c) $t^k \in r(s^k)$, $k = 1, 2, \ldots$ be a convergent companion sequence of points, and (d) t^0 be the limit of the $\{t^k\}_{k=1}^{\infty}$. $r(s)$ is upper semicontinuous if, for any such $\{s^k\}_{k=1}^{\infty}$, s^0, $\{t^k\}_{k=1}^{\infty}$, and t^0, the limit point of the companion sequence is in the image set of the limit of the original sequence. That is, $t^0 \in r(s^0)$. The lemma is proved by showing that, for arbitrary i, $t_i^0 \in r_i(s^0)$, and noting that the same argument applies for all $i \in N$.

Select an element of $r_i(s^0)$ and denote it t_i'. Then $P_i(s^0\backslash t_i^0) > P_i(s^0\backslash t_i')$ is impossible, for it contradicts the best-reply property of t_i'. If $P_i(s^0\backslash t_i^0) = P_i(s^0\backslash t_i')$, then $t_i^0 \in r_i(s^0)$; therefore, it remains to show that $P_i(s^0\backslash t_i^0) < P_i(s^0\backslash t_i')$ is impossible. To see this, assume for the moment that $P_i(s^0\backslash t_i^0) < P_i(s^0\backslash t_i')$ and let $P_i(s^0\backslash t_i') - P_i(s^0\backslash t_i^0) = \varepsilon > 0$. From the continuity of P_i, for any $\delta > 0$ there is finite k_δ such that, if $k > k_\delta$, $|P_i(s^k\backslash t_i^k) - P(s^0\backslash t_i^0)| < \delta$ and $|P_i(s^k\backslash t_i') - P_i(s^0\backslash t_i')| < \delta$. Choose $\delta < \varepsilon/4$. Then

$$P_i(s^k\backslash t_i') > P_i(s^0\backslash t_i') - \varepsilon/4 > P_i(s^0\backslash t_i') - 3\varepsilon/4 = P_i(s^0\backslash t_i^0) + \varepsilon/4 > P_i(s^k\backslash t_i^k) \tag{3.3}$$

which implies that $t_i^k \notin r_i(s^k)$ for $k > k_\delta$ because $P_i(s^k\backslash t_i') - P_i(s^k\backslash t_i^k) > \varepsilon/2$. But $t_i^k \notin r_i(s^k)$ is a contradiction; therefore, $P_i(s^0\backslash t_i^0) = P_i(s^0\backslash t_i')$ and r is upper semicontinuous. QED

2.4 Existence of an equilibrium point

In view of Lemmas 3.1 and 3.2 it is easy to prove that a noncooperative n-person game has an equilibrium point.

THEOREM 3.1 *Let* $\Gamma = (N, S, P)$ *be a game of complete information that satisfies Assumptions* 3.1, 3.2, *and* 3.3Q. Γ *has at least one equilibrium point.*

Proof. By Lemma 3.1, the set of equilibrium points of Γ is identical with the set of fixed points of the best-reply correspondence, $r(s)$; therefore, the theorem is proved if it can be shown that r has a fixed point (i.e., that there is some $s^* \in S$ for which $s^* \in r(s^*)$). This is established by the Kakutani fixed point theorem if $r(s)$ (a) is a correspondence whose domain, S, is compact and convex, (b) is defined for all $s \in S$, (c) has image sets, $r(s)$, that are contained in S for all s, (d) has convex image sets, and (e) is upper semicontinuous. (a) is satisfied by Assumption 3.1 and the fact that the Cartesian product of n convex sets is convex. To see (b) and (c), note that for any $s \in S$

$$r_i(s) = \{t_i \in S_i \mid P_i(s \backslash t_i) \geqslant P_o(s \backslash t_i') \text{ for all } t_i' \in S_i\} \tag{3.4}$$

Such t_i exist because they are the maximizers of a continuous function (P_i) defined over a compact set (S_i), and, by construction, they are in S_i. If $t \in r(s)$, then $t_i \in r_i(s)$, $i \in N$; therefore, $r(s)$ is defined for all $s \in S$ and $r(s) \subset S$. (d) follows easily from assumption 3.3Q. Suppose that t_i, $t_i' \in r_i(s)$. Then for $\lambda \in [0, 1]$, let $t_i^\lambda = \lambda t_i + (1-\lambda)t_i'$, and recall that concavity of P_i in s_i means that $P_i(s \backslash t_i^\lambda) \geqslant \lambda P_i(s \backslash t_i) + (1-\lambda)P_i(s \backslash t_i')$. Strict inequality implies that t_i, $t_i' \notin r_i(s)$, a contradiction; hence, equality holds and $t_i^\lambda \in r_i(s)$. Therefore, $r_i(s)$ is convex for $i \in N$ and $r(s)$, being the Cartesian product of convex sets, is also convex. (e) is established by Lemma 3.2, so the theorem is proved. Γ has an equilibrium point. QED

Theorem 3.1 is an existence theorem that sheds no light on the multiplicity of equilibrium, and, the possibility of multiple, separated equilibria is illustrated by the following infinite game. Let $N = \{1, 2\}$, $S_1 = S_2 = [-150, 150]$, $P_1(s) = 2{,}720{,}000s_1 - 33{,}600s_1s_2 - s_1^4$, and $P_2(s) = 2{,}720{,}000s_2 - 33{,}600s_1s_2 - s_2^4$. Equilibrium points of this game include (20, 80), (80, 20), (−59.139, 105.576), (105.576, −59.139), and (57.875, 57.875)[2]. Uniqueness of equilibrium is addressed in Section 4.

2.5 Examples of and comments on computing equilibrium points

The definition of equilibrium provides criteria for recognizing whether a given strategy combination is an equilibrium point and hints at techniques that can aid the computation of equilibria. Look again at the example in Section 2.4, which has at least five equilibrium points. An obvious technique for locating these equilibria is to start with the conditions

$$\frac{\partial P_1}{\partial s_1} = 2{,}720{,}000 - 33{,}600s_2 - 4s_1^3 = 0 \tag{3.5}$$

$$\frac{\partial P_2}{\partial s_2} = 2{,}720{,}000 - 33{,}600s_1 - 4s_2^3 = 0. \tag{3.6}$$

If the payoff functions are differentiable, then any *interior* equilibrium will satisfy conditions analogous to equations (3.5) and (3.6). A good procedure is thus to attempt to solve such equations. If a solution falls within the domain, S, then such a solution is an equilibrium. This particular game has another feature that can be exploited: symmetry. The symmetry of the payoff functions means that if (80, 20) is an equilibrium, then (20, 80) is also an equilibrium. Symmetry also invites searching for symmetric equilibria, such as (57.875, 57.875). This is done by setting $s_1 = s_2$ in equation (3.5) and then solving for s_1.

Consider now a variant of this game with the same players and payoff functions, but with $S_1 = S_2 = [-100, 90]$. Even though $(-59.139, 105.576)$ solves equations (3.5) and (3.6), it is not an equilibrium point because 105.576 is not in the strategy set of player 2. A good guess is that $s_2 = 90$ will work with player 1 choosing a best reply to it. The best-reply function of player 1 in this game is

$$\left.\begin{aligned} s_1 &= 90 \text{ for } s_2 \in [-100, -5.833) \\ &= (680{,}000 - 8{,}400 s_2)^{1/3} \text{ for } s_2 \in [-5.833, 90] \end{aligned}\right\} \tag{3.7}$$

The best reply for player 1 to $s_2 = 90$ is $s_1 = -42.358$. Conversely, the best-reply function for player 2 is

$$\left.\begin{aligned} s_2 &= 90 \text{ for } s_1 \in [-100, -5.833) \\ &= (680{,}000 - 8{,}400 s_2)^{1/3} \text{ for } s_1 \in [-5.833, 90] \end{aligned}\right\} \tag{3.8}$$

so $s^* = (-42.358, 90) = r(-42.358, 90)$ is an equilibrium point, as is $(90, -42.358)$. Note that the best-reply mapping is written here as a function because it is single valued for each $s \in S$.

For another example, suppose again that $N = \{1, 2\}$ and that the payoff functions are

$$P_1(s) = 100 s_{11} s_{22} - 200 s_{12} s_{22} - 15 s_{11}^2 + 2 s_{11} s_{12} s_{21} - 10 s_{12}^2 \tag{3.9}$$

$$P_2(s) = 50 s_{21} s_{12} + 100 s_{22} - 10 s_{21}^2 - 15 s_{21} s_{22} - 10 s_{22}^2 \tag{3.10}$$

and the strategy sets are

$$S_1 = \{s_1 \in R^2 \mid s_{11} \in [-20, 5] \text{ and } s_{12} \in [-10, 5]\} \tag{3.11}$$

$$S_2 = \{s_2 \in R^2 \mid s_{21} \in [-5, 12] \text{ and } s_{22} \in [-5, 8]\} \tag{3.12}$$

This game satisfies Assumptions 3.1, 3.2, and 3.3C. Compactness and convexity of S_1 and S_2 are self-evident, as is continuity of the payoff functions on S. To check concavity of P_1 on S_1 and of P_2 on S_2 we must examine the second derivatives. For player 1,

$$\left\| \frac{\partial^2 P_1}{\partial s_{11} \partial s_{1j}} \right\| = \begin{bmatrix} -30 & 2s_{21} \\ 2s_{21} & -20 \end{bmatrix} \tag{3.13}$$

P_1 is concave in s_1 if the matrix in equation (3.13) is negative semidefinite, which is implied by the principal diagonal elements being nonpositive and by the determinant being nonnegative. These conditions are all met: $-30 < 0$, $-20 < 0$, and $600 - 4s_{21}^2 > 0$ for $s_{21} \in [-5, 12]$. Note that concavity of P_1 depends on its domain. For example, if s_{21} took values anywhere in $[-5, 20]$ P_1 would no longer be concave everywhere on its domain. A similar exercise for P_2 shows it to be concave in s_2.

An equilibrium for this game can be found by simultaneously solving[3]

$$\frac{\partial P_1}{\partial s_{11}} = P_1^1 = 100s_{22} - 30s_{11} + 2s_{12}s_{21} = 0 \tag{3.14}$$

$$\frac{\partial P_1}{\partial s_{12}} = P_1^2 = -200s_{22} + 2s_{11}s_{21} - 20s_{12} = 0 \tag{3.15}$$

$$\frac{\partial P_2}{\partial s_{21}} = P_2^3 = 50s_{12} - 20s_{21} - 15s_{22} = 0 \tag{3.16}$$

$$\frac{\partial P_2}{\partial s_{22}} = P_2^4 = 100 - 15s_{21} - 20s_{22} = 0 \tag{3.17}$$

An equilibrium is $(s_1, s_2) = (1.45, 2.71, 6.89, -.18)$.

As a final example, let $N = \{1, 2\}$, let the payoff functions be given by equations (3.9) and (3.10), and let the strategy sets be $S_1 = \{s_1 \in R^2 \mid (5, 5) \leqslant s_1 \leqslant (10, 10)\}$ and $S_2 = \{s_2 \in R^2 \mid (5, 5) \leqslant s_2 \leqslant (10, 10)\}$. Equations (3.14) to (3.17) can be used to find an equilibrium even though the strategy combination $(1.45, 2.71, 6.89, -.18)$ is not in the new strategy space. From equation (3.14) it can be seen that $P_1^1 > 0$ for all possible s, which means that only $s_{11} = 10$ is compatible with equilibrium. This is most easily seen by choosing s_{11}, s_{12}, s_{21}, and s_{22} to minimize P_1^1 (i.e., choose $(10, 5, 5, 5)$). Doing the same with equations (3.15) and (3.17) shows that $P_1^2 < 0$ and $P_2^4 < 0$ for all $s \in S$; therefore, only $s_{12} = 5$ and $s_{22} = 5$ are compatible with equilibrium. Finally, s_{21} can be solved using equation (3.16) with the known values of s_{12} and s_{22}. This yields $s_{21} = 8.75$ and equilibrium is $s = (10, 5, 8.75, 5)$. This equilibrium is unique, because the values found for s_{11}, s_{12}, s_{21}, and s_{22} were shown by construction to be the only values compatible with equilibrium. The reader should realize that the processes sketched above to find equilibria involve some trial and error and are not guaranteed to locate equilibria in all games.

2.6 Other results on existence of equilibrium

This section is divided into two subsections, the first of which contains an existence theorem for equilbrium in *pseudogames*. A pseudogame is exactly like a game except that a player's payoff function will not, in general, be defined on all of S. In Section 2.6.2 other existence theorems are cited.

2.6.1 Equilibrium in pseudogames

Related to a game is something that could be termed a *pseudogame*, characterized by $(N, S, P, T_1, \ldots, T_n)$. N, S, and P have their familiar meanings and each $T_i \subset S$. T_i is the subset of S on which P_i is defined. A pseudogame is a game if $T_i = S$ for all $i \in N$. Debreu (1952) proves a corollary to Theorem 3.1 establishing existence of equilibrium for a class of pseudogames. The assumptions require some slight modification and a new condition must be added that place some restrictions on the T_i. Let $\tau_i(s) = \{t_i \in S_i \mid s\backslash t_i \in T_i\}$. Then P_i is defined for $s \in S$ if and only if $s_i \in \tau_i(s)$. Similarly, P is defined for $s \in S$ if and only if $s_i \in \tau_i(s)$ for all $i \in N$.

ASSUMPTION 3.2′ *$P_i(s)$ is defined, continuous, and bounded for all $s \in T_i \subset S$ and all $i \in N$. For each $s \in S$ and $i \in N$, the set $\tau_i(s)$ is nonempty, continuous, closed, and convex.*

Figure 3.4 illustrates the restrictions placed on the sets T_i. Each T_i is closed, and for any choice of $s_j \in S_j$ $(j \neq i)$, there is a (nonempty) set of strategies s_i from which player i can choose. This set is closed because T_i is closed, and it is also convex. Note, however, that T_i is not required to be convex, as Figure 3.4 illustrates. Assumption 3.3Q must be modified in the light of Assumption 3.2′.

ASSUMPTION 3.3Q′ *$P_i(s\backslash t_i)$ is quasiconcave with respect to $t_i \in \tau_i(s)$ for all $s \in S$ and all $i \in N$.*

COROLLARY *Let $\Gamma = (N, S, P, T_1, \ldots, T_n)$ be a pseudogame of complete information that satisfies Assumptions* 3.1, 3.2′, *and* 3.3Q. *Γ has at least one equilibrium point.*

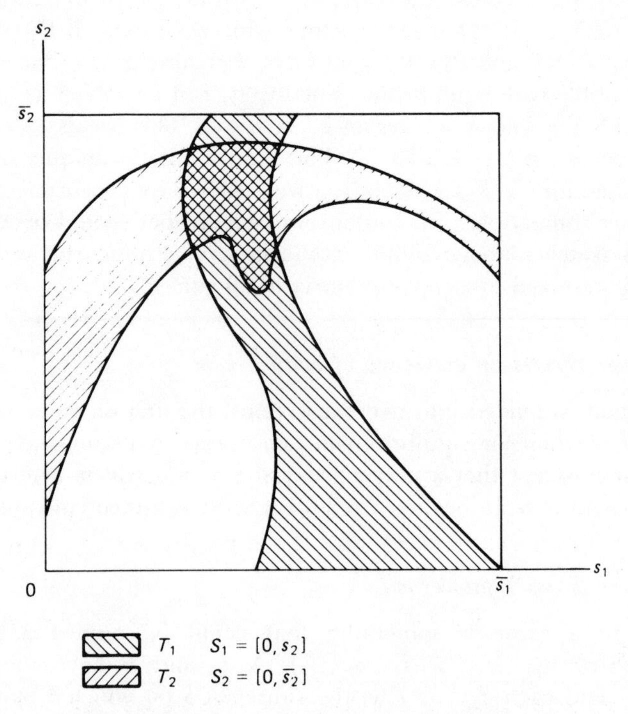

FIGURE 3.4 Strategy combinations for which payoff functions are defined

Proof. To prove the corollary, the best-reply mapping must be correctly defined. To that end, let

$$r_i(s) = \{t_i \in \tau_i(s) \mid P_i(s \backslash t_i) \geqslant P_i(s \backslash t_i') \text{ for all } t_i' \in \tau_i(s)\}, \; s \in S, \; i \in N \quad (3.18)$$

$r(s) = \times_{i \in N} r_i(s)$. $r(s)$ is defined for all s because $\tau_i(s)$ is nonempty and closed for all i and s; hence, a fixed point of r is an element of $\bigcap_{i \in N} T_i$. By argument parallel to that used in the proof of Theorem 3.1, $r(s)$ satisfies the Kakutani fixed-point theorem; hence, it has a fixed point that is an equilibrium point of the game. QED

Note that, of necessity, the equilibrium point, s^*, satisfies $s_i^* \in \tau_i(s^*)$, so that $s^* \in T_i$ for all i.

As a simple example of a pseudogame, imagine a Cournot duopoly in which the firms face the inverse demand function $p = 100 - q_1 - q_2$ and costs are nil for each firm. The inverse demand function is defined on $D = \{q \in R_+^2 \mid q_1 + q_2 \leqslant 100\}$. Then $P_i(q) = 100q_i - q_i^2 - q_1 q_2$ $(i = 1, 2)$, $S_1 = S_2 = [0, 100]$ (so that $S = \{q \in R_+^2 \mid q_1 \leqslant 100 \text{ and } q_2 \leqslant 100\}$ and $T_1 = T_2 = D$. It is natural for the strategy set of each player to be [0, 100], yet for some $q \in S$ payoffs will not be defined because a market clearing price is not defined when aggregate output exceeds 100. This pseudogame satisfies the conditions of the corollary and has an equilibrium.

This duopoly example can be modified to be a game by defining the inverse demand function more broadly as $\max\{0, p = 100 - q_1 - q_2\}$. Doing this implies a *free disposal* assumption for firms. From the standpoint of good modeling it is desirable to construct games rather than pseudogames whenever the underlying economics permits.

2.6.2 *Other existence theorems for games*

In Section 2.4 existence of equilibrium points was studied for a standard model satisfying Assumptions 3.1, 3.2, and 3.3Q. In addition, a model was examined for which the payoff functions was not defined on all of S. Incomplete information games are investigated in Section 5. Other interesting variations, not covered here, are games with an infinite number of players. Peleg and Yaari (1973) cover the case where the number of players is countably infinite and Schmeidler (1973) covers the uncountably infinite case. Another variation of the assumptions is obtained by relaxing the quasiconcavity requirement (see Nishimura and Friedman (1981)).

A final note here ties together Theorem 3.1 with some classic results. The first general existence theorem for equilibrium points in noncooperative games is that of Nash.

NASH (1951) THEOREM *Any finite game of complete information has an equilibrium point.*

It is left as an exercise for the reader to prove that Nash's theorem is a special case of Theorem 3.1.

3 Strictly competitive two-person games

Two-person zero-sum situations are rare in practice, since payoffs are utilities to the players and not money amounts. Consider a parlor game played for money in which the two players care only about how much money they win or lose and, apart from that, do not care about winning or losing as such. Suppose that the utility of each player is normalized so that $U_i(0)=0$, $i=1,2$, where $U_i(x_i)$ is the utility function of player i and x_i is the amount of money he wins. By hypothesis $x_2=-x_1$; however, it does not follow that $U_1(x_1)+U_2(-x_1)=0$. Assuming that $U_i(x_i)$ is strictly increasing in x_i, the salient feature of the game is that any change that helps one player will, of necessity, hurt the other. In this sense, the interests of the players are perfectly opposed. Such games are more easily imagined than zero-sum games. For example, in a two-candidate electoral race, each candidate wants to win, and by the largest margin possible; however, there is no reason to suppose the candidates' utilities are related by a zero- (or, equivalently, constant-) sum condition. Games of this variety, which include zero-sum games as a subclass, are called *strictly competitive games.*

3.1 Properties of equilibrium points for strictly competitive games

Two-person strictly competitive games are defined here in terms of Assumptions 3.1, 3.2, and 3.3Q. They are a subset of noncooperative games in which $n=2$ and $P_1(s')\geq P_1(s'')$ if and only if $P_2(s')\leq P_2(s'')$. Although some finite games are included among strictly competitive games, attention is not restricted to such games. After strictly competitive two-person games are formally defined, three of their interesting features are proved: (a) Any two equilibrium points, s' and s'', yield precisely the same payoffs. That is, $P_i(s')=P_i(s'')$, $i=1,2$. (b) If s' and s'' are equilibrium points, then (s'_1, s''_2) and (s''_1, s'_2) are also equilibrium points. Another way to state this condition is that each player i has a set of equilibrium strategies S_i^0 and any combination s for which $s_1\in S_1^0$ and $s_2\in S_2^0$ is an equilibrium point. Thus, the *set of equilibrium points* is $S^0=S_1^0\times S_2^0$. This is sometimes called the *interchangeability property*. (c) The set of equilibrium points is convex.

DEFINITION 3.6 *A* **strictly competitive game** *is a two-person game that satisfies Assumptions* 3.1, 3.2, *and* 3.3Q, *and in which, for all* $s, s'\in S$, $P_1(s)\geq P_1(s')$ *if and only if* $P_2(s)\leq P_2(s')$.

Theorem 3.2 establishes the properties (a) and (b) above: all equilibrium points yield the same payoffs and satisfy the interchangeability property.

THEOREM 3.2 *Let* $\Gamma=(N,S,P)$ *be a strictly competitive game of complete information and let* $S^0\subset S$ *be the set of equilibrium points of* Γ. *If* $s', s''\in S^0$ *then* $P_i(s')=P_i(s'')$, $i=1, 2$, $(s'_1, s''_2)\in S^0$, *and* $(s''_1, s'_2)\in S^0$.

Proof. Suppose that s', $s'' \in S^0$, and $P_1(s') \leqslant P_1(s'')$. Then $P_2(s'') \leqslant P_2(s')$ because the game is strictly competitive and $P_2(s_1'', s_2') \leqslant P_2(s'')$ because $P_2(s'') = \max_{s_2 \in S_2} P_2(s_1'', s_2)$. The latter inequality, along with the definition of a strictly competitive game, implies that $P_1(s_1'', s_2') \geqslant P_1(s'')$ and $P_1(s') = \max_{s_1 \in S_1} P_1(s_1, s_2')$ implies that $P_1(s') \geqslant P_1(s_1'', s_2')$; therefore, $P_1(s') \geqslant P_1(s'')$ and $P_1(s') \leqslant P_1(s'')$, which establishes that $P_i(s') = P_i(s'')$, $i = 1, 2$. Additionally, the latter, together with $P_1(s') \geqslant P_1(s_1'', s_2') \geqslant P_1(s'')$, implies that $(s_1'', s_2') \in S^0$. A parallel argument can be made to show that $(s_1', s_2'') \in S^0$. QED

Convexity of S^0, the set of equilibrium points, is established in Theorem 3.3. Lemma 3.3, which appears prior to the theorem, is used in its proof.

LEMMA 3.3 *Let* $\Gamma = (N, S, P)$ *be a strictly competitive game of complete information,* $s' \in S^0$ *and* $u \notin S^0$. *Then either* $\max_{s_1 \in S_1} P_1(s_1, u_2) > P_1(s')$ or $\max_{s_2 \in S_2} P_2(u_1, s_2) > P_2(s')$.

Proof. The lemma holds trivially for the case in which $P_1(s') \neq P_1(u)$, because either $P_1(s') < P_1(u)$ or $P_2(s') < P_2(u)$. Consider the remaining case, in which $P_i(s') = P_i(u)$, $i = 1, 2$. If $\max_{s_1 \in S_1} P_1(s_1, u_2) = P_1(s')$ and $\max_{s_2 \in S_2} P_2(u_1, s_2) = P_2(s')$, then $u \in S^0$, which is a contradiction; hence, the lemma follows. QED

THEOREM 3.3 *Let* $\Gamma = (N, S, P)$ *be a strictly competitive game of complete information whose set of equilibrium points is* S^0. *Then* S^0 *is convex.*

Proof. It suffices to show that, for any s', $s'' \in S^0$, $s' \neq s'$, and $\lambda \in [0, 1]$, that $s^\lambda = \lambda s' + (1 - \lambda)s''$ is an equilibrium point. Suppose that $s^\lambda \notin S^0$. Then, by Lemma 2.3, either $\max_{s_1 \in S_1} P_1(s_1, s_2^\lambda) = P_1(t_1, s_2^\lambda) > P_1(s') = P_1(s'')$ or a parallel condition holds for player 2. Without loss of generality, suppose the condition holds for player 1. Then $P_2(t_1, s_2^\lambda) < P_2(s') = P_2(s'')$. By quasiconcavity of P_2 in s_2, $P_2(t_1, s_2^\lambda) \geqslant \lambda P_2(t_1, s_2') + (1 - \lambda) P_2(t_1, s_2'')$; therefore, $P_2(t_1, s_2') \leqslant P_2(t_1, s_2^\lambda)$ or $P_2(t_1, s_2'') \leqslant P_2(t, s_2^\lambda)$. If the latter holds, then $P_1(t_1, s_2'') \geqslant P_1(t_1, s_2^\lambda) > P_1(s'')$, which implies that s'' is an not an equilibrium point and is a contradiction. If the former holds, then $P_2(t_1, s_2') \leqslant P_2(t_1, s_2^\lambda) < P_2(s')$, and a similar argument leads to a contradiction; thus S^0 is convex. QED

A lemma of Leininger, Thorlund-Petersen, and Weibull (1988) proves that any two-person strictly competitive game is equivalent to a two-person zero-sum game from a strategic standpoint. That is, for any $\Gamma = (N, S, P)$ that is two-person and strictly competitive there is an associated two-person zero-sum game $\Gamma_\phi = (N, S, P_1, \phi(P_2))$ having the property that s^* is an equilibrium point of Γ if and only if it is an equilibrium point of Γ_ϕ.

LEMMA 3.4 *Any strictly competitive game is equivalent to a two-person zero-sum game.*

Proof. For all $s \in S$ define $\phi(P_2(s)) = -P_1(s)$. Then $\phi(P_2(s)) \geqslant \phi(P_2(s'))$ if and only if $P_2(s) \geqslant P_2(s')$ because Γ is strictly competitive. Let s^* be an equilibrium point of Γ. Then $\phi(P_2(s^*)) \geqslant \phi(P_2(s_1^*, s_2))$ for all $s_2 \in S_2$. Conversely, if s^* were not an equilibrium point of Γ because s_2^* was not a best reply to s^* for player 2, then there would be some $s_2' \in S_2$ such that $P_2(s_1^*, s_2') > P_2(s^*)$ and $\phi(P_2(s_1^*, s_2')) > \phi(P_2(s^*))$ would also hold. QED

Theorem 3.3 is the von Neumann (1928) saddle point theorem, which can now be seen as a consequence of Theorems 3.1 and 3.2. From Theorem 3.1 existence of equilibrium is ensured. From Theorem 3.2, all equilibria of a finite two-person zero-sum game (N, S, P) must have identical payoffs; therefore, the value of such a game is well defined and is $v = P_1(s^*)$, where s^* is any equilibrium point of (N, S, P). The *saddle point* property of such an equilibrium results because a finite two-person game satisfies Assumption 3.3C and $P_2(s) = -P_1(s)$ for all s. Consequently, P_1 is concave in s_1 and convex in s_2, making any equilibrim s^* a saddle point of P_1.

3.2 Equilibrium characteristics of strictly competitive games that do not generalize

Taking Theorems 3.2 and 3.3 together, the equilibrium points of strictly competitive games (a) all have the same payoffs, (b) form a convex subset of the strategy space, and (c) have the interchangeability property (i.e., if s' and s'' are equilibrium points, then (s_1', s_2'') and (s_1'', s_2') are also). Interchangeability implies that S^0 is the Cartesian product of sets $S_1^0 \subset S_1$ and $S_2^0 \subset S_2$. Thus, each player i has a set of equilibrium strategies S_i^0 and, for any $t_1 \in S_1^0$ and $t_2 \in S_2^0$, the combination $t = (t_1, t_2)$ is an equilibrium point that affords exactly the same payoffs as any other equilibrium point. This remarkable fact means that no coordination is required for two players to choose equilibrium strategies. Contrast this with general n-person noncooperative games in which there may be multiple equilibrium points, but interchangeability does not generally hold. Then, for even a two-person noncooperative game, if $s', s'' \in S^0$ and the players choose (s_1', s_2''), the result need not be an equilibrium point. This suggests that where neither interchangeability nor uniqueness of equilibrium holds, the players must coordinate their strategy choices if an equilibrium point is to be achieved with certainty.

For two-person zero-sum games, Theorem 3.2 appears in Section 17 of von Neumann and Morgenstern (1944), and, although they do not mention Theorem 3.3, it follows easily from the linear structure of the game. For strictly competitive two-person games, Harsanyi (1977, page 170) asserts Theorem 3.2 as if it were well known. A proof of Theorems 3.2 and 3.3 is in Friedman (1984). That Theorems 3.2 and 3.3 do not extend to general two-person noncooperative games or to games of three or more persons that retain an appropriately modified condition of strict competitiveness can be illustrated by a pair of examples. Table 3.1 contains a two-person

TABLE 3.1 A two-person noncooperative game with two equilibrium points

		Strategies of player 2		
		1	2	3
Strategies of player 1	1	10, 8	4, 4	3, 2
	2	6, 10	14, 15	9, 20
	3	4, 10	8, 20	12, 25

noncooperative game that has two equilibrium points; (1, 1) and (3, 3). At the former, player 1 receives 10 and player 2 receives 8; at the latter, the payoffs are 12 and 25, respectively. These two equilibria are the only pure strategy equilibrium points of the game, their payoffs are not the same, they do not obey the interchangeability condition, and the set of equilibria is not convex. All of these assertions are easily verified. That the two equilibria yield different payoffs is readily seen from Table 3.1. The failure of interchangeability is verified by noting that (1, 3) is not an equilibrium point. To see that S^0, the set of equilibria, is not convex, recall that S_1 is the unit simplex in R^3; that is, $S_1 = \{s_1 \in R^3 \mid s_1 \geqslant 0, \sum_{i=1}^{3} s_{1i} = 1\}$. S_2 is similarly defined. Let $s' = [(1, 0, 0), (1, 0, 0)]$ and $s'' = [(0, 0, 1), (0, 0, 1)]$. These are, of course, the two pure strategy equilibrium points. Now consider $s^0 = s'/2 + s''/2$. The payoff to player 1 for s^0 is $(10 + 4 + 3 + 12)/4 = 7.25$. The combination $s^0 = [(\frac{1}{2}, 0, \frac{1}{2}), (\frac{1}{2}, 0, \frac{1}{2})]$. and player 1 would fare better with (0, 0, 1): the payoff to player 1 for $[(0, 0, 1), (\frac{1}{2}, 0, \frac{1}{2})]$ is 8.

The natural extension of strict competitiveness to three or more players is to stipulate that, for $s', s'' \in S$, if $P_i(s') > P_i(s'')$ for a player i, then there is at least one other player j for whom $P_j(s') < P_j(s'')$. This preserves the condition that, in comparing s' to s'', if some player i is better off at s', then some other player is better off at s''. Obviously n-person zero-sum games fulfill this condition. Table 3.2 illustrates a game in which there are just two pure strategy equilibrium points: (1, 1, 1) and (3, 3, 1). This game is zero sum; however, just like the game in Table 3.1, it satisfies none of the conditions proved in Theorems 3.2 and 3.3.

3.3 Strictly competitive games without quasiconcavity

Strictly competitive games are considered here in which Assumption 3.3Q is dropped and the players are restricted to using pure strategies. With Assumption 3.3Q holding, no player gains through the use of a mixed strategy. On the other hand, if Assumption 3.3Q does not hold, but a player i can select a probability distribution over elements of S_i, then the player's payoff will be a concave function of his mixed strategy. Because

TABLE 3.2 A three-person zero-sum game with two equilibrium points

Payoffs when player 3 chooses his strategy 1

Strategies of player 1		Strategies of player 2: 1	2	3
	1	8, 6, −14	2, 2, −4	1, 0, −1
	2	4, 8, −12	12, 13, −25	7, 18, −25
	3	2, 8, −10	6, 18, −24	10, 23, −33

Payoffs when player 3 chooses his strategy 2

Strategies of player 1		Strategies of player 2: 1	2	3
	1	10, 8, −18	4, 4, −8	3, 2, −5
	2	6, 10, −16	14, 15, −29	9, 20, −29
	3	4, 10, −14	8, 20, −28	12, 25, −37

mixed strategies reintroduce Assumption 3.3C by the back door they are concluded from this section.

Theorem 3.2 does not depend on the quasiconcavity condition; therefore, for games in which S^0 is not empty, any two equilibria yield the same payoffs and the interchangeability condition holds as well. Theorem 3.3 does not hold, as it depends critically on quasiconcavity.

Another aspect of these games is the security levels of the players. Denote by v_i the security level of player i. If the game has an equilibrium, then $v_i = P_i(s')$ for any $s' \in S^0$. Player 1 can assure herself at least v_1 by choosing s_1' and player 2 can keep her down to v_1 by choosing s_2'. If S^0 is empty, then there can be a difference between the payoff that player 1 can guarantee herself (v_i^m) and the payoff to which she can be held by player 2 (v_i^M). In general $v_i^M \geqslant v_i^m$ and when equilibrium exists, $v_i^M = v_i^m = v_i$. Player 1 can guarantee herself the highest payoff she can achieve under the rule that she pick s_1 and communicate her choice to player 2. Player 2 then picks s_2. For the s_1 that is chosen, player 2 selects s_2 to minimize $P_1(s_1, s_2)$. Forseeing this behavior, player 1 picks s_1 to attain the highest possible value for this minimum. Thus, $v_i^m = \max_{s_1} \min_{s_2} P_1(s_1, s_2)$. Player 1 can be kept down to $v_i^M = \min_{s_2} \max_{s_1} P_1(s_1, s_2)$. This corresponds to player 2 choosing s_2, communicating his choice to player 1, and player 1 then choosing s_1.

A simple example illustrates the possibility that $v_1^M > v_1^m$. Let $S_1 = S_2 = [-1, 1]$, $P_1 = s_1(s_1 + s_2)$, and $P_2 = -s_1(s_1 + s_2)$. $v_1^m = 0$ and $v_1^M = 1$. To see this, consider v_1^m first. If $s_1 = -1$, 0, or +1, the best player 2 can do is select $s_2 = -s_1$, which causes $P_1 = 0$. For other values of s_1 set $s_2 = -1$ or +1, with the sign selected to be opposite of the sign of s_1. Then $P_1 < 0$. On the other hand, if $s_2 = 0$, player 1 can set $s_1 = 1$ and achieve $P_1 = 1$. For other values of s_2, set $s_1 = -1$ or $+1$, choosing s_1 to have the same sign as s_2. Then $P_1 = 1 + |s_2| > 1$. So player 1 can only guarantee 0 to herself, but can only be held to 1.

4 Uniqueness of equilibrium points

Multiple equilibria are troublesome from more than one standpoint. First, it is not clear that players can be expected to coordinate to play an equilibrium strategy combination when there is more than one equilibrium point; if they have no means of communication, even if each one selects a strategy associated with an equilibrium point, the resulting combination may not be an equilibrium. Were the players able to communicate, then they could agree on a particular equilibrium point, but we are then left with the problem of figuring out which equilibrium point they would be expected to choose. The situation is much happier if equilibrium is unique. Then it is reasonable to view all players as being able to calculate the equilibrium and to presume the equilibrium is the only reasonable outcome. Furthermore, in many applications, the particular restrictions that imply a unique equilibrium are reasonable assumptions to make; however, that must always be judged case by case.

4.1 Some additional restrictions needed to ensure uniqueness

Two theorems on uniqueness are given in this section. Neither is a special case of the other, so both are useful to have in one's arsenal. Both require that Assumption 3.3Q be modified so that the best-reply mapping is a (single-valued) function. The first theorem requires, furthermore, that the best-reply function be a contraction. The second theorem, by contrast, does not restrict the best-reply function in this way, but it requires differentiability and places some additional restrictions that are explained later. Taken as given in this section are Assumptions 3.1 and 3.2, complete information, and

ASSUMPTION 3.3Q″ *$P_i(s \backslash t_i)$ is strictly quasiconcave with respect to $t_i \in S_i$ for all $i \in N$.*

Strict quasiconcavity means that if $s \in S$, $t_i \in S_i$, $t'_i \in S_i$, $t_i \neq t'_i$, $P_i(s \backslash t_i) = P_i(s \backslash t'_i)$, and $\lambda \in (0, 1)$ then

$$P_i(s \backslash \lambda t_i + (1 - \lambda) t'_i) > P_i(s \backslash t_i) \tag{3.19}$$

The key feature of using Assumption 3.3Q″ in place of Assumption 3.3Q is stated in Lemma 3.5

LEMMA 3.5 *For a game* $\Gamma = (N, S, P)$ *satisfying Assumptions* 3.1, 3.2, *and* 3.3Q″, *the set*

$$\{t_i \in S_i \mid P_i(s \backslash t_i) \geq P_i(s \backslash t_i') \text{ for all } t_i' \in S_i\} \tag{3.20}$$

consists of exactly one element for each $S \in S$.

Proof. Suppose that t_i^0 and t_i^1 both maximize $P_i(s \backslash t_i)$. Then, by Assumption 3.3Q, if $t_i^0 \neq t_i^1$ and $t_i' = (t_i^0 + t_i^1)/2$,

$$P_i(s \backslash t_i') > P_i(s \backslash t_i^0) = P_i(s \backslash t_i^1) \tag{3.21}$$

however, the latter inequality contradicts the optimality of t_i^0 and t_i^1. Thus $t_i^0 = t_i^1$. QED

For the remainder of this section, $r(s)$ is understood to be single valued; hence, the best-reply function can be stated as $t = r(s)$.

LEMMA 3.6 *The best-reply function,* $r(s)$, *is continuous.*

Proof. This is implied by the upper semicontinuity of the best-reply correspondence. A proof could be given directly by writing a proof parallel to the proof of Lemma 3.2. QED

4.2 Uniqueness of equilibrium: contraction mapping formulation

Preparing for the first uniqueness theorem, we must define the *distance between two points of* R^m and *contraction.*

DEFINITION 3.7 *Let* $x, y \in R^m$. *The* **distance from** x **to** y, *denoted either* $d(x,y)$ *or* $\|x - y\|$, *is* $d(x,y) = \max_i |x_i - y_i|$.

DEFINITION 3.8 *Let* $f(x)$ *be a function with domain* $A \subset R^m$ *and range* $B \subset R^n$. *If there is a positive scalar* $\lambda < 1$ *such that for any* $x, x' \in A$, $d(f(x), f(x')) \leq \lambda d(x, x')$, *then* $f(x)$ *is a* **contraction.**

Simply put, a contraction leaves the images of two points closer than the original points themselves were. When $f(x)$ is a differentiable function, the contraction condition in Definition 3.8 is fulfilled if $\sum_{i=1}^{m} |\partial f_j(x)/\partial x_i| \leq \lambda$ for each component $f_j(x)$ of $f(x) = (f_1(x), \ldots, f_n(x))$. For example, $y = .3x_1 - .2x_2 + .4x_3$ satisfies the contraction condition for $\lambda = .9$.

THEOREM 3.4 *Let* $\Gamma = (N, S, P)$ *be a game of complete information that satisfies Assumption* 3.1, 3.2, *and* 3.3Q″ . *If the best-reply function,* $r(s)$, *is a contraction, then* Γ *has exactly one equilibrium point.*

Proof. From Theorem 3.1, the game is known to have at least one equilibrium point. Suppose that $s', s'' \in S$ are equilibrium points (i.e., $s' = r(s')$ and $s'' = r(s'')$). Then $d(r(s'), r(s'')) = d(s', s'')$, but r being a contraction means that $d(r(s'), r(s'')) \leq \lambda d(s', s'')$ for $\lambda < 1$. The latter can hold only for $s' = s''$, which implies that the equilibrium point is unique. QED

4.3 Uniqueness of equilibrium: the univalent mapping approach

The second uniqueness theorem is not restricted to games in which $r(s)$ is a contraction, but it does require differentiability of the payoff functions. Let $\mathring{S}$ denote the *interior* of S.

DEFINITION 3.9 *$\Gamma = (N, S, P)$ is a* **smooth game** *if the following derivatives exist and are continuous on $\mathring{S}$: $\partial P_i/\partial s_{jk}$, $k = 1, \ldots, m$, $j \in N$ and $\partial^2 P_i/\partial s_{ik}\partial s_{jl}$, k, $l = 1, \ldots, m$, $i, j \in N$.*

DEFINITION 3.10 *$\Gamma = (N, S, P)$ is a* **strictly smooth game** *if it is a smooth game, and for each s' on the boundary of S (i.e., $s' \in S$, $s' \notin \mathring{S}$)*

$$\frac{\partial P_i(s')}{\partial s_{ik}} = \lim_{t \to s'} \frac{\partial P_i(t)}{\partial s_{ik}}, \qquad k = 1, \ldots, m, \quad i \in N \tag{3.22}$$

$$\frac{\partial^2 P_i(s')}{\partial s_{ik}\partial s_{jl}} = \lim_{t \to s'} \frac{\partial^2 P_i(t)}{\partial s_{ik}\partial s_{jl}}, \qquad k, l = 1, \ldots, m, \quad i, j \in N \tag{3.23}$$

exist for all sequences of points in $\mathring{S}$ converging to s' on the boundary of S.

The second partial derivatives that exist everywhere in S for a strictly smooth game are precisely the derivatives that appear in the Jacobian of the system

$$\frac{\partial P_i}{\partial s_{ik}} = 0, \qquad k = 1, \ldots, m, \quad i \in N \tag{3.24}$$

This Jacobian, denoted $J(s)$, is a square matrix with $m \times n$ rows and columns. Its elements are $\partial^2 P_i/\partial s_{ik}\partial s_{jl}$ ($i, j \in N$ and k, l, $= 1, \ldots, m$). It is the Jacobian of the implicit form of the best-reply function, which must fulfill a special condition everywhere on its domain, S. Recall that *a square symmetric matrix A is* **negative definite** *if all principal minor subdeterminants of odd order are negative and those of even order are positive.* A related condition for nonsymmetric square matrices is *quasinegative definiteness.*

DEFINITION 3.11 *Let A be an $m \times m$ matrix. A is* **negative quasidefinite** *if $B = A + A^T$ is negative definite.*

At a glance, it may appear that a negative quasidefinite matrix differs only trivially from a negative definite matrix. An example can illuminate the difference. Suppose that

$$A = \begin{bmatrix} -1 & a \\ a & -1 \end{bmatrix} \tag{3.25}$$

Then the determinant of A is $|A| = 1 - a^2$ and $|A| > 0$ is required for negative definiteness. This, in turn, means that $|a| < 1$. Now consider

$$A^* - \begin{bmatrix} -1 & a \\ b & -1 \end{bmatrix} \tag{3.26}$$

and

$$B = A^* + A^{*T} = \begin{bmatrix} -2 & a+b \\ a+b & -2 \end{bmatrix} \tag{3.27}$$

Then $|B| = 4 - (a+b)^2$ and negative quasidefiniteness requires $(a+b)^2 < 4$. a and b are not constrained to be less than 1 in absolute value, but the sum of their absolute values must be less than 2. So $a = 98.5$ and $b = -100$ satisfies negative quasidefiniteness. To see the connecting link between these concepts from another angle, let A be an $m \times m$ matrix and x be a vector in R^m. A is negative quasidefinite if, for all $x \neq 0$, $xAx^T < 0$; however, A is negative definite if A is both negative quasidefinite and symmetric.

The uniqueness theorem requires that $J(s)$ be negative quasidefinite for all $s \in S$. The theorem allowing the proof of uniqueness is the following:

GALE–NIKAIDO (1965) UNIVALENCE THEOREM *Let $f(x)$ be a function from a convex set $X \subset R^m$ to R^m. If the Jacobian of f is negative quasidefinite for all $x \in X$, then f is one to one. (That is, if $f(x') = y'$, then, for all $x \notin x'$, $f(x) \neq y'$.)*

THEOREM 3.5 *Let $\Gamma = (N, S, P)$ be a smooth game of complete information that satisfies Assumptions 3.1, 3.2, and 3.3Q. Assume that $J(s)$, the Jacobian of the implicit form of the best-reply function, is negative quasidefinite for all $s \in \mathring{S}$, and that, for any $s \in S$, $r(s) \in \mathring{S}$. Then Γ has a unique equilibrium point.*

Proof. The theorem is a direct consequence of Theorem 3.1 and the Gale–Nikaido univalence theorem. From the former, equilibria are known to exist, and, by the condition that $r(s) \in \mathring{S}$, there are no equilibrium points on the boundary of S. Thus, all equilibrium points must satisfy conventional first-order conditions: $\partial P_i(S^*)/\partial s_{jk} = 0$, $k = 1, \ldots, m$, i, $j \in N$. Denote the system of first derivatives of all the P_i with respect to the s_{ik} by $P'(s)$.[3] An equilibrium point s^* corresponds to $P'(s) = 0$. $P'(s)$ is a function from $\mathring{S} \subset R^{nm}$ to R^{nm}. Its Jacobian is $J(s)$, which is negative quasidefinite; therefore $P'(s)$ is univalent and takes on the value $P'(s) = 0$ only once, establishing uniqueness of equilibrium. QED

A stronger result is the Rosen uniqueness theorem:

ROSEN (1965) UNIQUENESS THEOREM *Let $\Gamma = (N, S, P)$ be a strictly smooth game of complete information that satisfies Assumptions 3.1, 3.2, and 3.3Q″. If $J(s)$ is negative quasidefinite for all $s \in S$, then Γ has a unique equilibrium point.*

Unlike Theorem 3.5, the Rosen uniqueness theorem is based on an extension of the differentiability and quasidefiniteness conditions from $\mathring{S}$ to S. This permits a global uniqueness theorem that is not hampered by restriction to models in which all equilibria must be in $\mathring{S}$. These two theorems are closer to each other than either is to Theorem 3.4. Neither Theorem 3.4 nor the Rosen uniqueness theorem is a generalization of the other, because each requires an assumption that is stronger than the other theorem requires. Theorem 3.4 does not require differentiability of the payoff functions, while the other two theorems require continuous second partial derivatives. But, for differentiable payoff functions, the contraction condition of Theorem 3.4 is much more confining than the negative quasidefiniteness of the Jacobian, $J(s)$. If $r(s)$ is a differentiable contraction, then $J(s)$ is always negative quasidefinite and fulfills the dominant diagonal condition $P_i^i(s) + \sum_{j \neq i} |P_j^i(s)| < 0$, $i \in N$, but negative quasidefiniteness does not imply the dominant diagonal condition, as the 2×2 example with $a = 99.5$ and $b = -100$ clearly demonstrates. In practice it is useful to have all of these uniqueness theorems available.

4.4 *Examples of games with unique equilibrium points*

These theorems on uniqueness are easily illustrated with simple examples. Suppose $S_1 = [-10, 0]$, $S_2 = [-3, 0]$,

$$P_1(s) = 10s_1 + 7s_1s_2 - s_1^2 \tag{3.28}$$

$$P_2(s) = 15s_2 + 5s_1s_2 - s_2^2 \tag{3.29}$$

The first-order conditions for an interior equilibrium point are

$$10 + 7s_2 - 2s_1 = 0 \tag{3.30}$$

$$15 + 5s_1 - 2s_2 = 0 \tag{3.31}$$

Equations (3.30) and (3.31) are solved at $(-125/31, -80/31)$. In addition to this interior equilibrium, there is a boundary equilibrium, $(0, 0)$. Checking for negative quasidefiniteness, the Jacobian of equations (3.30) and (3.31) is

$$\begin{bmatrix} -2 & 7 \\ 5 & -2 \end{bmatrix} = J(s) \tag{3.32}$$

and $J(s) + J^T(s)$ is

$$\begin{bmatrix} -4 & +12 \\ +12 & -4 \end{bmatrix} \tag{3.33}$$

The determinant of equation (3.33) is $16 - 144 = -128$. To satisfy negative quasidefiniteness the diagonal elements must be negative, which they are, and the determinant must be positive. Furthermore, the best-reply mapping is not a contraction. Solving for $r(s)$ from equations (3.30) and (3.31) yields

$$s_1 = 5 + 3.5s_2 \tag{3.34}$$

$$s_2 = 7.5 + 2.5s_1 \tag{3.35}$$

Neither equation (3.34) nor (3.35) is a contraction. For example, $\partial s_1/\partial s_2 = 3.5 > 1$.

Consider a slight variant of this example. S_1, S_2, and P_2 are as before, but

$$P_1(s) = 10s_1 - 7s_1s_2 - s_1^2 \tag{3.36}$$

The first-order conditions for an interior equilibrium are equation (3.31) and

$$10 - 7s_2 - 2s_1 = 0 \tag{3.37}$$

These are not simultaneously satisfied anywhere on S; however, there is a unique equilibrium at (0, 0). The Jacobian of this system is

$$\begin{bmatrix} -2 & -7 \\ 5 & -2 \end{bmatrix} \tag{3.38}$$

and $J(s) + J^T(s)$ is

$$\begin{bmatrix} -4 & -2 \\ -2 & -4 \end{bmatrix} \tag{3.39}$$

This matrix is negative definite, thus the quasinegative definiteness conditions are met: $-4 < 0$ and the determinant of equation (3.39) is $16 - 4 = 12 > 0$. If the example were changed again by letting $S_2 = [-3, +3]$ with S_1, P_1, and P_2 unchanged, there would be a unique equilibrium at $(-85/39, 80/39)$. It can be easily verified that the best-reply function is not a contraction for the second and third examples as well. Thus, the uniqueness theorems in Section 4 give sufficient, but not necessary, conditions for uniqueness. It is left as an exercise to the reader to suggest modifications to one of the examples that change the best-reply function into a contraction.

5 Noncooperative games under incomplete information

It is possible to specify an equilibrium concept for some incomplete information games by making assumptions about the way that incompleteness enters into the game. The following treatment is based on Harsanyi (1967, 1968a, 1968b). It is assumed that each player i knows the set N of players and each of the strategy sets $S_j, j \in N$; however, players do not know the correct payoff functions for the game. Each player knows her own payoff function and knows the set from which the payoff function of any other player is drawn. For each player i there is a finite number v of associated *types*. Each player i knows her own type, $b_i \in V = \{1, \ldots, v\}$ and has a probability distribution over the types of each of the remaining players. If the beliefs of the players (i.e., their respective probability distributions over the types of their rivals) are consistent in the sense spelled out presently in Assumption 3.5, then the incomplete information game can be analyzed as if it were a complete information game with $n \cdot v$ players in which nature has the first move. Thus, each type of each player is regarded as a distinct player. Nature chooses a vector of player types, $\beta \in V^n$, that determines which of the $n \cdot v$ players will be active. The result of that random draw is not revealed beyond type β_i of player i knowing that she is active and the remaining players learning that they are inactive. Type β_i of player i does not know any of the other selected β_j.

Thus, even knowing a player's type does not totally determine that player's payoff function. $\beta = (\beta_1, \ldots, \beta_n) \in V^n$ indexes the set of possible payoff functions for all players in that, given some $\beta \in V^n$, the payoff functions $P_{i\beta}$, $i \in N$, are precisely determined. A vector β will be called a *state*.

DEFINITION 3.12 *$P_i^* = \{P_{i\beta}\}_{\beta \in V^n}$ is the* **set of potential payoff functions for player** *$i \in N$.*

DEFINITION 3.13 *$P^* = \times_{i \in N} P_i^*$ is the* **set of potential payoff functions** *for an incomplete information game Γ_I. $P^* = \{(P_{1\beta}, \ldots, P_{n\beta}) \mid P_{i\beta} \in P_i^*, \ i \in N, \beta \in V\}$.*

An element of P^* is a vector of payoff functions $(P_{1\beta}, \ldots, P_{n\beta}) = P_\beta$.

ASSUMPTION 3.4 *$P_{i\beta}(s)$ is defined, bounded, and continuous on $s \in S$, and concave in s_i, for all $i \in N$ and all $\beta \in V^n$.*

From Assumption 3.4, the strategy sets are identical for all possible payoff functions of the ith player, and the payoff functions satisfy Assumptions 3.2 and 3.3C.

5.1 *Definition of a game of incomplete information*

Differing information among the players arises because each player i is assumed to know the true value of β_i, and to have a subjective probability distribution over $\beta^i = (\beta_1, \ldots, \beta_{i-1}, \beta_{i+1}, \ldots, \beta_n)$, the values of the β_j pertaining to the other players.

DEFINITION 3.14 *π_i is a* **subjective probability distribution for player** *i. It satisfies $\pi_i(\beta^i) \geqslant 0$ for each $\beta^i \in V^{n-1}$ and $\sum_{\beta^i \in V^{n-1}} \pi_i(\beta^i) = 1$.*

Letting $\pi = (\pi_1, \ldots, \pi_n)$, a game of incomplete information is denoted $\Gamma_I = (N, S, V, P^*, \pi)$.

DEFINITION 3.15 *A* **noncooperative game of incomplete information** *$\Gamma_I = (N, S, V, P^*, \pi)$ satisfies Assumptions 3.1 and 3.4, and is a game in which each player i knows (a) all the strategy sets S_j, $j \in N$, (b) all the possible payoff functions $P_{j\beta}$, $j \in N$, $\beta \in V^n$, (c) the true value of β_i, denoted b_i, and (d) has her own subjective probability distribution π_i over $\beta^i \in V^{n-1}$. Each player knows the nature of the information in the possession of all players and is aware that all players are thus informed.*

Note that the players are unable to calculate equilibrium points on their own if each player i is ignorant of the subjective probability distributions of the others. It is not reasonable to suppose they would get together and exchange this information, for, even if they could communicate, there is no reason to think that a player would be truthful or that he would be believed. By contrast, in a complete information game, each player has the information needed to calculate all of the equilibrium points. To define *equilibrium points for* Γ_I, we assume that each player does have some information concerning the subjective probability distributions of the others, although this information is not entirely complete. Then it is shown that the incomplete information game has a mathematical structure identical to a complete information game, and the equilibrium points of the complete information game are defined to be the equilibrium points of the incomplete information game. This identification of equilibrium points is done in a natural way.

5.2 Consistency of players' subjective beliefs

The true state is denoted b. Let $(\beta^i, b_i) = (\beta_1, \ldots, \beta_{i-1}, b_i, \beta_{i+1}, \ldots, \beta_n) = \beta \backslash b_i$. For a player i, the possible states are therefore $\{(\beta^i, b_i) \mid \beta^i \in V^{n-1}\}$ because his own type, b_i, is known to him with certainty. The subjective beliefs of the players, π, are consistent with one another if they are all appropriate conditional distributions of some specific common distribution θ defined on V^n. That is, π is consistent if there exists some θ defined on V^n such that $\pi_i = \theta(\beta^i \mid b_i)$ for each $i \in N$. Thus,

ASSUMPTION 3.5 *Suppose $b \in V^n$ is the true state and θ is a probability distribution over V^n that is known to all players and that satisfies*

$$\pi_i(\beta^i) = \theta(\beta^i \mid b_i) = \frac{\theta(\beta^i, b_i)}{\sum_{k^i \in V^{n-1}} \theta(k^i, b_i)} \tag{3.40}$$

Each player i believes that the subjective beliefs of player j, type β_j, are $\theta(\cdot \mid \beta_j)$.

Although technically $\theta(\cdot)$ is a probability distribution over V^n, it implies a distribution over P^*, and it should not cause confusion to refer to it as a distribution over P^*. In Assumption 3.5, $\theta(\cdot \mid \beta_i)$ is the conditional distribution of β^i given β_i, and $\pi^i(\beta^i)$ is required to equal this conditional distribution for $\beta_i = b_i$. In addition, θ embodies the beliefs that each player holds concerning the subjective probability distributions of the others. Thus, if player i believes the correct value of β_j (for $j \neq i$) is β'_j, then player i supposes that the subjective probability distribution of player j is $\theta(\cdot \mid \beta'_j)$.

DEFINITION 3.16 *A* **consistent game of incomplete information** $\Gamma_I = (N, S, V, P^*, b, \theta)$ *is a game of incomplete information that satisfies Assumption* 3.5.

5.3 A complete information companion game

It is now possible to look for equilibrium points in consistent games of incomplete information. These games satisfy Assumptions 3.1, 3.4, and 3.5, as well as the information conditions stated in Definition 3.15. The technique is to specify a mathematically equivalent complete information game and define the equilibrium points of the complete information game to be the equilibrium points of the incomplete information game. To define the complete information game, we suppose that there are nv players in a game in which the players fall into n groups, with v players in each group. Player (i, β_i) is the β_ith player (type) in group i, $i \in N$, $\beta_i \in V$. The first move in this game is made by nature and is the selection of a $\beta \in V^n$, which designates the active players. Thus, there will be n active players, $(1, \beta_1), \ldots, (n, \beta_n)$. To be active means that the player (i, β_i) will achieve the payoff $P_{i\beta}(s_{1\beta}, \ldots, s_{n\beta}) = P_{i\beta}(s_\beta)$. An inactive player receives a payoff of 0. Following nature's move, each player i is informed whether she is active; however, a player (i, β_i) does not learn whether any given player (j, β_j), $(j \neq i)$ is active. Naturally if (i, β_i) knows she is active then she knows that (i, β_i') is not active (for $\beta_i \neq \beta_i'$). Although the active player will achieve the payoff $P_{i\beta}(s_\beta)$, the payoff function as perceived by player (i, β_i) is not $P_{i\beta}$ because she is ignorant of β^i. An active player has the payoff function $\bar{P}_{i\beta_i}$, and any player (j, k_j) who is not active has a payoff of 0. Technically, the payoff function of a player must depend on the strategy choices of all nv players. The random mechanism that selects β has the distribution $\theta(\beta)$. The set of players is $N \times V$ and a single player is denoted (i, β_i). Let $\sigma_{i\beta_i}$ denote a strategy selected by player (i, β_i). The strategy space of this player is the same for all $\beta_i \in V$ and is S_i. The joint strategy space of the nv player game is $\times_{(i,\beta_i) \in N \times V} S_i = \times_{i \in N} S_i^v = S^v$. An element of S^v is denoted σ. Let $\sigma_\beta = (\sigma_{1\beta_1}, \ldots, \sigma_{n\beta_n}) \in S$ and let $\delta_{k_i\beta_i}$ be the Kronecker delta (i.e., $\delta_{k_i\beta_i} = 1$ if $k_i = \beta_i$ and $\delta_{k_i\beta_i} = 0$ if $k_i \neq \beta_i$). Then the payoff function of active player (i, β_i) is

$$\bar{P}_{i\beta_i}(\sigma) = \sum_{k \in V^n} \delta_{k_i\beta_i} \theta(k) P_{ik}(\sigma_k) = \sum_{k^i \in V^{n-1}} \theta(k \mid \beta_i) P_{i,k^i,\beta_i}(\sigma_{k^i\beta_i}). \tag{3.41}$$

Letting $\bar{P} = (\bar{P}_{11}, \ldots, \bar{P}_{1v}, \ldots, \bar{P}_{n1}, \ldots, \bar{P}_{nv})$, the complete information game corresponding to $\Gamma_I = (N, S, V, P^*, \theta)$ is $\Gamma_C = (N \times V, S^v, \bar{P})$, and Γ_C is called the *companion game of* Γ_I.

THEOREM 3.6. *The complete information game $\Gamma_C = (N \times V, S^v, \bar{P})$, which is the companion game to the consistent incomplete information game $\Gamma_I = (N, S, V, P^*, b, \theta)$, has a noncooperative equilibrium point.*

Proof. It suffices to show that Γ_C satisfies Assumptions 3.1, 3.2, and 3.3C. Clearly, each S_i is compact and convex. Therefore, S^v is compact, because it is a product of compact sets S. $\bar{P}_{i\beta_i}$ is continuous and bounded because it is the sum of a finite number of continuous and bounded functions. Similarly, $\bar{P}_{i\beta_i}$ is concave in $\sigma_{i\beta_i}$ because it is a sum of concave functions. QED

5.4 Existence of equilibrium points for incomplete information games

The equilibrium points of Γ_I are defined to be those of Γ_C. Recall that in Γ_I each player knows which type she is. That is, there is $b \in V$, which is the true player specification. Now suppose that player i selects s_i to maximise her expected payoff, and, although player i does know b_i, she does not know b_j $(j \neq i)$. Indeed, $\pi_i(\beta^i)$ is her subjective probability distribution over the possible payoff functions of the other players. Player i recognizes that another player j will choose s_j differently according to the true value of β_j, so player i will think of player j actually choosing $u_j = (u_{j1}, \ldots, u_{jv}) \in S_j^v$, where $u_{j\beta_j}$ is the strategy choice of player j contingent on $\beta_j = b_j$. Thus, player i views her payoff function as the expected payoff

$$F_{ib_i}(s_i, u') = \sum_{\beta^i \in V^{n-1}} \pi_i(\beta^i) P_{i,\beta^i,b_i}(u_{1\beta_1}, \ldots, u_{i-1,\beta_{i-1}}, s_i, u_{i+1,\beta_{i+1}}, \ldots, u_{n\beta_n}) \tag{3.42}$$

Note that equations (3.40) and (3.41) are equivalent.

DEFINITION 3.17 *$s^* \in S$ is an* **equilibrium point relative to *b*** *for the consistent noncooperative incomplete information game $\Gamma_I = (N, S, V, P^*, b, \theta)$ if there is $u^* \in S^v$ such that $u^*_{ib_i} = s^*_i$ and*

$$\max_{s_i \in S_i} F_{ib_i}(s_i, u^{*i}) = F_{ib_i}(s^*_i, u^{*i}) \tag{3.43}$$

for all $i \in N$.

Under Definition 3.17, each player i has a system of expectations π^i concerning what the true payoff functions will be for the other players and concerning the strategy choices the other players will make as a function of the various payoff functions. These expectations are consistent in the sense that all the $\pi^i = (\pi_1, \ldots, \pi_{i-1}, \pi_{i+1}, \ldots, \pi_n)$ are conditional distributions for a common $\theta(\beta)$ and all players $i \neq j$ expect that player j will select $u^*_{j\beta_j}$ if her payoff function is $\bar{P}_{j\beta_j}$.

DEFINITION 3.18 *$u^* \in S^v$ is* **an equilibrium point for the consistent noncooperative incomplete information game** *$\Gamma_I = (N, S, V, P^*, b, \theta)$ if*

$$\max_{s_i \in S_i} F_{i\beta_i}(s_i, u^{*i}) = F_{i\beta_i}(u^*_{i\beta_i}, u^{*i}) \tag{3.44}$$

for all $i \in N$ and $\beta_i \in V$.

Definition 3.18 specifies a u^* that satisfies Definition 3.17 for all possible $k \in V^n$.

THEOREM 3.7 *Let $\Gamma_C = (N \times V, S^v, P)$ be the companion game to $\Gamma_I = (N, S, V, P^*, b, \theta)$. Then $\sigma^* \in S^v$ is an equilibrium point of Γ_C if and only if it is an equilibrium point of Γ_I.*

Proof. The games and the definitions of their respective equilibrium points are mathematically equivalent. QED

A generalization of this model can be obtained easily by allowing the strategy set of player i to be different for each value of β_i, giving each player (i, β_i) her own distinct strategy space.

TABLE 3.3 A game of incomplete information

		States for player 2	
		1	2
States for player 1	1	.1	.4
	2	.3	.2

Conditional distributions (subjective probability assessments)

For player 1

	$\beta_2=1$	$\beta_2=2$
If $\beta_1=1$	.2	.8
If $\beta_1=2$	.6	.4

For player 2

	$\beta_1=1$	$\beta_1=2$
If $\beta_2=1$	.25	.75
If $\beta_2=2$	$\frac{2}{3}$	$\frac{1}{5}$

Strategy spaces $S_1=[-15,25]$ $S_2=[-100,80]$

Payoff functions for each player, conditional on the state

State	For player 1	For player 2
(1, 1)	$P_{111}(u_{11},u_{21})=100u_{11}-u_{11}^2+u_{21}^2-5u_{11}u_{21}$	$P_{211}(u_{11},u_{21})=50u_{21}-u_{21}^2+10u_{11}u_{21}$
(1, 2)	$P_{112}(u_{11},u_{22})=-u_{11}^2-u_{22}^2-u_{11}u_{22}$	$P_{212}(u_{11},u_{22})=30u_{11}-u_{22}^2-10u_{11}u_{22}$
(2, 1)	$P_{121}(u_{12},u_{21})=-u_{12}^2+3u_{12}u_{21}$	$P_{221}(u_{12},u_{21})=-u_{21}^2-5u_{12}u_{21}$
(2, 2)	$P_{122}(u_{12},u_{22})=30u_{12}-u_{12}^2+u_{12}u_{22}$	$P_{222}(u_{12},u_{22})=-u_{22}^2+20u_{12}u_{22}$

5.5 A numerical example

An example of a consistent noncooperative incomplete information game is presented in Table 3.3 for two players and two states per player. The payoff functions for the complete information companion game are

$$F_{11}(u)=.2P_{111}+.8P_{112}=20u_{11}-u_{11}^2+.2u_{21}^2-u_{11}u_{21}-.8u_{22}^2-.8u_{11}u_{22} \tag{3.45}$$

$$F_{12}(u)=.6P_{121}+.4P_{122}=12u_{12}-u_{12}^2+1.8u_{12}u_{21}+.4u_{12}u_{22} \tag{3.46}$$

$$F_{21}(u)=.25P_{211}+.75P_{221}=12.5u_{21}-u_{21}^2+2.5u_{11}u_{21}-3.75u_{12}u_{21} \tag{3.47}$$

$$F_{22}(u)=\tfrac{2}{3}P_{212}+\tfrac{1}{3}P_{222}=20u_{11}-u_{22}^2-6\tfrac{2}{3}u_{11}u_{22}+6\tfrac{2}{3}u_{12}u_{22} \tag{3.48}$$

An equilibrium point is a solution to the equation system

$$\frac{\partial F_{11}}{\partial u_{11}}=20-2u_{11}-u_{21}-.8u_{22}=0 \tag{3.49}$$

TABLE 3.4 Probability distribution over states for a game of incomplete information

		States of player 2	
		1	2
States of player 1	1	a	$4a$
	2	$3a$	$1-8a$

$$\frac{\partial F_{12}}{\partial u_{12}} = 12 - 2u_{12} + 1.8u_{21} + .4u_{22} = 0 \tag{3.50}$$

$$\frac{\partial F_{21}}{\partial u_{21}} = 12.5 - 2u_{21} + 2.5u_{11} - 3.75u_{12} = 0 \tag{3.51}$$

$$\frac{\partial F_{22}}{\partial u_{22}} = -2u_{22} - 6\tfrac{2}{3}u_{11} + 6\tfrac{2}{5} - u_{12} = 0 \tag{3.52}$$

The equilibrium point is $u^* = (12.05, 8.49, 5.41, -11.89)$ and the associated expected payoffs are 37.98, 72.10, 29.34, and 381.82.

It is interesting to consider a variant of the example in Table 3.3. Suppose that the true state is $\beta_1 = 1$ and $\beta_2 = 1$. Of course, player 1 is unaware of the true value of β_2, just as player 2 is unaware of the true value of β_1. Say that the two players' subjective probability assessments are as shown in Table 3.3. That is, player 1 places a probability of .2 on the event $\beta_2 = 1$ and .8 on the event $\beta_2 = 2$; while player 2 believes $\beta_1 = 1$ with probability .25 and $\beta_1 = 2$ with probability .75. A large family of probability distributions θ are consistent with these subjective distributions; that is, all the distributions shown in Table 3.4 for $0 \leqslant a \leqslant .125$. In Table 3.3, $a = .1$. Table 3.5 displays the equilibrium strategies and payoffs for several values of a. Clearly, both u^* and equilibrium payoffs are very sensitive to the value of a, which is not surprising. Many complete

TABLE 3.5 Equilibrium strategies and payoffs for a game of incomplete information

a	u_{11}	u_{12}	u_{21}	u_{22}	F_{11}	F_{12}	F_{21}	F_{22}
0	11.30	−3.75	24.70	−37.50	−847.2	14.06	807.6	1406.3
.001	11.36	−3.77	27.53	−37.82	−1551.0	−270.1	814.1	−4286.4
.05	20.53	−7.87	46.67	−84.67	−4878.0	61.89	228.6	7325.9
.1	12.05	8.49	5.41	−11.89	37.98	72.10	29.34	381.8
.125	−10.98	−2.94	−1.96	54.90	−2289.9	8.64	−51.05	2684.6

information games are compatible with this incomplete information game in the sense that they can be companion games, and, although each complete information game has a unique Nash equilibrium, the equilibrium changes rapidly as a is changed. Merely imposing mutually consistent beliefs on the players does not bring determinateness to the outcome.

6 Applications of noncooperative games

Three examples of noncooperative games are sketched in this section. The first is the Cournot (1838) oligopoly, the classic instance of a noncooperative game in economics. The second is a model of general equilibrium, and the third is the free rider problem. These examples are by no means exhaustive, but they illustrate the underlying principles of noncooperative games, bring out some interesting special problems, and give some indication of the scope of noncooperative games for economics.

6.1 Cournot oligopoly

Suppose a single market in which n firms offering a homogeneous good sell to a very large number of consumers whose willingness to purchase the product is summarized by the inverse demand function $p = f(Q)$, where p is the market price and Q is the total output in the industry. The set of firms is $N = \{1, \ldots, n\}$ and the output of firm $i \in N$ is q_i. Thus, $Q = \sum_{i \in N} q_i$. Firms produce output, all of which is sold at whatever price will clear the market. Each firm faces a cost of production, $C_i(q_i)$; hence, letting $q = (q_1, \ldots, q_n)$, the profit to firm i is $\pi_i(q) = q_i f(Q) - C_i(q_i)$.

Common, although not universal, assumptions placed on the model are as follows

CONDITION 3.1 *The inverse demand function is finite valued, nonnegative, defined for all* $Q \in [0, \infty)$, *and twice continuously differentiable wherever* $f(Q) > 0$. *In addition,* $f(0) > 0$, *and if* $f(Q) > 0$, *then* $f'(Q) < 0$.

CONDITION 3.2 $C_i(q_i)$ *is defined for all* $q_i \in [0, \infty)$, *nonnegative, convex, twice continuously differentiable, and* $C_i'(q_i) \geqslant \varepsilon > 0$.

CONDITION 3.3 $Qf(Q)$ *is bounded and is strictly concave for all* q *such that* $f(Q) > 0$.

For the most part, these assumptions have natural economic interpretations. Condition 3.1 stipulates a demand function that is downward sloping, starting from a point on the price axis. Condition 3.2 specifies nonnegative fixed cost ($C_i(0) \geqslant 0$) and positive, nondecreasing marginal cost ($C_i'(q_i) > 0$). These stem from C_i being nonnegative, increasing, and convex. Condition 3.3 asserts concavity of $Qf(Q)$, total industry revenue. This, in turn, implies concavity of $q_i f(\sum_{j \in N} q_j)$ with respect to q_i. Therefore, $\pi_i(q)$ is concave in q_i for all q such that $f(Q) > 0$, because it is a sum of two concave functions, $q_i f(Q)$ and $-C_i(q_i)$. On the Cournot model, see Cournot's own description (1838, Chapter 7), and more recent expositions, such as Fellner (1949, Chapter 2) and Friedman (1977, Chapter 2; 1983, Chapter 2).) In time-honored fashion, Cournot discussed stability in a way that helped confuse later readers about whether he was dealing with a strictly static, one-shot model or a multiperiod model. For our present purpose, I treat the model as strictly one period. Thus, each firm knows the inverse demand function and the cost functions of all firms. For each firm i, strategy is simply an output level, q_i. In principle, the firm's strategy set can be $[0, \infty)$; however, in practice, a finite upper bound can be invoked. The market operates very simply: The n firms simultaneously choose output levels $q_1, \ldots, q_n$, which determine the profits of all firms. There is no past history of behavior to guide the firms, nor is there any future to concern them. Cournot's concept of equilibrium in this market is characterized by an output vector q^c such that no single firm could have greater profits by having selected an output level different from q_i^c, given the output levels of the others. That is, $\pi_i(q^c) \geqslant \pi_i(q^c \backslash q_i)$ for all admissible q_i and all $i \in N$. Clearly, this is the Nash equilibrium in a game in which each firm is a player, the π_i are the payoff functions, q_i is the strategy of player i, and some subset of $[0, \infty)$ is each firm's strategy set.

From the foregoing description, it may not be evident that this model actually has a Nash equilibrium. To verify existence of equilibrium, it is sufficient to prove that the model satisfies the assumptions of an existence theorem such Theorem 3.1. To this end, the firms' strategy spaces must be carefully defined. The strategy space of each firm i will consist of an interval $[0, q_i^0]$, where $q_i^0 < \infty$. Although a firm can, in principle, select any nonnegative output level, there are output levels so high that they cannot, under any circumstances, be best replies. Thus, q_i^0 must be chosen so that any larger output level would never be a best reply. By Condition 3.3, $QF(Q)$ is bounded above; therefore, let z^* be its least upper bound. Now let q_i^0 be defined by $z^* = q_i^0 \varepsilon$ or $q_i^0 = z^*/\varepsilon$, where ε is the lower bound on marginal cost. Then clearly for $q_i > q_i^0$, the firm's revenue must fall short of its total variable cost and the firm would never, in equilibrium, choose such a large output. Because z^* is finite and $\varepsilon > 0$, q_i^0 is finite.

Some demand functions touch the quantity axis. That is, there may be a finite output level, Q^*, for which $f(Q^*) = 0$ and $f'(Q) > 0$ if $Q < Q^*$. Clearly, $q_i^0 \leqslant Q^*$. If $f(Q) > 0$ for all positive Q, then $Q^* = \infty$.

Therefore, it is reasonable to take $S_i = [0, q_i^0]$ as the strategy set of player i and to let $S = \times_{i \in N} S_i$. $\pi_i(q)$ is defined for all $q \in S$. Assumptions 3.1, 3.2, and 3.3Q are satisfied: $S_i = [0, q_i^0]$ is compact and convex, $\pi_i(q)$ is defined, continuous, and bounded for all $q \in S$. Finally, $\pi_i(q \backslash t_i)$ is quasiconcave in t_i. Therefore, from Theorem 3.1, the market has a Cournot equilibrium.

As a numerical example, let $n = 2$, $p = 100 - 4Q + 3Q^2 - Q^3$, $C_1(q_1) = 4q_1$, and $C_2(q_2) = 2q_2 + .1q_2^2$. Then

$$\pi_1(q) = 96q_1 - 4q_1(q_1 + q_2) + 3q_1(q_1 + q_2)^2 - q_1(q_1 + q_2)^3 \tag{3.53}$$

$$\pi_2(q) = 98q_2 - 4q_1q_2 - 4.1q_2^2 + 3q_2(q_1 + q_2)^2 - q_2(q_1 + q_2)^3 \tag{3.54}$$

As the reader can verify, the Cournot equilibrium occurs at the solution of the equations

$$\frac{\partial \pi_1}{\partial q_1} = 96 - 8q_1 - 4q_2 + 6q_1(q_1 + q_2) + 3(1 - q_1)(q_1 + q_2)^2 - (q_1 + q_2)^3 = 0 \quad (3.55)$$

$$\frac{\partial \pi_2}{\partial q_2} = 98 - 4q_1 - 8.2q_2 + 6q_2(q_1 + q_2) + 3(1 - q_2)(q_1 + q_2)^2 - (q_1 + q_2)^3 = 0 \quad (3.56)$$

which is $q' = (2.028, 2.081)$.

6.2 General competitive exchange equilibrium

Now a Walrasian model of a general competitive exchange economy is analyzed as a noncooperative pseudogame. In brief, a Walrasian pure trade model is made up of a set of consumers $N' = \{1, \ldots, n\}$ in which each consumer is endowed with a bundle of consumption goods $w^i \in R^m_+$, and markets in which consumers can trade, selling any portion of their endowments in exchange for whatever they wish, at fixed prices $p \in R^m_{++}$. Let x^i denote the final consumption bundle of consumer i. Trading is then limited by the conditions $x^i \geqslant 0$ and $pw^i = px^i$. The former requires that quantities be nonnegative, and the latter is the conventional budget constraint. Equilibrium is characterized by a price vector p^* such that, at p^*, $\sum_{i \in N'} w^i = \sum_{i \in N'} x^i$, and the consumption bundle x^i maximizes the utility of each consumer $i \in N'$, relative to his budget constraint (i.e., markets clear in equilibrium, and each consumer has optimized).

Approaching this model as a noncooperative pseudogame is natural in one respect, yet peculiar in another. In a competitive economy, each agent is assumed to choose his action (strategy) from his set of available alternatives so as to maximize his objection function (payoff). This is clearly noncooperative. A pillar of competitive economics—indeed, the element defining the term *competitive*—is that the actions of one agent have no effect on the opportunities facing any other agent. In a pure trade model, such as that examined here, having no effect on the opportunities of others means that (a) the endowment of a consumer is fixed, independently of the actions of others, and (b) no single agent can affect prices. Thus, the player interactions that generally take place in noncooperative games, which many regard as a distinguishing characteristic of such games, are absent. Nonetheless, a competitive economy can be regarded as a noncooperative pseudogame, although in certain respects, a simple one. The model set out here and the game theoretic approach are based on the classic article by Arrow and Debreu (1954).

Approaching a competitive economy as a noncooperative pseudogame provides another means of proving existence of equilibrium; however, in taking this approach, a means must be built into the pseudogame for price determination. This is done by the addition of another player, player 0, whose objective function and strategy spaces are described presently. Thus, the set of players becomes $N = \{0, 1, \ldots, n\}$. Before describing player 0, the consumers' payoff functions and strategy sets are described. The preferences of consumer i are represented by the utility functions $u(x^i)$, where $u_i(x^i)$ is scalar valued and $x^i \in R^m_+$ is a commodity bundle. The utility function u_i is defined for all $x^i \in R^m_i$ and, if $x^i, y^i \in R^m_+$ then $u_i(x^i) \geqslant u_i(y^i)$ means that x^i is preferred or indifferent to y^i. Suppose, further, that u_i is quasiconcave, continuous, and finite valued for finite x^i. The vector $w^i \in R^m_+$ is the endowment of player i, and $w^0 = \sum_{i \in N'} w^i$ is the total endowment of the economy. The individual consumer, facing prices $p \in R^m_{++}$, is regarded as having a budget set $B(p, w^i) = \{x^i \in R^m_+ \mid px^i \leqslant pw^i\}$. Thus, $B(p, w^i)$ is the set of commodity bundles costing no more than the value of player i's endowment when prices are p.

Player 0, the price chooser, is fictitious in the sense that he does not correspond to an actual maximizing economic agent in the usual Walrasian model, but he is a useful artifact. Player 0 chooses a price vector $p \in \Phi = \{p \in R^m_{++} \mid \sum_{i \in N'} p_i = 1\}$ and wishes to minimize the extent to which the economy is out of equilibrium. If $(x^1, \ldots, x^n)$ are chosen by the consumers, $x^0 = \sum_{i \in N'} x^i$ is aggregate demand. Excess supply of good j in the economy is $w^0_j - x^0_j$. The sum of the value of (positive) excess supplies in the economy is $\sum_{j=1}^m p_j \max\{w^0_j - x^0_j, 0\}$. It is this value that player 0 seeks to minimize. An alternative version of the objective function of player 0 is $\sum_{j=1}^m p_j \min\{(x^0_j - w^0_j), 0\}$. This is the negative of the preceding formulation (i.e., it is the negative of the value of aggregate excess supplies), and player 0 would wish to maximize it.

Returning to the consumers, it is well known that for any $p \gg 0$, there is a solution to the following problem: $\max u_i(x^i)$ subject to $x^i \in B(p, w^i)$. If u_i is strictly quasiconcave, the x^i that achieves the maximum is unique. If weak quasiconcavity holds, then, for a given (p, w^i), there is a nonempty, convex set of consumption bundles, denoted $d^i(p, w^i)$, which maximize utility. $d^i(p, w^i)$ is called the *demand correspondence*. Thus, $d^i(p, w^i) \subset R^m_+$ and $y^i \in d^i(p, w^i)$ if and only if $u_i(y^i) = \max_{x^i \in B(p, w^i)} u_i(x^i)$.

The purpose of the following formulation is to use the noncooperative equilibrium concept and the corollary to Theorem 3.1 as tools to prove existence of a competitive equilibrium. The pseudogame is defined by Conditions 3.4 to 3.6.

CONDITION 3.4 *$u_i(x^i)$ is a quasiconcave and strictly increasing utility function defined on R^m_+ and satisfying nonsatiation. Let $d^i(p, w^i)$ be the demand correspondence for consumer i, defined for all $p \in \Phi$, and let $z^i_j(p, w^i) = \inf\{y_j \mid y \in d^i(p, w^i)\}$. $z^i_j(p, w^i)$ is the greatest lower bound on the demand for good j by consumer i when prices are p and endowment is w^i. For any $\{p^l\} \in \Phi$ such that $p^l \to p^0$ and $p^0_j = 0$, $z^i_j(p^l, w^l) \to +\infty$ as $l \to \infty$.*

This condition establishes that the payoff of player i ($\neq 0$) is continuous in all strategies and concave in the strategy of player i (the continuity with respect to other players' strategies is trivial because these other strategies do not enter the payoff function) and that the demand for a good goes to infinity as the price of that good goes to 0, which ensures that individual prices close to 0 need not be considered. They can never arise in equilibrium because markets cannot clear. This is important in ensuring compact strategy sets, as seen in Condition 3.6. Let $z_j^0(p, w) = \sum_{i \in N'} z_j^i(p, w^i)$, where $w = (w^1, \ldots, w^n)$ and let $\varepsilon > 0$ satisfy $z_j^0(p, w) > w_j^0$ if $p \in \Phi$ and $p_j < \varepsilon$, for $j = 1, \ldots, m$. That is, if $p_j < \varepsilon$ then the aggregate demand for good j must exceed the aggregate endowment of that good, and the market for good j cannot clear. Condition 3.4 guarantees that such an ε can be found.

CONDITION 3.5 *The strategy set of player* 0 *is* $S_0 = \{p \in \Phi \mid p_j \geqslant \varepsilon, j = 1, \ldots, m\}$, *and his payoff function is* $\sum_{j=1}^m p_j \min\{(x_j^0 - w_j^0), 0\}$.

CONDITION 3.6 *The strategy set of player* $i > 0$ *is* $S_i = \bigcup_{p \in S_0} B(p, w^i)$. *The payoff function of player* $i > 0$, $P_i(p, x^1, \ldots, x^n)$, *is defined on the set* $T_i = \{(p, x^1, \ldots, x^n) \mid p \in S_0, x^i \in B(p, w^i)\}$, *and* $P_i(p, x^1, \ldots, x^n) = u_i(x^i)$.

A pseudogame satisfying Conditions 3.4 to 3.6 also satisfies Assumptions 3.1, 3.2′, and 3.3Q′; hence the corollary to Theorem 3.1 can be applied, and an equilibrium point to this game is a competitive equilibrium. The converse is also true: A competitive equilibrium in this model is an equilibrium point to the game. To see the latter, suppose that $(p, y^1, \ldots, y^n)$ is a competitive equilibrium. Then, $u_i(y^i) = \max_{x^i \in B(p, w^i)} u_i(x^i)$ for $i \in N$. This may also be stated in the form $y^i \in d^i(p, w^i)$, $i \in N$. The demand correspondence d^i is, of course, the best-reply mapping of player $i > 0$. Because markets clear at a competitive equilibrium, $x_i^0 = w_j^0$ for all j and $\sum_{j=1}^m p_j \min\{(x_j^0 - w_j^0), 0\} = 0$, which is clearly the maximum value of player 0's payoff function. Now, to see that an equilibrium point is a competitive equilibrium, suppose that $(p, y^1, \ldots, y^n)$ is an equilibrium point. Then, clearly $y^i \in d^i(p, w^i)$, otherwise, consumer i would not be choosing a best reply. Let $y^0 = \sum_{i=1}^n y^i$, and suppose that $\sum_{j=1}^m p_j \min\{(y_j^0 - w_j^0), 0\} < 0$. Then, for at least one value of j, say j^*, $y_{j^*}^0 < w_{j^*}^0$. This follows from Walras' law, which states that $\sum_j p_j(y_j^0 - w_j^0) = \sum_j(\sum_i y_j^i - \sum_i w_j^i) = \sum_i[\sum_j (p_j y_j^i - p_i w_j^i)] = 0$. Note that if $y_j^0 = w_j^0$ does not hold for all j, then there must be at least one market j^* in which $y_{j^*}^0 < w_{j^*}^0$ and at least one j^{**} in which $y_{j^{**}}^0 > w_{j^{**}}^0$. Furthermore, $p_{j^*} > \varepsilon$ due to Condition 3.4. Therefore, p is not a best reply to $(y^1, \ldots, y^n)$. The price vector p' yields a higher value of the payoff function of player 0 than does p, where $p_j' = p_j$ for $j \neq j^*, j^{**}$, $p_{j^*}' = \varepsilon$, and $p_{j^{**}}' = p_{j^*} + p_{j^{**}} - \varepsilon$. Thus, if $(p, y^1, \ldots, y^n)$ is an equilibrium point, it is also a competitive equilibrium. For a general treatment of competitive equilibrium see Hildenbrand and Kirman (1976) or Debreu (1959).

As a numerical example, imagine an economy with two commodities, apples and oranges, and 10 consumers. Table 3.6 gives the utility functions and endowments for the consumers. The typical consumer, holding w_1^i of apples and w_2^i of oranges, seeks to maximize the utility function $u_i(x_1^i, x_2^i) = x_1^i(x_2^i)^a$ subject to the budget constraint $p_1 w_1^i + p_2 w_2^i = p_1 x_1^i + p_2 x_2^i$. From the

TABLE 3.6 A competitive economy

Consumer	Utility	Endowment		Demand at $(\frac{1}{3}, \frac{2}{3})$	
		Apples	Oranges	Apples	Oranges
1	$x_1x_2^{10}$	5	3	1	5
2	$x_1x_2^8$	6	6	2	8
3	$x_1x_2^5$	8	2	2	5
4	$x_1x_2^3$	6	1	2	3
5	$x_1x_2^{3/2}$	16	2	8	6
6	$x_1x_2^{2/3}$	16	2	12	4
7	$x_1x_2^{1/3}$	2	7	12	2
8	$x_1x_2^{1/5}$	8	5	15	1.5
9	$x_1x_2^{1/8}$	4	7	16	1
10	$x_1x_2^{1/10}$	9	1	10	.5

budget constraint, $x_1^i = w_1^i + (w_2^i - x_2^i)p_2/p_1$. Using this in the utility function, consumer i wishes to maximize

$$[w_1^i + (w_2^i - x_2^i)p_2/p_1][x_2^i]^a \tag{3.57}$$

with respect to x_2^i. Solving this gives

$$x_1^i = \frac{1}{1+a}(w_1^i + p_2 w_2^i/p_1) \tag{3.58}$$

$$x_2^i = \frac{a}{1+a}(p_1 w_1^i/p_2 + w_2^i) \tag{3.59}$$

It is easily verified that $p = (\frac{1}{3}, \frac{2}{3})$ is an equilibrium price vector. Demands at this price vector are shown in the last two columns of Table 3.6. At $p = (\frac{1}{3}, \frac{2}{3})$ the objective function of player 0 is equal to 0, but, away from an equilibrium, her objective function would be strictly negative.

6.3 The free rider problem

The free rider problem concerns a conflict between individual incentives and Pareto optimality. This conflict arises when a group of people is concerned with the provision of a public good, and each person knows only her own preferences. No one knows the preferences of any other person; hence, each is dependent on the information that others voluntarily provide concerning the value to them of the public good and concerning their willingness to pay toward its cost.

Consider the following example: Twenty neighbors jointly own a piece of land on which they may build a swimming pool. Suppose that the pool must, of necessity, be open for all 20 to use (exclusion is impossible), and that it is sufficiently large that there are no congestion effects. Assume that the pool will cost \$200,000 and that each person i has a value V_i that he attaches to the pool. That is, person i is indifferent between (a) his original income and no pool, and (b) a diminution in his income of V_i and the pool. Clearly, if $\sum_{i=1}^{n} V_i > \$200,000$, it would be possible to build the pool with person i contributing w_i toward its cost, leaving all the residents better off than they would have been without the pool. This requires that $w_i < V_i$ for all i and $\sum_{i=1}^{n} w_i = 200,000$. Each person knows that $V_i \geqslant 0$ for all persons, and w_i must be nonnegative.

Now imagine that each person actually values the pool at $V_i = \$15,000$, that each is asked to state a valuation, w_i, and that the pool is built if and only if $\sum_i w_i \geqslant \$200,000$. Thus, the total true valuation of \$300,000 greatly exceeds its cost. Imagine now the position of one person i. If $\sum_{j \neq i} w_j \geqslant \$200,000$, then the pool is built no matter what value player i announces for w_i, and the optimal value for player i is $w_i = 0$. Similarly, if $\sum_{j \neq i} w_j < \$185,000$ one person cannot cause the pool to be built unless he commits to a contribution high enough to make himself worse off, and $w_i = 0$ is still optimal. If $\$185,000 \leqslant \sum_{j \neq i} w_j < 200,000$, then the ideal amount for person i to announce is $w_i = \$200,000 - \sum_{j \neq i} w_j$, just enough to ensure that the pool is built. In general, this amount is less than V_i, thus the narrowly selfish incentive of each person is almost surely to understate the true value of the pool to him.

Suppose person i regards the total bids of the others ($\sum_{j \neq i} w_j$) as a random variable that has a rectangular distribution over the interval from a to $a + b$. Suppose, too, that the cost of the pool is $a + kb$, where $k \in (0, 1)$, and the value of the pool to person i, V_i, is independent of his income level. Then person i's subjective probability estimate that the pool will be built, as a function of his own bid, w_i, is $[(1 - k)b + w_i]/b$. If u_i^* is the utility he achieves with no pool and $u_i^* + V_i - w_i$ his utility if the pool is built and he pays w_i, then person i's expected utility is

$$u_i^* + \frac{(1-k)bV_i + [V_i - (1-k)b]w_i - w_i^2}{b} \tag{3.60}$$

If $V_i \leqslant (1 - k)b$, the optimal value of w_i is 0. Otherwise it is $w_i = [V_i - (1 - k)b]/2$. Note that $(1 - k)b$ is the difference between the cost of the pool $(a + kb)$ and the largest aggregate amount that person i believes the others would bid $(a + b)$. Thus, even with $k = 1$, the highest bid that would be rational for person i is $V_i/2$. This is merely a simple example, not a general theory, it illustrates the force of the free rider problem.

Ideally, a solution to this problem would be a means, or mechanism, of decision that would encourage each person to state his valuation of the pool truthfully and, along with this, devise a schedule of contributions (taxes) t_i

such that $t_i \leqslant V_i$ for all i and $\sum_{i=1}^{20} t_i = 200{,}000$. All this is too much to ask; however, some improvement can be made.

Imagine the following variant of the swimming pool situation: $N = \{1, \ldots, n\}$ is the set of economic agents (e.g., the neighbors). Each person i announces a valuation $s_i \in S_i = R$ for a specific proposal consisting of the swimming pool plus a system of taxes $t = (t_1, \ldots, t_n)$ to pay for it. The aggregate tax, $\sum_{i \in N} t_i$ is just equal to the cost of the public good. Supposing that V_i is the value of the pool to person i, then the value of the proposal (pool with tax plan t) to him is $v_i = V_i - t_i$. Clearly, the situation here is a noncooperative game of incomplete information. It remains to describe the payoff functions of the players. These will specify actual payments and a decision on the public project as functions of the strategies chosen by the players. These payoff functions, along with the decision and tax rules, embody a *Groves mechanism,* which is a structure under which telling the truth is always a best choice. Consequently, announcing v_i will maximize the utility of player i, no matter what other players select. (On Groves mechanisms and incentive compatibility, see Green and Laffont (1979, Part II), Tideman and Tullock (1976), or Groves and Ledyard (1977).) Thus, $v = (v_1, \ldots, v_n) \in S$ is a Nash equilibrium for the game no matter how each player perceives the payoff functions of the others. That is, there are special features of the game that make it irrelevant to the analysis that there is incomplete information. In essence, this is achieved by making the payoff of a player dependent on the way his actions affect the others.

Two caveats concerning Groves mechanisms must be stated. First, the sum of revenues collected for a project need not equal the cost of the project. The dominant strategy property (that truth telling is best) is lost if budget balance is forced to occur. Second, truth telling need not remain optimal if two or more players can collude. Specifically, let $f_i(s)$ be a function such that, for all $s \in S$ and all $s_i' \in S_i$, $f_i(s) = f_i(s \backslash s_i')$. In other words, f_i depends on all the s_j $(j \neq i)$, but does not vary with s_i. The project is adopted if and only if $\sum_{j \in N} s_j \geqslant 0$; hence the payoff function of player i is

$$\begin{aligned} P_i(s) &= f_i(s) + v_i + \sum_{j \neq i} s_j \qquad \text{if } \sum_{j \in N} s_j \geqslant 0 \\ &= f_i(s) \qquad\qquad\qquad\quad \text{if } \sum_{j \in N} s_j < 0 \end{aligned} \tag{3.61}$$

The game, then, is described by (N, S, P). Interestingly, v_i is a best reply to any $s \in S$. That is, $P_i(s \backslash v_i) \geqslant P_i(s)$ for all $s \in S$ and all $i \in N$. Such a strategy is called a *dominant strategy*. To see that v_i is a best reply under all circumstances, consider the best-reply mapping of player i:

$$\begin{aligned} r_i(s) &= \left\{ s_i' \in S_i \,\middle|\, s_i' \geqslant -\sum_{j \neq i} s_j \right\} \qquad \text{if } v_i + \sum_{j \neq i} s_j \geqslant 0 \\ &= \left\{ s_i' \in S_i \,\middle|\, s_i' < -\sum_{j \neq i} s_j \right\} \qquad \text{if } v_i + \sum_{j \neq i} s_j < 0 \end{aligned} \tag{3.62}$$

Clearly, $v_i \in r_i(s)$ in either case, $v \in S$ is an equilibrium point of the game.

This model fails to satisfy the usual assumptions for noncooperative games in several ways: the strategy sets are not compact, (b) the payoff functions P_i are discontinuous at $s_i = -\sum_{j \neq i} s_j$, and (c) each player knows nothing about the payoff functions of the other players. These three differences do not prevent the model from having at least one equilibrium point due to the special structure of the model.

It is interesting to examine more closely the utility functions appearing in the free rider game just discussed. To do this, a slightly more general approach is now taken. Suppose that there are m private goods, $1, \ldots, m$, and that good 0 is the swimming pool. Let $x' = (x_1^i, \ldots, x_n^i)$ so that the complete consumption bundle for consumer i is (x_0, x^i), where $x_0 = 1$ or 0 according to whether the pool is $(x_0 = 1)$ or is not $(x_0 = 0)$ obtained. Suppose $p = (p_1, \ldots, p_m)$, the prices of private goods, are unaffected by the swimming pool decision. In general, the representation of preferences would be by means of a utility function, $U_i(x_0, x^i)$ that depended on the commodity bundle that a player actually consumed; however, the requirement that the value of the pool to person i is independent of the rest of his consumption bundle places additional restrictions on the form of U_i. To make these restrictions clear, an indirect utility function is first derived. Suppose x_0, p, and income M_i is given. The consumer then seeks to maximize $U_i(x_0, x^i)$ with respect to x^i and subject to a conventional budget constraint: $M_i = px^i$. A result of this process is that, for each (x_0, p, M_i) there is a maximum attainable utility level summarized in the indirect utility function $u_i(x_0, p, M_i)$. Supposing U_i to be strictly concave and letting $d^i(x_0, p, M_i)$, denote the consumer's demand system. Then the functions U_i and u_i are related by

$$U_i(x_0, d^i(x_0, p, M_i)) = u_i(x_0, p, M_i) \tag{3.63}$$

The value of the swimming pool to consumer i, given prices p and an income M_i, where M_i is the income that the consumer will have available for spending on private goods if there is no pool, is $\phi_i(p, M_i)$, defined by

$$u_i(0, p, M_i) = u_i(1, p, M_i - \phi_i(p, M_i)) \tag{3.64}$$

On the left in equation (3.64) is the utility level that the consumer will achieve in the absence of a pool. On the right is the same utility level, which is achieved at an income of $M_i - \phi_i(p, M_i)$. Thus, if the consumer pays $\phi_i(p, M_i)$ for the pool, he is indifferent between the two situations. Suppose that the form of u_i is

$$u_i(x_0, p, M_i) = a_i x_0 + M_i \theta_i(p) \tag{3.65}$$

Then the equation defining $\phi_i(p, M_i)$ is

$$M_i \theta_i(p) = a_i + [M_i - \phi_i(p, M_i)]\theta_i(p) \tag{3.66}$$

or

$$\phi_i(p, M_i) = \frac{a_i}{\theta_i(p)} \tag{3.67}$$

If ϕ_i is to be independent of income, which is all we require, then equation (3.65) puts sufficient restrictions on the utility functions. Utility in the form of equation (3.65) is a rather strong assumption. Since prices are constant throughout this discussion, let $a_i/\theta_i(p) = V_i$, where V_i is the value of the pool. Recalling that a utility function will continue to represent the same preferences when divided by a positive constant, we can divide equation (3.65) by $\theta_i(p)$, leaving $u_i(x_0, p, M_i) = V_i x_0 + M_i$ as the utility function.

Now consider a proposal to build the swimming pool and to finance it with payments $t = (t_1, \ldots, t_n)$. The sum $\sum_{i \in N} t_i$ is the cost of the pool. Each person i is asked to state an amount w_i with the understanding that the project will be undertaken if $\sum_{i \in N} w_j \geq 0$, and that each person i will receive $\sum_{j \neq i} w_j$ if the project is undertaken. Thus, the utility function of person i is

$$\begin{array}{ll} M_i & \text{if } \sum_{j \in N} w_j < 0 \\ M_i + (V_i - t_i) + \sum_{j \neq i} w_j & \text{if } \sum_{j \in N} w_j \geq 0 \end{array} \tag{3.68}$$

That is, the status quo is maintained if $\sum_{j \in N} w_j < 0$ and the project is not undertaken; however, if the project is undertaken, person i receives the benefit of it (V_i) minus his preassigned tax (t_i) plus another sum determined by the strategies announced by the other players.

The interesting feature of this scheme is that player i can do no better than to announce $w_i = V_i - t_i$, which is the true net benefit to him of the project. To see this, note that player i's utility does not depend on w_i, except with respect to whether the project is undertaken. That is, for all w_i such that $\sum_{j \in N} w_j \geq 0$, the utility is $M_i + (V_i - t_i) + \sum_{j \neq i} w_j$. Player i benefits from having the project undertaken if and only if $M_i + (V_i - t_i) + \sum_{j \neq i} w_j > M_i$ or $V_i - t_i > -\sum_{j \neq i} w_j$. If w_i is chosen larger than $V_i - t_i$, then the project might be undertaken under circumstances that might be worse for player i than no project. Conversely, if $w_i < V_i - t_i$, the project could fail to be adopted in circumstances where player i would have been better off with it being carried out. Only $w_i = V_i - t_i$ is foolproof.

7 Concluding comments

The basic analytical tool of this chapter is the noncooperative equilibrium, which is defined for games in which each player knows the payoff functions and strategy spaces of all players. In this setting, equilibrium is characterized by a strategy combination under which each player is maximizing his payoff given the strategies of the other players.

The assumption of complete information is used virtually throughout the chapter; even in the section on incomplete information, the incomplete information equilibrium is found by defining a complete information game whose equilibria have obvious analogues in the incomplete information game. Then these analogue-equilibria are taken to be the equilibrium points of the incomplete information game. Advantages of these incomplete information equilibria are that they are clearly related to the non-cooperative equilibrium (for complete information games) and the players have, at equilibrium, beliefs that are mutually consistent. Yet there may be other ways of defining such equilibria that appear equally appealing.

The complete information condition itself may be criticized as being unrealistic; however, complete information is a natural starting point for the investigation of noncooperative games. It is easier to deal with conceptually than incomplete information, the condition is bound to be met in some instances of interest, and it provides a natural benchmark against which to measure equilibria in incomplete information games.

Finally, if the assumption of complete information is accepted, at least provisionally, one can question whether the noncooperative equilibrium is appealing in such games. A clear point in its favor is that it is defined by a condition of internal consistency: If the players in a game choose a noncooperative equilibrium point combination, then no player, ex post, will see that she could have gotten a larger payoff by selecting a different strategy. This is good, and, indeed, in many particular games, it may be an overriding concern. But note that the attractiveness of the equilibrium rests, in part, on the confidence that each player must have that every other player will (a) approach the game in the same way and (b) believe that all players believe that they will all approach the game this way. The Nash approach looks less appealing if players have reason to think that other players may violate either (a) or (b). This issue is addressed in Chapter 5.

Another issue arises, even when noncooperative equilibrium behavior is accepted as reasonable, if a game lacks a unique equilibrium point. Supposing that the players cannot meet and discuss which equilibrium to select, there is no reason to suppose they would select a noncooperative equilibrium strategy combination; each player may choose a strategy assuming a different equilibrium point as the outcome. This problem is absent when the equilibrium is unique or when the game is a strictly competitive two-person game; however, many games fail to fall into either of these categories.

In summary, although the noncooperative equilibrium is not free from disadvantages, it has appealing characteristics. In this, it is like most of what is best in economic theory: it is a useful and interesting tool that provides genuine, important insights, but, at the same time, leaves considerable scope for improvement or supplementation.

Exercises

1. For the following two-person zero-sum games, find the value and the equilibrium strategies.

a.

5	0
2	4

b.

5	0	2
2	4	8

c.

5	0	2
2	4	3

2. The 'prisoners' dilemma' is a game in which each of two players has two pure strategies, confess (C) and maintain innocence (I). The two players are criminals who have committed a crime together and who are being separately grilled by the police. Thus, they are moving simultaneously in a game having one move per player. If both choose I, their payoffs are 15 each each (in utility terms), corresponding to conviction on a minor charge. If both confess, they receive 5 utility units each, corresponding to conviction on a more serious charge. If one confesses and the other does not, the payoffs are 20 and 0, respectively. In this case, the police will let the confessor off free and throw the book at the other. Make a game tree for this game.
3. In the following bimatrix game, player 1 chooses the row and receives the first payoff in a payoff pair. Find an equilibrium point for the game.

5, 0	0, 8
2, 6	4, 5

4. Suppose a game in strategic form is given by (N, S, P), where $N = \{1, 2\}$, $S_1 = [0, 100]$, $S_2 = [0, 100]$, $P_1(s) = 25s_1 - 4s_1^2 + 15s_1s_2$, and $P_2(s) = 100s_2 - 50s_1 - s_2^2 - s_1s_2$. What is the best-reply mapping? Find an equilibrium point for the game.
5. Find an equilibrium point for the game described in problem 4 if $S_1 = [0, 30]$.
6. Suppose a game (N, S, P) given by $N = \{1, 2\}$, $S_1 = [10, 20]$, $S_2 = [0, 15]$, $P_1(s) = 40s_1 + 5s_1s_2 - 2s_1^2$, and $P_2(s) = 50s_2 - 3s_1s_2 - s_2^2$. What is the best-reply mapping? Find an equilibrium point.
7. Can any of the theorems on the uniqueness of noncooperative equilibrium be used to prove uniqueness in problems 4, 5, or 6?
8. Suppose a two-person game in which the strategy spaces are $S_1 = [-100, 100]$ and $S_2[10, 50]$, and let the payoff functions be $P_1(s) = 2s_2^2 - s_1^2$ and $P_2(s) = 4s_1^2s_2^2 - s_1^4 - 4s_2^4$. Find an equilibrium point for this game. Is this game strictly competitive? Determine whether Assumptions 3.1, 3.2, 3.3Q, and 3.3C are satisfied. Suppose the game is changed by letting $S_1 = [-10, 10]$.

Notes

1. It goes without saying that one may contemplate noncooperative games in which (a) N is not finite, (b) the $P_i(s)$ are defined on only a (proper) subset of S, (c) the P_i are not quasiconcave in s_i, or (d) the game is one of incomplete information. These relaxations of assumptions are discussed in Section 2.6.
2. I do not know whether this game has equilibrium points other than the five listed here; I strongly suspect that it does not.

3. That is,

$$P'(s) = \begin{bmatrix} \partial P_1/\partial s_{11} \\ \vdots \\ \partial P_1/\partial s_{1m} \\ \vdots \\ \partial P_n/\partial s_{n1} \\ \vdots \\ \partial P_n/\partial s_{nm} \end{bmatrix}$$

The notation P_i^j will also be used to denote the partial derivative of P_i with respect to its jth argument. P_i^{jk} will denote an analogously defined second partial derivative.

4

Multiperiod noncooperative games without time dependence

The strategic form of a noncooperative game tends to obscure interesting aspects of how the game is played, aspects that are of paramount importance in many applications. For example, in an infinite horizon oligopoly, economists want to understand what determines the individual choices of the firms within each time period. While it is important to know that equilibria exist, it is also important to understand characteristics of equilibrium behavior. Indeed, our usual conception of a game is something such as, say, a single hand of poker, where a sequence of moves is played, the game ends, and payoffs are realized. It is a short step from a single hand of poker to a sequence of hands. The sequence itself should be viewed as a game, or supergame, whose components parts are games, each having its own structure of moves and payoffs. Naturally, an astute poker player will adopt a strategy that takes account of the connections between distinct poker hands. How one plays in a particular hand (game) can influence the way opponents play in subsequent hands. Thus one's play in one hand can affect one's payoffs in later hands.

Two points emerge from this illustration. First, it is not desirable to analyze each hand as if it were totally independent of all others. Second, the optimal play of a single hand will differ according to whether the single play is the whole game or is one of a long sequence of plays.

1 Introduction

The games studied in this chapter and in Chapter 5 are often called *supergames*. The word is intended to suggest a sequence of games, finite or infinite in number, played by a fixed set of players. A *repeated game* is a supergame in which the same (ordinary) game is played at each iteration. It is central to the study of supergames that all players realize they will play a sequence of games and know that this is common knowledge. Various classes of supergames are investigated in the sections that follow and in Chapter 5. It is clearly possible to model such a sequence in a strategic form and obscure the underlying structure of individual component games; however, it is interesting to retain explicit knowledge of these component games and of the choices made within them. Thus, we are left with a *semiextensive* form, in which the move structure is not totally obscured.

The difference between extensive and semiextensive forms lies in the way simultaneous moves are modeled. In the extensive form, players are modeled as if no two players move at the same moment and information sets are used to preserve the appropriate information conditions. In the semiextensive form, the simultaneous moves of the players are modeled as a game in strategic form. Thus, there is a succession of points in time ($t = 0, 1, 2, \ldots$). At each point each player makes a choice. The simultaneous choices at one such time are represented within a (component) game in strategic form and the (super) game is the sequence of these games.

The most fascinating feature of supergames is that the set of equilibria in them is greatly and interestingly enlarged as compared with the single play of a component game. In particular, the set of non-cooperative equilibria of, say, a repeated game, will often include realizations on the payoff possibility frontier which are, in general, unobtainable as equilibrium outcomes in the ordinary game that is being repeated. Luce and Raiffa (1957) noticed this in connection with the prisoner's dilemma.

1.1 The repeated prisoners' dilemma

Table 4.1 shows a prisoners' dilemma game in which the players are the robbers Bonnie and Clyde. After each robbery, the police bring the two in for questioning. The table shows their utilities associated with each possible pair of actions they can take. As a game to be played just once, there is a unique equilibrium point: Both players choose *confess* and each receives a payoff of 5. Now consider the repeated game based on Table 4.1, repeated infinitely many times, and suppose each player discounts future payoffs using a discount parameter of .9. Always to confess is still an equilibrium point, yielding a supergame payoff of $5(1 + .9 + .9^2 + \ldots) = 50$ to each player. Contrast this with a strategy for Bonnie under which she refuses to confess at the first iteration and continues to refuse in each later period if both have in the past also always refused. But if Clyde fails even once to cooperate, Bonnie will revert forever to the safe policy of confessing. If Cylde adopts a parallel strategy, then they can reap 10 units per period and

TABLE 4.1 A prisoners' dilemma game

Bonnie \ Clyde	Confess	Not confess
Confess	5, 5	15, 0
Not confess	0, 15	10, 10

obtain a supergame payoff of 100 each. If Clyde takes advantage of Bonnie's willingness to cooperate, he can get 15 in the first period and 5 in each later period for a supergame payoff of $15 + 5(.9 + .9^2 + \ldots) = 60$. Clearly, it is not in Clyde's interest to follow the latter strategy.

Note that the ability in period t to condition choice on the observed actions of past periods allows Bonnie and Clyde to achieve in a noncooperative equilibrium an outcome normally associated with cooperation (i.e., with collusion and binding agreements). In principle, it is not even necessary that they discuss plans in advance. If each presumes the other will have the good sense to act in this quasicooperative way, their presumptions will be correct, they will be at a noncooperative equilibrium, and they will each receive a payoff of 10 per period.

The first systematic study of these equilibria was done by Aumann (1959, 1961), who proved that *core* (see Chapter 7) outcomes are attainable as noncooperative equilibrium points in repeated games. An earlier qualitative insight into the driving force behind these equilibria comes from Hume (1739–1740), who writes that

> we can better satisfy our appetites in an oblique and artificial manner, than by their headlong and impetuous motion. Hence I learn to do a service to another, without bearing him any real kindness; because I forsee, that he will return my service, in expectation of another of the same kind, and in order to maintain the same correspondence of good offices with me or with others. And accordingly, after I have serv'd him, and he is in possession of the advantage arising from my action, he is induc'd to perform his part, as foreseeing the consequences of his refusal.

And in describing why people are quite reliable about keeping promises when no legal sanction requires them to do so, and when keeping them is inconvenient or costly, Hume comments

> After these signs [i.e., promises] are instituted, whoever uses them is immediately bound by his interest to execute his engagements, and must never expect to be trusted any more, if he refuse to perform what he promis'd.

1.2 Trigger strategy equilibria and the 'folk theorem' for repeated games

The equilibrium in the repeated prisoners' dilemma is an example of both a *trigger strategy equilibrium* and an instance of the *folk theorem for repeated games*. A trigger strategy is a strategy under which a player uses two single-shot actions, s_i^* and s_i^c. The player will begin by choosing s_i^* and will have in mind some action combination s^*. If all players $j \in N$ choose s_j^* in each iteration of the game, then player i continues to choose s_i^*; however, if any player j ever deviates from s_j^* then, as soon as that deviation is detected by player i she switches to choosing s_i^c and continues to choose s_i^c in all future iterations, no matter what choices are made by others. When s^c is an equilibrium of the single-period game (such as *confess, confess* in the prisoners' dilemma), s^* yields a larger payoff to each player than does s^c, and the players' discount parameters are sufficiently large, then the trigger strategy combination is a subgame perfect noncooperative equilibrium.

The *folk theorem for repeated games* states that any attainable payoff vector in a single-shot game can be the realized equilibrium payoff in each period of a repeated game if the payoff vector gives more to each player than does the lowest payoff to which the player could be forcibly held. In the repeated prisoners' dilemma a player can be held down to a payoff of 5 per period. According to the folk theorem, any feasible outcome that gives at least 5 per period to each player can be achieved as the outcome of a non-cooperative equilibrium strategy combination in the infinitely repeated game. It is a special feature of the prisoners' dilemma that the lowest payoff to which a player can be held is also the single-shot noncooperative equilibrium payoff. In general, a player can be held to a lower payoff than that. The folk theorem applies to games in which players do not discount their future receipts, and says nothing about subgame perfection. The origin of the folk theorem is apparently not known; however, Luce and Raiffa (1957) discuss the repeated prisoners' dilemma as a means of attaining the efficient (*not confess, not confess*) outcome by noncooperative means.

The folk theorem and trigger strategy equilibria have an essential element in common. Both are concerned with the attainment of Pareto efficient outcomes in repeated games via noncooperative play. Both sorts of equilibria address the larger question of what outcomes can be supported by noncooperative equilibrium strategy combinations. Trigger strategies focus on single-shot equilibria as the *threat*. Consequently, a payoff vector cannot be supported by a trigger strategy combination unless it gives at least as much to each player as does some single-period equilibrium point. The folk theorem and the theorems closely related to it (the *extended folk theorems*) shows that a payoff may be supported by a noncooperative equilibrium strategy combination if it gives at least as much to each player as the player could be held to by the others. In general, then, these theorems show that more outcomes can be supported by equilibrium strategy combinations that can be supported by trigger strategies. Thus, trigger strategy equilibria are a special case of the extended folk theorems, which raises the question of why anyone should bother with trigger strategies. Trigger strategy equilibria are studied because they have a much simpler form than do the strategies supporting outcomes that trigger strategies cannot support. The simpler form is appealing and likely to be easier to implement.

1.3 Outline of the chapter

Section 2 sets up the class of models that is used in most of the chapter. It also defines subgames for these models and defines subgame perfection. some basic results, due mainly to Abreu (1988), on subgame perfect equilibria in repeated games are developed. Abreu has shown that attention can be restricted to a relatively simple class of strategies. Section 3 takes up theorems on the existence of equilibrium in infinite-horizon repeated games, including subgame perfect trigger strategy equilibria in games with discounting; the folk theorem; the folk theorem extended to subgame perfection; and the folk theorem extended to subgame perfection when the players discount. Section 4 adapts many of the results developed in Section 3 to finite-horizon repeated games. Some comparisons are made in Section 5 between trigger strategy equilibria and the equilibria accounted for in the sophisticated descendants of the folk theorem. Section 6 examines economic applications, and Section 7 includes some concluding remarks.

2 The formulation of repeated games, subgame perfection, and simple strategy combinations

The connection between *games* and *supergames,* and the relationship between *extensive* and *semiextensive* forms requires clarification. The names *supergames* and *repeated games* arose out of historical accident. The accident is that certain one-move-per-player situations, such as the Cournot oligopoly, have long been regarded as games, where as multiperiod versions of the same situation have only recently been formally studied. The result is that the single move version has come to be thought of as "a game," and the infinite period counterpart has come to be labeled as "an infinitely repeated game," or a "supergame." In fact, the single-move Cournot market is one game, and the countably infinite Cournot market is merely another game.

Developments in what are called supergames are very important to economists, because the supergame models have often been inspired by, and developed with an eye toward, interesting economic applications. Furthermore, specific results have been obtained that go beyond existence of equilibrium points and specify some characteristics of equilibrium.

The desire to know more than merely whether equilibrium exists has led to the use of a *semiextensive form* of a game. Recall that the extensive form details the game in a move by move way, so that each individual move appears to take place at a different point in time. No two players move simultaneously; however, the device of information sets allows an extensive form to be logically equivalent to a game in which some moves are simultaneous. Recall, too, that the extensive form is not usually helpful for games that are not very small and simple. The semiextensive form utilized here is based on stating strategies in terms of individual moves and on writing payoff functions in terms of individual moves. In the material that follows in this chapter and Chapter 5, the *individual move of player i at time t* is the same as the *ordinary game strategy of player i at time t.* Reference is made to *time t,* because the games under study are assumed to have an explicit temporal structure. In each time period, the n players simultaneously select moves. These moves can be interpreted as strategies in a game confined to the current time period. That is, each time period has associated with it a payoff function for each player and a strategy set (set of available moves) for each player. A *repeated game* is a game in which the circumstances of the initial time period (payoff functions and sets of available moves) repeat themselves identically in each succeeding period. Of course, the player does not merely consider each occurrence in isolation; he is interested in his overall payoff over the whole time horizon, and he considers strategies that direct the choices of each individual period's action from this global perspective. These strategies are called supergame strategies to distinguish them from the single-period actions.[1]

This section is divided into three subsections. In Section 2.1 a structure for infinitely repeated games is developed and an *antifolk theorem* is presented to show the kind of information flow that is needed to make the study of repeated games interesting. In Section 2.2 subgames and subgame perfect equilibrium are defined. Section 2.3 defines and develops results concerning a class of simple strategy combinations for repeated gams.

2.1 Repeated game definitions and the antifolk theorem

The main purpose of this section is to introduce basic definitions used in supergames. Four categories of supergames are distinguished, all of which are supergames without time-dependent structures. The absence of *structural time dependence* means that the payoff associated with a particular time period depends only on the actions of that time period, as in the Bonnie and Clyde example. Note that the absence of structural time dependence does not preclude strategic time dependence. *Strategic time dependence* is present when the action taken in a time period by a player depends on the history of actions in the game to that time.[2] Again, this is illustrated in the Bonnie and Clyde example, where the *action* taken at any time can depend on the past actions of all players. The four categories of supergames are the combinations obtainable from finite versus infinite horizons, and from repeated versus nonrepeated games. Clearly, infinite-horizon nonrepeated supergames contain the other three categories as special cases. Throughout this section, only infinite-horizon repeated games will be covered.

2.1.1 Information flow over time and the antifolk theorem

Suppose that $\Gamma = (N, S, P)$ is a game in strategic form satisfying Assumptions 3.1, 3.2, and 3.3Q and imagine that the players will engage in this game in each of the time periods $t = 0, 1, 2, \ldots,$. Assume, too, that each player $i \in N$ discounts the future using the discount parameter $\alpha_i = 1/(1 + r_i)$, where $r_i \geqslant 0$ is the discount rate. Letting s_{it} be the action chosen by player i in the tth play of the game, and $s_t = (s_{1t}, \ldots, s_{nt})$, the discounted payoff stream to player i is

$$\sum_{t=0}^{T} \alpha_i^t P_i(s_t), \qquad i \in N \tag{4.1}$$

Allowing the possibility that $T = \infty$, the number of plays can be finite or infinite.

DEFINITION 4.1 *A* **repeated game** *is* $\Gamma = (N, S, P, \alpha)$ *where* (N, S, P) *satisfies Assumptions* 3.1, 3.2, *and* 3.3Q, $\alpha = (\alpha_1, \ldots, \alpha_n)$ *satisfies* $\alpha_i \in (0, 1]$ *for all* $i \in N$.

In the repeated game $\Gamma = (N, S, P, \alpha)$, the game (N, S, P) that is repeated is called the *constituent game*, the *stage game*, or the *single-shot game*.

Something must be specified regarding the flow of information over the time horizon. The player is not actually committed to s_{it} until time t occurs. That is, even if a sequence, $s_{i0}, s_{i1}, s_{i2}, \ldots$ is decided at time 0, the player retains the right at any time t to select *at that moment* any element of S_{it}. Two natural alternative assumptions are: (a) At each play of the game, after s_{it} is chosen for all $i \in N$, then all players are informed of s_t. (b) s_{it} must be chosen for all $i \in N$ and $t = 0, 1, \ldots$ before any player is informed of the choices made by the other players. These two assumptions are stated as Rules 4.1 and 4.1′, respectively.

RULE 4.1 *At each time t, the* s_{it} $(i \in N)$ *are chosen simultaneously; however, for* $t > 0$, s_t, $\tau = 0, 1, \ldots, t-1$ *is known to all players.*

RULE 4.1′ *At each time t, the* s_{it} $(i \in N)$ *are chosen simultaneously. For* $t > 0$, *player i knows* $s_{i\tau}$ *but does not know* $s_{j\tau}$ $(j \neq i)$, $\tau = 0, 1, \ldots, t-1$. *The realized values of* $P_i(s_t)$ *are not revealed to the players until after all choices are made.*

Rules 4.1 and 4.1′ have different implications regarding the way the players' strategy spaces should be modeled. Under Rule 4.1′, a player accumulates no information as time passes, therefore, her choice at any time t cannot be a function of the previous actions of the other players. Thus, her strategy space is naturally $\mathscr{S}_i = \times_{t=0}^{\infty} S_{it}$. Let elements of $\mathscr{S}_i$ be denoted σ_i and let $\sigma = (\sigma_1, \ldots, \sigma_n) \in \mathscr{S} = \times_{i \in N} \mathscr{S}_i$. Under Rule 4.1, the information revealed to each player is the past constituent game pure strategy combination that was chosen; however, in most games of this chapter, the players do not choose mixed strategies. The following theorem, found in Kaneko (1982), gives conditions under which an equilibrium in a repeated game can only be a succession of equilibria of the constituent game. It is an antifolk theorem in the sense that it gives conditions under which a folk-type theorem cannot hold.

THEOREM 4.1 *A repeated game* $\Gamma = (N, S, P, \alpha)$ *satisfying Rule* 4.1′ *has a noncooperative equilibrium.* $\sigma^* = (s_0^*, s_1^*, \ldots, s_T^*)$ *is a noncooperative equilibrium of* Γ *if and only if* s_t^* *is a noncooperative equilibrium for the game* (N, S, P), $t = 0, \ldots, T$.

Proof. (N, S, P) has an equilibrium point; the theorem is therefore proved if it is shown that σ^* is an equilibrium point of Γ if and only if s_t^* is an equilibrium point of (N, S, P) for all t.

Suppose that s_t^* is an equilibrium point of (N, S, P), $t = 0, \ldots, T$. Let $\sigma_i' \in \mathcal{S}_i$, $\sigma_i' \neq \sigma_i^*$. Then $P_i(s_t^* \backslash s_{it}') \neq P_i(s_t^*)$ only if $s_{it}' \neq s_{it}^*$. Because the individual iterations of (N, S, P) are structurally independent (i.e., the action of one period s_t does not enter into the payoff functions of any period other than period t), $P_i(s_t^* \backslash s_{it}') \neq P_i(s_t^*)$ means $P_i(s_t^* \backslash s_{it}') < P_i(s_t^*)$. Thus, σ^* is an equilibrium point of Γ.

Suppose now that σ^* is an equilibrium point of Γ. If s_t^* were not an equilibrium point of (N, S, P) for some value of t, then there would be a player i who could choose σ_i' with $s_{i\tau}' = s_{i\tau}^*$ for $\tau \neq t$, $P_i(s_t^* \backslash s_{it}') > P_i(s_t^*)$, and $P_i(s_\tau^* \backslash s_{i\tau}') = P_i(s_\tau^*)$ for all $\tau \neq t$. Thus, if s_t^* were not an equilibrium point for (N, S, P) for all t, then σ^* could not be an equilibrium point for (N, S, P, α). QED

This theorem is certainly not surprising; indeed, it is trivial. In a supergame composed of independent plays of a game, and with players receiving no information as the game progresses, it is clear that a supergame equilibrium must, of necessity, be a sequence of plays of individual game equilibria. Note that s_t^* need not equal $s_{t'}^*$ as long as both are equilibrium points of (N, S, P). It is interesting to note that an equilibrium point for a game, such as Γ in Theorem 4.1, is still an equilibrium point for an otherwise identical game in which Rule 4.1′ is replaced with Rule 4.1. To see why this is true, it is first necessary to contemplate strategy spaces for games under Rule 4.1.

2.1.2 *Strategies and strategy spaces for repeated games*

The most appropriate way to define strategy spaces for the players takes into account all the information the players will have at each decision point and assumes this information will be used. Under Rule 4.1, the players will accumulate information on the history of actual period by period choices. Let h_i be the *history of the game* at time t. Then, $h_t = (s_0, s_1, \ldots, s_{t-1})$, where s_τ, $\tau = 0, \ldots, t-1$, is understood to be the actual game move combination chosen in time τ. Then a player i can select her own action in period t as a function of the history of the game to that time. Any *decision function* $\nu_{it}(h_t)$ is suitable as long as it is defined for all $h_t \in S^t = \times_{\tau=0}^{t-1} S$ and has values in S_i. The set of histories at time 0, which is empty, will be denoted S^0. This is done to permit simpler statements. Thus, saying that ν_{i0} is any function from S^0 to S_i is a rather fancy way of saying that ν_{i0} can be any element of S_i.

DEFINITION 4.2 *The* **set of decision functions for player** *i at time* $t \geq 0$ *is* $V_{it} = \{\nu_{it} \mid \nu_{it} : S^t \to S_i\}$.

DEFINITION 4.3 *A* **complete memory repeated game strategy** $\sigma_i = (v_{i0}, v_{i1}, v_{i2}, \ldots)$ *is a strategy that satisfies* $v_{i0} \in S_i$, *and* $v_{it} \in V_{it}$, $t = 1, \ldots$. *The set of such strategies for player i is denoted* $\mathcal{S}_i^*$.

Most of the models of Chapters 4 and 5 are based on assumptions under which mixed strategies need not be considered. Thus, it is reasonable to assume that s_{jt} is observable, after it is played, by players $i \neq j$. If mixed strategies were used, then it would be appropriate to define h_t as the history of actual pure moves that were played and to assume that the underlying distributions governing the selection of moves were not observable by other players. The virtue of the space $\mathcal{S}_i^*$ is that it is the natural (pure) strategy space under the information conditions specified in Rule 4.1. In period 0, there is no past history; hence, all a player can do is choose some $s_i \in S_i$. In a later period $t > 0$, a player has seen $(s_0, \ldots, s_{t-1}) = h_t$. Clearly, any function associating h_t with an element of S_i, defined for all $h_t \in S^t$, is imaginable and should not be ruled out. That is, no such function should be thought technically beyond the player's reach.

The elements of the strategy space space $\mathcal{S}_i^*$ are often called *closed-loop strategies* to connote that the action taken in a period will, in general, depend on information that has become available after the beginning of play. This contrasts with elements of $\mathcal{S}_i$, which are called *open-loop strategies,* under which the action chosen by a player i for some period t does not depend on any information that might become available as the game progresses from period 0 to period $t-1$. Of course, under Rule 4.1′ no such information accumulates; however, it is possible to contemplate both open- and closed-loop strategies under Rule 4.1.

DEFINITION 4.4 σ_i *is a* **closed-loop strategy** *if* $\sigma_i = (v_{i0}, \ldots, v_{iT})$, *where* $v_{it} \in V_{it}$ *for* $t = 0, 1, 2, \ldots$.

Note that $\mathcal{S}^*$ contains $\mathcal{S}$ and is the largest set of closed-loop strategies compatible with Rule 4.1. It is clearly possible to contemplate smaller sets of closed-loop strategies that are subsets of $\mathcal{S}^*$ but are larger than $\mathcal{S}$, the set of open-loop strategies.

DEFINITION 4.5 σ_i *is an* **open-loop strategy** *if* $\sigma_i = (v_{i0}, \ldots)$ *and* $v_{it}(h_t) = v_{it}(h_t')$ *for all* $h_t, h_t' \in S^t$ *and* $t = 0, 1, 2, \ldots$.

Open-loop strategies are defined here as a special (degenerate) case of closed-loop strategies.

If a specific strategy combination σ is played starting from $t = 0$, then it will cause one particular sequence of single-period actions to be selected. This sequence is called a *path.* A path is necessarily an element of $\mathcal{S}$ and any element of $\mathcal{S}$ may be regarded as a path. Thus, a strategy combination, $\sigma = (v_0, v_1, \ldots)$, *induces a path* $u(\sigma) = (u_0(\sigma), u_1(\sigma), \ldots) \in \mathcal{S}$, where $u_0(\sigma) = v_0$ and $u_t(\sigma) = v_t(u_0(\sigma), \ldots, u_{t-1}(\sigma))$ for $t \geq 1$. The payoffs in the repeated game are derived from the payoffs received in each time period and these, in turn, depend upon the particular action combination selected in that period. Thus, repeated game payoffs associated with σ must be defined in terms of the path induced by σ. The repeated game payoff function is

$$G_i(\sigma) = \sum_{t=0}^{\infty} \alpha_i^t P_i(u_t(\sigma)) \tag{4.2}$$

Letting $G = (G_1, \ldots, G_n)$, a repeated game can be denoted as in Definition 4.6.

DEFINITION 4.6 $\Gamma^* = (N, \mathcal{S}^*, G)$ *is a* **repeated game with closed-loop strategies** *under Rule* 4.1.

On the face of it, restricting attention to strategies in $\mathcal{S}$ (i.e., to open-loop strategies) when Rule 4.1 holds seems unreasonably restrictive. The player is being prevented from taking explicit advantage of information that will come into his possession; however, for games satisfying Assumptions 3.1, 3.2, and 3.3Q and Rule 4.1, a noncooperative equilibrium with the strategy space restricted to $\mathcal{S}$ is still a noncooperative equilibrium when the enlarged strategy space $\mathcal{S}^*$ is considered.

COROLLARY *Let* $\Gamma = (N, \mathcal{S}, G)$ *and* $\Gamma^* = (N, \mathcal{S}^*, G)$ *satisfy Assumptions* 3.1, 3.2, *and* 3.3Q. *Let* Γ *satisfy Rule* 4.1′ *and* Γ^* *satisfy Rule* 4.1. *Then if* $\sigma^* \in \mathcal{S}$ *is an equilibrium point of* Γ, *it is also an equilibrium point of* Γ^*.

Proof. It suffices to note that, given the σ_j^* $(j \neq i)$, player i could not alter her strategy from σ_i^* and increase her payoff, $i \in N$. QED

It is crucial to this corollary that the players know in advance what the exact payoff functions will be in each future period, and that the intended strategy of each player can be carried out precisely as each plans. If, in each period, a player might face either P_i^0 or P_i^1 as a single-period payoff function, with the realizations being randomly determined and announced at the start of each period before that period's action is selected, then an open-loop strategy combination would not, in general, be an equilibrium point in a game allowing closed-loop strategies. A player could improve on an open-loop strategy by selecting a strategy that directed, say, $s_{it} = s_i^0$ when P_i^0 is the tth-period payoff function and $s_{it} = s_i^1$ when P_i^1 is the tth-period payoff function.

2.1.3 A note on discount parameters and continuation probabilities

Suppose the players in a repeated game do not discount their payoffs, that is, they value a dollar received in 10 years the same as a dollar received today. However, suppose that the repeated game is finite with a positive probability $p \in (0, 1)$ that it will continue into time $t + 1$ given that it has reached time t. In such a model the payoff function remains equation (4.2).

The only change is that $\alpha_i = p$ for all i and the interpretation of α_i is changed. From an analytical standpoint nothing is altered. Note that for any $t > 0$ there is a strictly positive probability (p^t) of reaching this period.

Both discounting and a stopping probability can fall within the model. Suppose player i discounts at the rate $r_i > 0$ and the continuation probability is p. Then $\alpha_i = p/(1 + r_i)$.

2.2 Subgames and subgame perfection

Subgame perfection is easily brought into the repeated game framework and is very useful. Recalling the definition of *subgame* from Chapter 2, it is easy to see that what constitutes a subgame here is the play of the repeated game from any given time period, t, onward, together with h_t, the history of the game at time t. To put it another way, to name a specific h_t is to specify a subgame. The subgame begins at time t with no player having chosen an action for time t. It is not possible to have a subgame at time t that starts when some players have chosen time t actions, but others have not, because the actions of any time period are chosen simultaneously; therefore, such a starting point is an information set containing multiple nodes. The only single-node information sets occurring at time t are at points when no player has made a time t choice. This is illustrated in Figure 4.1 for a two-person game. The game begins with player 1 choosing an action for period 0. The next information set, where player 2 selects an action for $t = 0$, contains an uncountable infinity of nodes. The nodes following the move of player 2 are all in distinct information sets and a subgame begins

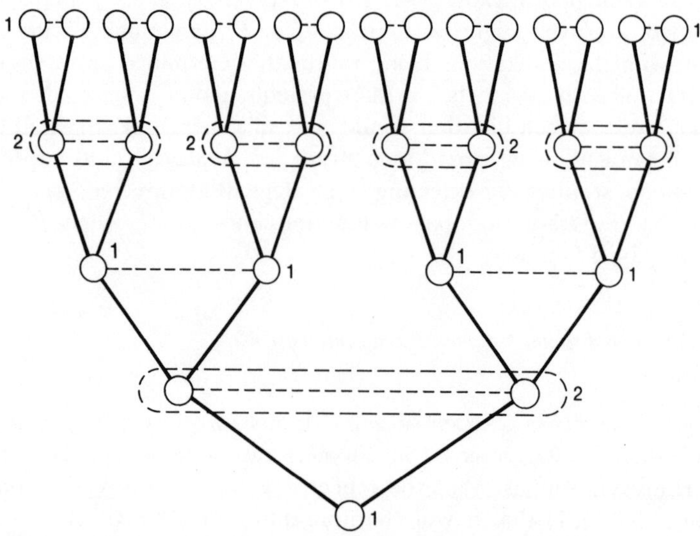

FIGURE 4.1 A schematic game tree for a two-person repeated game.

at each of them. Following the second move of player 1, note that player 2 will know what player 1 chose at $t = 0$, but not what she chose at $t = 1$.

Again, given a game $(N, \mathcal{S}^*, G)$ under Rule 4.1, specifying a history h_t specifies a subgame. The set of strategies in the subgame available to player i is based on the decision functions available to the player. These, in turn, are defined on histories that begin with h_t. Thus, a decision function for time t is really just an element of S_i. A decision function for some later time $t+k$ is a function of any history h_{t+k} that is identical with h_t in its first t move combinations.

DEFINITION 4.7 *The* **set of decision functions for player** i **at time** $t+k$ **in the subgame specified by** h_t *is*

$$V_{i,h_{t+k}} = \{v_{i,t+k} \mid v_{i,t+k}: \{h_t\} \times S^k \rightarrow S_i\},\ k = 0, 1, \ldots$$

$V_{h_{t+k}} = \times_{i \in N} V_{i,h_{t+k}}$ and $v_{t+k} = (v_{1,t+k}, \ldots, v_{n,t+k})$.

A strategy of player i in the subgame specified by the history h_t is denoted $\sigma_{i,h_t} = (v_{it}, v_{i,t+1}, \ldots)$. The strategy space for player i in this subgame, is denoted $\mathcal{S}^*_{i,h_t}$.

DEFINITION 4.8 *The* **strategy space of player** i **in the subgame specified by** h_t *is*

$$\mathcal{S}^*_{i,h_t} = \{\sigma_{i,h_t} = (v_{it}, v_{i,t+1}, \ldots) \mid v_{i,t+k} \in V_{i,h_{t+k}},\ k = 0, 1, \ldots\}.$$

The joint strategy space is $\mathcal{S}^*_{h_t} = \times_{i \in N} \mathcal{S}^*_{i,h_t}$ and an element of $\mathcal{S}^*_{h_t}$ is denoted σ_{h_t}. σ_{h_t} *induces* a *path* $u(\sigma_{h_t}) \in \mathcal{S}$ in the obvious way; namely, $u(\sigma_{h_t}) = (u_t(\sigma_{h_t}), u_{t+1}(\sigma_{h_t}), \ldots)$, where $u_t(\sigma_{h_t}) = v_t(h_t)$ and, for $k > 0$, $u_{t+k}(\sigma_{h_t}) = v_{t+k}(h_t, u_t(\sigma_{h_t}), \ldots, u_{t+k-1}(\sigma_{h_t}))$. It is important to note that, in general, $u_{t+k}(\sigma) \neq u_{t+k}(\sigma_{h_t})$. This is because it is permissible to consider any conceivable history h_t and not merely the history that would be generated by following σ from the beginning. The subgame payoff function for player i is

$$G_{ih_t}(\sigma_{h_t}) = \sum_{\tau=t}^{\infty} \alpha_i^{\tau-t} P_i(u_\tau(\sigma_{h_t})) \tag{4.3}$$

DEFINITION 4.9 *Let* $\Gamma = (N, \mathcal{S}^*, G)$ *be a repeated game with a closed-loop strategy set and for* $t \geqslant 0$ *let* h_t *be the history of the game through period* t. *Then the* **subgame of** Γ **at time** t **with history** h_t *is described by* (a) *the set of players* N, (b) *the strategy spaces* $\mathcal{S}^*_{ih_t}$, $i \in N$, (c) *the payoff functions* G_{ih_t}, *and* (d) *the history* h_t. *This subgame is denoted* $\Gamma_{h_t} = (N, \mathcal{S}^*_{h_t}, G_{h_t})$.

DEFINITION 4.10 $\sigma^* \in \mathcal{S}^*$ *is a* **subgame perfect equilibrium point** *of* Γ *if* σ^* *is an equilibrium point of* Γ_{h_t} *for* $t = 0, 1, \ldots, \infty$ *and for all* $h_t \in S^t$.

Under Definition 4.10, σ^* is subgame perfect if it is an equilibrium point for any possible subgame of the original game. That is, the strategy combination induced on Γ_{h_t} by σ^* must be an equilibrium point in each subgame Γ_{h_t}, even if the subgame Γ_{h_t} would never be encountered when σ^* is actually played.

2.3 The sufficiency of simple strategies

Many interesting strategy combinations can be described by a small number of paths together with rules stating when to switch from one path to another. Such strategy combinations, formally defined here, are called *simple strategy combinations*. An example examined earlier is the trigger strategy combination in the repeated prisoners' dilemma. Bonnie and Clyde start out on the path under which they each choose *not confess* in each period. If either deviates from this path, they switch to a path under which they each choose *confess* in each period. A simple strategy combination will cause a specific path to be followed if no player deviates from his strategy and it will cause any deviating player to be punished by switching to a path that gives the deviator a low payoff. Extensions of the folk theorem, discussed in Sections 3 and 4, are proved by the construction of subgame perfect simple strategy combinations. Simple strategy combinations are precisely defined in Section 2.3.1. The punishment possibilities inherent in repeated game strategies are explored in Section 2.3.2, where it is found that simple strategy combinations can punish as effectively as any strategy combinations. It is then shown in Section 2.3.3 that any outcome (path) supportable by a subgame perfect strategy combination can be supported by a simple strategy combination that is subgame perfect.

2.3.1 *Simple strategy combinations*

In general, the simple strategy combinations are characterized by $n+1$ paths. The first path, μ^0, is followed at the beginning and continues to be followed until a deviation from it occurs. Frequently μ^0 can be interpreted as cooperative behavior. In the context of a simple strategy combination, this cooperative behavior will be supported by a subgame perfect strategy combination. This is so in the repeated prisoners' dilemma, where the *not confess* path represents cooperative behavior that is supported by the trigger strategy combination. The remaining paths, $\mu^1, \ldots, \mu^n$ are called *punishment paths*, with μ^i being the punishment path of player i. In the repeated prisoners' dilemma, there is just one punishment path, the *confess* path, which is used to punish deviations by either player. The rules governing path choice under a simple strategy combination are that, at any time, there is a current path to be followed. The current path is followed forever if no player ever deviates from it. If some player i deviates, then that player's punishment path μ^i becomes the current path and it will be followed forever unless a later deviation from μ^i occurs. In general, whenever a deviation occurs, the deviator's punishment path becomes the current path and it is followed from the start of that path. Thus, if player i would deviate when μ^i is the current path, then that very same path μ^i remains the current path, but it is restarted from its beginning. Because the μ^i generally punish more severely in the earlier phase than the later, deviation from μ^i by player i will tend to be costly. Deviations by two or more players at the same time are strategically unimportant because collusion between players is not considered; however, the strategies must specify what happens if there are multiple deviations. The rule followed here is that when players i, j, k deviate at the same time the punishment path associated with $\min\{i, j, k\}$ is followed. This is arbitrary and any clear rule will suffice.

DEFINITION 4.11 *A* **simple strategy combination,** $\sigma = (\mathbf{v}_0, \mathbf{v}_1, \ldots) \in \mathcal{S}^*$, *is characterized by* $\mu^i \in \mathcal{S}$, $i = 0, 1, \ldots, n$ *according to the following rules*:

a) *At any time* t, *one of the* μ^i *is the current path. At* $t = 0$ *the current path is* μ^0 *and* $\mathbf{v}_0 = \mu_0^0$.

b) *If* $h_t = (\mu_0^0, \ldots, \mu_{t-1}^0)$ *then* $\mathbf{v}_t(h_t) = \mu_t^0$ *and* μ^0 *is the current path.*

c) *If* $s_t = \mathbf{v}_t(h_t) = \mu_\tau^i$ *then the current path does not change and* $s_{t+1} = \mathbf{v}_{t+1}(h_t, \mu_\tau^i) = \mu_{\tau+1}^i$.

d) *If* $s_{it} \neq \mathbf{v}_{it}(h_t)$ *and* $s_{jt} = \mathbf{v}_{jt}(h_t)$ *for all* $j < i$, *then at time* $t + 1$ *the current path is* μ^i *and* $\mathbf{v}_{t+1}(h_{t+1}) = \mu_0^i$.

In view of the preceding definition simple strategy combinations may be written $\sigma^\mu = (\mu^0, \mu^1, \ldots, \mu^n)$. As an exercise the reader might want to write out the definition of σ_i^μ, a simple strategy for player i.

2.3.2 *Optimal penal codes*

A punishment for player i is more than a path, μ^i, on which player i fares poorly; it is a strategy combination the players will have the proper incentives to follow and for which the realized path is μ^i. Thus, to avoid noncredible threats, a punishment for player i must be subgame perfect. A *penal code* is a collection of n punishments, one for each player. Attention is restricted here to penal codes that are optimal, which means that they punish the players as severely as possible. Let $\mathcal{S}^P \subset \mathcal{S}^*$ be the subgame perfect equilibria of $(N, \mathcal{S}^*, G)$.

LEMMA 4.1 *Let* $(N, \mathcal{S}^*, G)$ *be a repeated game. Then* $\mathcal{S}^P$ *is not empty.*

Proof. Let (N, S, P) be the constituent game of $(N, \mathcal{S}^*, G)$, let s^c be an equilibrium point of (N, S, P), and let $\mu^i = (s^c, s^c, \ldots)$ for $i = 0$ and all $i \in N$. Then $\sigma^\mu = (\mu^0, \mu^1, \ldots, \mu^n) \in \mathcal{S}^P$. QED

Let $\underline{G}_i = \min_{\sigma \in \mathcal{S}^P} G_i(\sigma)$, $i \in N$. $\underline{G}_i$ is the lowest payoff that player i could possibly receive under a subgame perfect strategy combination. A penal code is optimal if, for each player i, her punishment holds her to $\underline{G}_i$.

DEFINITION 4.12 *Let* $\underline{\sigma}^i \in \mathcal{S}^P$, $i \in N$. *Then* $(\underline{\sigma}^1, \ldots, \underline{\sigma}^n)$ is an **optimal penal code** *if* $G_i(\underline{\sigma}^i) = \underline{G}_i$.

An optimal penal code $(\underline{\sigma}^1, \ldots, \underline{\sigma}^n)$ is *simple* if each of the $\underline{\sigma}^i$ is a simple strategy combination.

DEFINITION 4.13 *If* $G_i(\underline{\mu}^i) = \underline{G}_i$ *for* $\mu^i \in \mathcal{S}$ *and* $i \in N$, *and* σ^i *is a simple strategy combination given by* $(\mu^i, \underline{\mu}^1, \ldots, \mu^n)$, *then* $(\underline{\sigma}^1, \ldots, \underline{\sigma}^n)$ *is a* **simple optimal penal code.**

In view of Definition 4.13 a simple optimal penal code can be characterized by the n paths that it utilizes, $(\mu^1, \ldots, \mu^n)$.

2.3.3 *Supporting attainable outcomes*

This section contains several important results concerning simple strategy combinations. The first, Theorem 4.2, gives necessary and sufficient conditions for a simple strategy combination to be subgame perfect. Later results establish the existence of optimal penal codes and then give necessary and sufficient conditions characterizing the set of paths that can be supported by optimal penal codes.

THEOREM 4.2 *The simple strategy combination $\sigma^\mu = (\mu^\sigma, \mu^1, \ldots, \mu^n)$ is a subgame perfect equilibrium point if and only if*

$$\sum_{\tau=t}^{\infty} \alpha_i^{\tau-t} P_i(\mu_\tau^j) \geqslant P_i(\mu_t^j \backslash s_i) + \alpha_i G_i(\mu^i) \tag{4.4}$$

for all $s_i \in S_i$, $j \in \{0\} \cup N$, $t \geqslant 0$, and $i \in N$.

Proof. First it is proved that if σ^μ is subgame perfect, then equation (4.4) must hold. Suppose that equation (4.4) is violated for some i, j, t, and s_i. There must be a time t' and a history $h_{t'}$ such that μ^j is the current path at time t' and has been the current path for exactly t immediately previous periods. Then $h_{t'}$ identifies a subgame for which $\sigma^\mu_{h_{t'}}$ is not an equilibrium because the subgame strategy of player i is not a best reply. Therefore, if σ^μ is subgame perfect, then equation (4.4) must hold.

Now it is shown that if equation (4.4) holds for all i, j, t, and s_i then σ^μ must be subgame perfect. That equation (4.4) holds means that no single deviation from σ_i^μ by player i can increase his payoff on any subgame. It is necessary to see whether a sequence of deviations by player i can increase his payoff. Suppose there is such a sequence. Then there is some strategy σ_i' and history h_t such that

$$G_{ih_t}(\sigma^\mu_{h_t} \backslash \sigma'_{ih_t}) - G_{ih_t}(\sigma^\mu_{h_t}) = \varepsilon > 0 \tag{4.5}$$

Because the P_i are bounded and players discount future payoffs, equation (4.5) implies that there is a finite sequence satisfying equation (4.5) for $\frac{1}{2}\varepsilon$. The reason is that there is a finite time T such that payoffs received in the subgame from T onward, discounted to t, must be less than $\frac{1}{2}\varepsilon$. Therefore, following a strategy that is equivalent to using σ'_{ih_t} until T and then using $\sigma^\mu_{ih_t}$ must be superior to σ^μ_{i,h_t} on the subgame by at least $\frac{1}{2}\varepsilon$. Consequently, σ'_{i,h_t} may as well be viewed as prescribing only a finite number of deviations. Suppose the times at which these deviations occur are $t_1, \ldots, t_k$, where $t_l < t_{l+1}$. Consider the last deviation, taking place at t_k. This one cannot increase the subgame payoff of player i from time t_k onward, nor will the choice made by player i at time t_k affect the choices made by any player at any earlier time. Consequently, the deviation at t_k can be replaced with nondeviating behavior and the payoff of player i will either increase or remain the same. It cannot fall. The same argument can be applied successively to the deviations to times $t_{k-1}, \ldots, t_1$. Thus, all the deviations can be eliminated with no decrease in payoff on the subgame, which means that no sequence of deviations, finite or infinite, can be found to increase the payoff of player i on any subgame if no single such deviation can be found. This completes the proof. QED

To summarize the content of Theorem 4.2, under $\sigma^\mu = (\mu^0, \mu^1, \ldots, \mu^n)$, $\mu^0 = u(\sigma^\mu)$ is the path that will be followed under σ^μ if no player deviates. μ^i is the punishment path for player i, which means that if player i deviates at time t and no player deviates thereafter, then player i will receive a payoff of $G_i(\mu^i)$ from period $t+1$ onward, discounted to $t+1$. σ^μ is subgame perfect if and only if no player at any time t ever has a profitable single deviation from σ_i^μ given that, following the deviation, her payoff will be $G_i(\mu^i)$.

The worst possible punishment path payoff for player i is $G_i(\mu^i) = \underline{G}_i$. Lemma 4.2 is the unsurprising result that, along the path of a subgame perfect equilibrium, single deviations under which $G_i(\mu^i) = \underline{G}_i$ do not pay.

LEMMA 4.2 *If $\sigma \in \mathcal{S}^P$ and $u(\sigma)$ is the path induced by σ, then*

$$\sum_{\tau=t}^{\infty} \alpha_i^{\tau-t} P_i(u_\tau(\sigma)) \geqslant P_i(u_t(\sigma)\backslash s_i) + \alpha_i \underline{G}_i \tag{4.6}$$

for all $s_i \in S_i$, $t = 0, 1, \ldots$, and $i \in N$.

Proof. Note first that player i cannot be held to a payoff lower than $\underline{G}_i$ on any subgame, given that a subgame perfect equilibrium is played, because $\underline{G}_i$ is the smallest payoff of player i that is consistent with subgame perfect equilibrium play. Suppose that equation (4.6) fails for some i, s_i, and t. Then there would be a subgame on which player i could achieve a larger payoff by deviating from σ_i than she receives without deviating, in which case σ would not be subgame perfect. Therefore, if σ is subgame perfect, equation (4.6) must hold for all $s_i \in S_i$, $t = 0, 1, \ldots$, and $i \in N$. QED

Let $M^P = \{u(\sigma) \mid \sigma \in \mathcal{S}^P\}$. $M^P \subset \mathcal{S}$ and is the set of paths generated by subgame perfect strategy combinations. Letting γ be the maximum of the discount parameters, α_i, a norm and a distance can be defined on $\mathcal{S}$ as follows. For μ, $\mu' \in \mathcal{S}$, the norm is given by $\|\mu\| = \sum_{t=0}^{\infty} \gamma^t \|\mu_t\|$ and the distance is $d(\mu, \mu') = \|\mu - \mu'\|$.

Lemma 4.3 establishes that any player i can be held to a repeated game payoff of $\underline{G}_i$ by means of a subgame perfect simple strategy combination.

LEMMA 4.3 *A simple optimal penal code exists.*

Proof. Let $\{\sigma^{il}\}_{l=1}^{\infty} \subset \mathcal{S}^P$ be a sequence for which $\{G_i(\sigma^{il})\}_{l=1}^{\infty}$ converges to $\underline{G}_i$, and let $u(\sigma^{il}) = \mu^{il}$. Because $\mathcal{S}$ is compact, the sequence $\{\mu^{il}\}_{l=1}^{\infty}$ has a cluster point, μ^i, for which $G_i(\mu^i) = \underline{G}_i$. Repeat the preceding for each $i \in N$. It will now be proved that the simple strategy combination $\sigma^i = (\mu^i, \mu^1, \ldots, \mu^n)$ is subgame perfect. If σ^i is not subgame perfect, then, by Theorem 4.2, there are k, j, t', and s_k such that

$$G_{k,h_t}(\sigma^i_{h_{t'}}) < P_k(\mu^j_t \backslash s_k) + \alpha_k G_k(\mu^k) = P_k(\mu^j_t \backslash s_k) + \alpha_k \underline{G}_k \tag{4.7}$$

Letting $\sigma^{il} = (\mu^{il}, \mu^{1l}, \ldots, \mu^{nl})$, by continuity of G_k, for large l

$$G_{k,h_{t'}}(\sigma^{il}_{h_{t'}}) < P_k(\mu^{jl}_t \backslash s_k) + \alpha_k \underline{G}_k \tag{4.8}$$

which contradicts $\mu^{jl} \in M^P$. Therefore, σ^i is a subgame perfect. This can be repeated for each $i \in N$, which completes the proof. QED

As the corollary below shows, any optimal penal code has associated with it a simple optimal penal code. The latter is found by looking for the paths that would be followed under each of the n punishments specified by the code.

COROLLARY *Let $(\sigma^1, \ldots, \sigma^n)$ be an optimal penal code. Then $(\mu^1, \ldots, \mu^n)$ is a simple optimal penal code where $\mu^i = u(\underline{\sigma}^i)$, $i \in N$.*

The proof is left to the reader.

Theorem 4.2 and Lemma 4.2 imply Theorem 4.3, which characterizes the simple strategy combinations that are subgame perfect. $\sigma^\mu = (\mu^0, \mu^1, \ldots, \mu^n)$ is subgame perfect if and only if $(\mu^1, \ldots, \mu^n)$ is an optimal penal code and no single deviation from μ^0 is profitable for any player at any time.

THEOREM 4.3 *Let $(\mu^1, \ldots, \mu^n)$ be an optimal penal code. Then $\mu^0 \in M^P$ and $\sigma^\mu = (\mu^0, \mu^1, \ldots, \mu^n) \in \mathscr{S}^P$ if and only if*

$$\sum_{\tau=t}^{\infty} \alpha_i^{\tau-t} P_i(\mu^0_\tau) \geqslant P_i(\mu^0_t \backslash s_i) + \alpha_i \underline{G}_i \tag{4.9}$$

for all $i \in N$ and $s_i \in S_i$.

3 Equilibria in infinite-horizon repeated games

In Section 3.1 trigger strategy equilibria in games with discounting are examined. The classic folk theorem in repeated games without discounting and its strengthening to subgame perfection are discussed in Section 3.2. In Section 3.3 the subgame perfect folk theorem is proved for games with discounting. Finally, trigger strategy equilibria in general nontime-dependent supergames are examined in Section 3.4.

3.1 Trigger strategy equilibria in repeated games with discounting

3.1.1 Characterizing grim trigger and finite reversion trigger strategy combinations

Trigger strategy combinations are a special case of simple strategy combinations. Under a trigger strategy combination $(\mu^0, \mu^1, \ldots, \mu^n)$ is specialized in two ways. First, $\mu^i = \mu^j$ for all $i, j \in N$. That is, all punishment paths are the same, no matter who deviates. Second, μ^0 is a constant path so that $\mu^0 = (s^*, s^*, s^*, \ldots)$ for some $s^* \in S$. The punishment paths show considerable regularity as well. At one extreme are the *grim trigger strategies,* under which $\mu^i = (s^c, s^c, s^c, \ldots)$ for all $i \in N$ and for some $s^c \in S$. If a trigger strategy is not grim, then it incorporates *finite reversion.* This means that $\mu^i_t = s^c$ for $t = 1, \ldots, T$ and $\mu^i_t = s^*$ for $t > T$. Thus, under finite reversion, the punishment lasts for only T periods, while under the grim trigger, punishment lasts forever.

Interesting trigger strategies are restricted to those for which s^c is an equilibrium point of (N, S, P) and $P(s^*) \gg P(s^c)$. In view of the simplicity of trigger strategies, grim trigger strategy combinations can be described by (s^*, s^c) or (s^*, s^c, ∞) and finite reversion trigger strategy combinations by (s^*, s^c, T).

In thinking about trigger strategies, recall the switching rules for simple strategy combinations. As applied to grim trigger strategies, player i chooses s_i^* in period 0 and chooses s_i^* at any later time t if and only if $s_\tau = s^*$ for $\tau = 0, \ldots, t-1$. That is, s_i^* continues to be chosen as long as the observed action combination has been s^* in all past periods. Otherwise, s_i^c is chosen. If any player deviates from s^c, s_i^c continues to be chosen.

As applied to finite reversion trigger strategy combinations, player i chooses s_i^* in period 0 and chooses s_i^* at any later time t if $s_\tau = s^*$ for $\tau = 0, \ldots, t-1$. That is, s_i^* continues to be chosen as long as the observed action combination has been s^* in all past periods. If any player deviates from s^* in some period t, then s_i^c should be chosen for periods $t+1, \ldots, t+T$, and at $t+T+1$ player i should return to s_i^*. If any player deviates from s^c at some time t when it is supposed to be chosen, the clock should be turned back and the punishment cycle restarted. This means that s^c is to be chosen by the players for periods $t+1, \ldots, t+T$, after which play reverts to s^*. The basic rule here is to punish any deviation by any player from any prescribed action by T successive periods of s^c.

3.1.2 An example of a cooperative outcome supported by noncooperative equilibrium strategies

Trigger strategy equilibria explain how people who place no value on each other's welfare, each caring only about himself, can cooperate to attain Pareto efficient outcomes when binding agreements among them are not possible. For example, imagine a lake around which are located rental cottages. The cottages are in five clusters, with each cluster having a single owner. Each summer season is a single time period. At the start of each season the five cottage owners choose rental rates for their cabins that must remain in effect for the duration of the season. Let s_{it} be the (scalar) rental rate chosen by owner i for season t, and suppose that

$$P_i(s_t) = 180s_{it} - 6s_{it}^2 + s_{it} \sum_{j=1}^{5} s_{jt}, \qquad i = 1, \ldots, 5 \tag{4.10}$$

It is easily seen that if each period is regarded as a separate noncooperative game, the choice of $s_{it} = 30$ for all i will be an equilibrium point, and each player's payoff will be 4,500 per period. If all players choose trigger strategies with $s_{it} = 90$ being selected in each period, and with the understanding that the players will switch to $s_{i,t+1} = 30$ if, in some period t, a choice other than 90 is observed, then the payoff per period will be 8,100.

The choice of $s_{it} = 30$ by all players may be thought to yield very high occupancy rates, while $s_{it} = 90$ yields only a moderate occupancy rate coupled with a very high price per unit. Under a trigger strategy, a single firm can take advantage of the high prices of the others and, for one period only, attain an extraordinary payoff. Given $s_{jt} = 90$ for the others, owner i will maximize her single-period payoff by choosing $s_{it} = 54$ and will receive 14,580. The fact that one owner can obtain 14,580 by choosing 54 when all others choose 90 shows that the choice of 90 by all owners is not an equilibrium point for the single-period game. Now suppose that all owners except owner i are using trigger strategies, and consider the alternatives facing owner i. If she also uses a trigger strategy she receives 8,100 in each period or $8{,}100/(1-\alpha_i)$ for the supergame, where α_i is her discount parameter. If she maximizes her first-period gain, then she gets 14,580 in period 1 and the best she can do in the subsequent periods is 4,500. The discounted value of this course of action is $14{,}580 + 4{,}500\alpha_i/(1-\alpha_i)$. It is easily seen that the trigger strategy is superior if $\alpha_i > 9/14$, it is inferior if $\alpha_i < 9/14$, and the two strategies yield equal supergame payoffs if $\alpha_i = 9/14$. When the trigger strategies form a noncooperative equilibrium, the extra gain available by maximizing against $s_{jt} = 90$ ($14{,}580 - 8{,}100 = 6{,}480$) is smaller than the future discounted losses suffered by reverting to the single-period noncooperative equilibrium ($[8{,}100 - 4{,}500]\alpha_i/[1-\alpha_i]$).

3.1.3 *Conditions for subgame perfect trigger strategy equilibria*

Trigger strategy equilibria may be considered self-enforcing agreements where the enforcement mechanism is a threat made by all players that, in the event that the agreement is violated, they will change their actions to actions that correspond to a single-period noncooperative equilibrium whose associated payoffs are worse for every player than are the agreed-on payoffs. Because the threat involves playing noncooperative equilibrium strategies, it is perforce credible. If all owners say they will choose 30, then no single owner can find a better alternative. The two pillars supporting the trigger strategy equilibrium are (a) a credible threat and (b) the inability of each player to deviate from the trigger strategy and increase his own supergame payoff, given that the other players follow their trigger strategies. Condition (b) is the requirement that the trigger strategies are a noncooperative equilibrium, and condition (a) is the further stipulation that the equilibrium is subgame perfect. In the remainder of this section, results on trigger strategy equilibria are formally developed.

DEFINITION 4.14 *$\phi_i(s)$ is the* **best-reply payoff of player i relative to** *s in the game (N, S, P) if $\phi_i(s) = \max_{s_i' \in S_i} P_i(s \backslash s_i')$.*

Clearly the function ϕ_i is defined everywhere on S.

Theorem 4.4 gives conditions on s^*, s^c, and the discount parameters under which grim trigger strategy equilibria exist. This theorem is a special case of Theorem 4.2.

THEOREM 4.4 *Let $\Gamma = (N, S, P, \alpha)$ be a repeated game satisfying Rule 4.1, $s^c \in S$ be an equilibrium point of (N, S, P), and $(s^*, s^c) \in S \times S$ be a grim trigger strategy combination. If and only if*

$$\alpha_i \geq \frac{\phi_i(s^*) - P_i(s^*)}{\phi_i(s^*) - P_i(s^c)}, \qquad i \in N \tag{4.11}$$

then (s^*, s^c) is a subgame perfect equilibrium point of Γ.

Proof. For the present model, equation (4.4) from Theorem 4.2 can be rewritten

$$\frac{P_i(s^*)}{1 - \alpha_i} \geq \phi_i(s^*) + \frac{\alpha_i P_i(s^c)}{1 - \alpha_i} \tag{4.12}$$

for all $i \in N$. Equation (4.12) is equivalent to equation (4.11) QED

It is obvious that equation (4.12) cannot be satisfied for player i and $\alpha_i \in (0, 1)$ if $P_i(s^*) \leq P_i(s^c)$. A converse to this statement, stated in Corollary 1, is that if $P(s^*) \gg P(s^c)$, then there are discount parameters for which grim trigger strategy equilibria exist. A second corollary addresses existence of equilibrium for finite reversion trigger strategies.

COROLLARY 1 *Under the conditions of Theorem 4.4 if $P(s^*) \gg P(s^c)$ then there are $\alpha_i \in (0, 1)$, $i \in N$, such that (s^*, s^c) is a trigger strategy noncooperative equilibrium.*

Proof. Because $\phi_i(s^*) \geq P_i(s^*) > P_i(s^c)$, it is immediate that

$$\alpha_i^* = \frac{\phi_i(s^*) - P_i(s^*)}{\phi_i(s^*) - P_i(s^c)} \in (0, 1), \qquad i \in N \tag{4.13}$$

Choosing $\alpha_i \in [\alpha_i^*, 1)$, $i \in N$, completes the proof. QED

COROLLARY 2 *Under the conditions of Theorem 4.4, if $P(s^*) \gg P(s^c)$, then there are discount parameters $\alpha \in (0, 1)^n$ and finite times T such that the finite reversion trigger strategy (s^*, s^c, T) is a subgame perfect equilibrium.*

Proof. Define α_i^* as in equation (4.13) for all players and choose each $\alpha_i \in (\alpha_i^*, 1)$. Equilibrium requires

$$\frac{P_i(s^*)}{1 - \alpha_i} \geq \phi_i(s^*) + \frac{\alpha_i P_i(s^c)(1 - \alpha_i^T)}{1 - \alpha_i} + \frac{\alpha_i^{T+1} P_i(s^*)}{1 - \alpha_i} \tag{4.14}$$

for all players. Equation (4.14) is equivalent to

$$\alpha_i \geq \frac{\phi_i(s^*) - P_i(s^*)}{\phi_i(s^*) - P_i(s^c)} + \alpha_i^{T+1}(P_i(s^*) - P_i(s^c)) \tag{4.15}$$

With $\alpha_i > \alpha_i^*$, there will be a finite T_i such that equation (4.15) will hold for all $T > T_i$; therefore, equation (4.14) will hold for all $i \in N$ for $T > \max_{i \in N} T_i$. QED

Aumann (1959, 1961) provides the first systematic study of cooperative outcomes in repeated games that are achieved by means of noncooperative behavior.[3]

Figure 4.2 illustrates the relationship between s^* and s^c for a two-player example. Any payoff in the shaded region, denoted $H(\Gamma)$, above and to the right of $P(s^c)$ could be $P(s^*)$, given suitable discount parameters, α. Depending on the values of the players' discount parameters, there are a large number of potential equilibria; it is therefore appealing to look for ways to select among them. One obvious technique is to eliminate

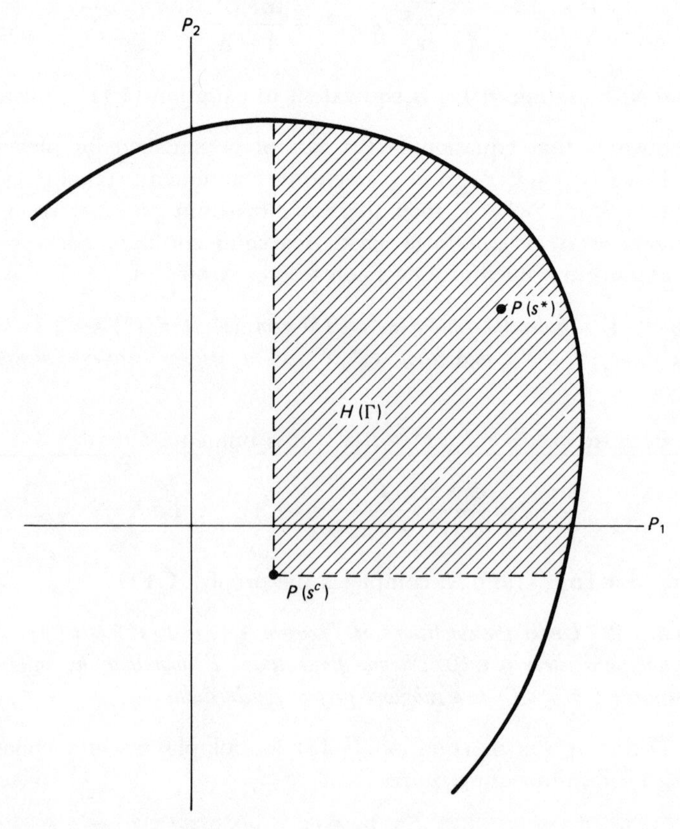

FIGURE 4.2 Repeated game outcomes that dominate the single-shot Nash equilibrium.

dominated equilibria. That is, let $E(\Gamma)$ denote the set of trigger strategies, (s^*, s^c), for the game $\Gamma = (N, S, P, \alpha)$. That $(s^*, s^c) \in E(\Gamma)$ means that s^c is an equilibrium of the single-shot game (N, S, P) and equation (4.11) is satisfied. The *locally efficient* elements of $E(\Gamma)$, denoted $E^*(\Gamma)$, provide the outcomes associated with undominated equilibria. These are

$$E^*(\Gamma = \{(s^*, s^c) \in E(\Gamma) \mid x \gg P(s^*) \text{ implies } x \notin H(\Gamma)\} \qquad (4.16)$$

The locally efficient payoff vectors, $\{P(s^*) \in H(\Gamma) \mid (s^*, s^c) \in E^*(\Gamma)\}$, may lie underneath the efficient frontier of the repeated game. That local efficiency does not imply global efficiency may be seen by examining Figure 4.3. Note that $P(s^*)$ is locally efficient but the discounted payoffs of both players can be increased by selecting s' in odd-numbered periods and s'' in even-numbered periods. The trigger strategy specification must be restated to require this alternation, along with reversion to s^c if a player violates the prescribed alternation between s' and s''. Supposing the payoffs to be those shown in Figure 4.3 and assuming the discount parameters of the two players to be .9, the trigger strategy based on s^* gives a discounted payoff of 54 to each player, while the trigger strategy based on alternating between s' and s'' gives player 1 a payoff of 57.89 and player 2, 62.11. If $(s^*, s^c) \in E(\Gamma)$ and $P(s^*)$ lies on the payoff possibility frontier, then supergame payoffs under (s^*, s^c) will certainly be Pareto optimal if the payoff possibility frontier is concave. This condition is violated in Figure 4.3. A weaker condition is that $P(s^*)$ lie on the upper right boundary of the convex hull of $H(\Gamma)$. Figure 4.4 repeats Figure 4.3 with the boundary of the

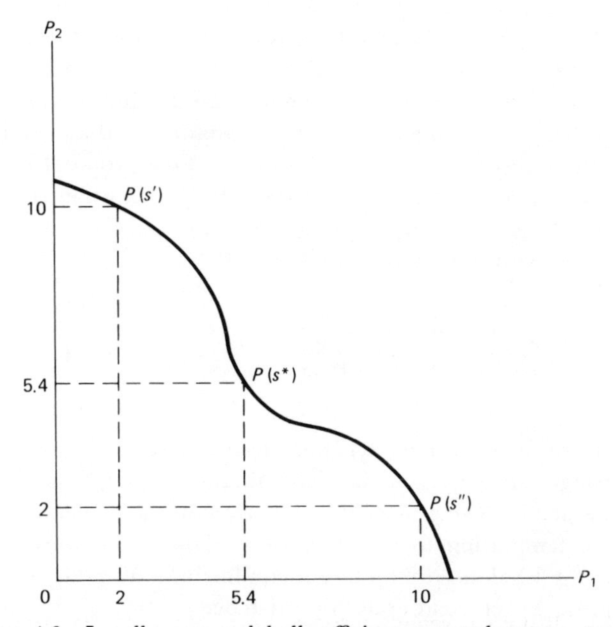

FIGURE 4.3 Locally versus globally efficient repeated game outcomes.

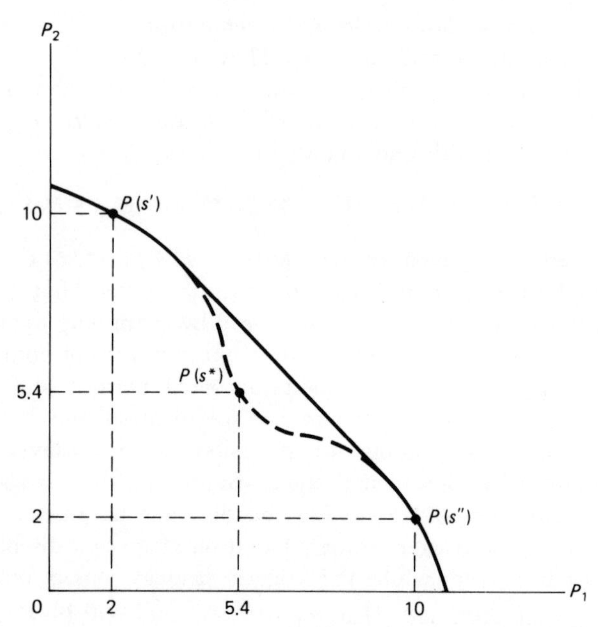

FIGURE 4.4 In a repated game the set of attainable payoffs is approximately convex.

convex hull drawn solid and the rest of the boundary of $H(\Gamma)$ drawn as a broken curve.

Even restricting attention to globally efficient trigger strategies, it is clear that there are a great many possible equilibrium points, particularly if the discount parameters are close to unity. While there is no a priori reason to eliminate any of these efficient subgame perfect equilibrium points, it is nonetheless desirable to single out a small number of these equilibria if an appealing criterion can be found for doing so. One proposal is the *balanced temptation equilibrium,* found in Friedman (1971). Basically, the idea is to select a point s^* such that α_i^* in equation (4.13) has the same value for all i.. This can be expressed equally well by

$$\frac{\phi_i(s^*) - P_i(s^*)}{P_i(s^*) - P_i(s^c)} = \frac{\phi_j(s^*) - P_j(s^*)}{P_j(s^*) - P_j(s^c)}, \qquad i,j \in N \tag{4.17}$$

The numerator in equation (4.17) is the single-period gain to defecting from the trigger strategy and the denominator is the per period loss that will be sustained following defection. The ratio of the two can be thought to measure the temptation to defect, with a higher ratio meaning a larger temptation. At a balanced temptation equilibrium, all players' temptations are equalized. Exact conditions for existence of this equilibrium can be found in Friedman (1971) or (1977, Chapter 8).

3.2 The folk theorem for repeated games without discounting and its extension to subgame perfection

The theorems discussed in this section and in Section 3.3 deal with equilibria that support *individually rational* outcomes of the constituent game. These are outcomes at which each player receives a payoff in each period that exceeds the minimum payoff to which that player can be held by the remaining players in the constituent game. Denote such a payoff by v_i and let $v = (v_1, \ldots, v_n)$. The folk theorem for repeated games states that a constant path μ^0 can be supported as the outcome of a noncooperative equilibrium strategy combination, σ, in a repeated game without discounting as long as $P(\mu_t^0) \gg v$ for all t. Of course, $u(\sigma) = \mu^0$. Aumann and Shapley (1976) and, independently, Rubinstein (1979) have proved (see Theorem 4.6) that the strategy combination σ can be selected so that it is subgame perfect.

DEFINITION 4.15 *Let* (N, S, P) *be a game,* $v_i = \min_{s \in S} \phi_i(s)$, *and* $v = (v_1, \ldots, v_n)$. *Then* v_i *is the* **minimax payoff of player** i *in* (N, S, P).

The minimax payoff, v_i, is defined for games satisfying Assumptions 3.1 and 3.2; a quasiconcavity condition is not necessary. First, ϕ_i is a continuous function; for, even though the best-reply mapping $(r_i(s))$ need not be single valued or continuous, the maximum attainable payoff varies continuously with s. Second, v_i is the minimum value of ϕ_i. This minimum is attained due to the continuity of ϕ_i and the compactness of S. Denote by $s^i \in S$ the strategy combination that attains v_i. That is, $P_i(s^i) = v_i$ and $P_i(s \backslash s_i^i) \geq P_i(s^i)$ for all $s \in S$.

DEFINITION 4.16 *In the game* (N, S, P), $P(s)$ *is an* **individually rational payoff vector** *if* $P(s) \gg v$. *The* **set of individually rational payoff vectors** *is* $H^*(N, S, P) = \{P(s) \gg v \mid s \in S\}$.

The infinitely repeated game without discounting is denoted $\Gamma = (N, S, P, 1) = (N, \mathcal{S}^*, G)$. Payoffs are, of course, unbounded; therefore, the repeated game payoff functions are not the same as under discounting. Instead

$$G_i(\sigma) = \liminf \frac{1}{t+1} \sum_{\tau=0}^{t} P_i(u_\tau(\sigma)) \tag{4.18}$$

for $i \in N$.

THEOREM 4.5 (*The folk theorem for repeated games*) *Let* $\Gamma = (N, S, P, 1) = (N, \mathcal{S}^*, G)$ *be a repeated game without discounting and let* $x \in H^*(N, S, P)$. *Then* Γ *has an equilibrium point* σ *such that* $P(u_t(\sigma)) = x$ *for all* t.

Proof. An equilibrium will be constructed that is a simple strategy combination $\sigma^\mu = (\mu^0, \mu^1, \ldots, \mu^n)$. First, denote by s^* an action combination for which $P(s^*) = x$. Let $\mu^0 = (s^*, s^*, s^*, \ldots)$ and $\mu^i = (s^i, s^i, s^i, \ldots)$ for $i \in N$. As each $\phi_i(s^*)$ is bounded and $x_i > v_i$, it cannot pay any player to deviate from σ_i^μ, which completes the proof. QED

The equilibrium strategy combination constructed above is extremely simple. It specifies that if a player i deviates from choosing s_i^*, then everyone switches to choosing s^i, which will drive the deviator to a payoff of v_i per period. No attention is paid here to the incentives of players other than i to participate in the punishment of player i. That is, σ^μ is not necessarily subgame perfect. Subgame perfection is addressed next. Let $\bar{z}_i = \max_{s \in S} P_i(s)$ and $\underline{z}_i = \min_{s \in S} P_i(s)$. $\bar{z}_i$ is the largest possible single period payoff that player i could receive and $\underline{z}_i$ is the smallest. Both are finite because P_i is a continuous function and S is a compact set.

THEOREM 4.6 *Let $\Gamma = (N, S, P, 1) = (N, \mathcal{S}^*, G)$ be a repeated game without discounting and let $x \in H^*(N, S, P)$. Then Γ has a subgame perfect equilibrium point σ such that $P(u_t(\sigma)) = x$ for all t.*

Proof. The theorem is proved by constructing a strategy combination, σ^* that consists of an infinite sequence of paths. $\sigma^* = (\mu^0, \mu^{11}, \ldots, \mu^{n1}, \mu^{12}, \ldots, \mu^{n2}, \mu^{13}, \ldots, \mu^{n3}, \ldots)$ is similar to a simple strategy combination except that the number of paths comprising σ^* is large and the switching rules are a little different. These paths define hierarchies of punishments. If no player ever deviates, then μ^0 is followed. That is, $u(\sigma^*) = \mu^0$. Suppose that player i deviates when the current path is μ^{jk}. Then the current path becomes $\mu^{i,k+1}$. The paths are defined as follows: For μ^0, $\mu_t^0 = s^*$ for all t. For μ^{jk}, $\mu_t^{jk} = s^i$ for $t = 1, \ldots, T_k$ and $\mu_t^{jk} = s^*$ for $t > T_k$. Deviation from μ^0 is a *first level deviation* and causes a *first level punishment*. Deviation from $\mu^{j,k-1}$ is a *k-th level deviation* and causes a kth-level punishment. It remains to determine the values of the T_k.

Consider first-level deviation by player i. The payoffs on the paths μ^0 and μ^{i1} are identical beginning T_1 periods following the period of deviation. Therefore, deviation is inferior to staying on μ^0 if

$$\bar{z}_i + T_1 v_1 < (T_1 + 1)x_i \tag{4.19}$$

Equation (4.19) can be rewritten $T_1 > (\bar{z}_i - x_i)/(x_i - v_i)$; therefore, T_1 is the smallest integer satisfying equation (4.19) for all i. Proceeding inductively to determine T_k, given that T_{k-1} has been found, two sorts of deviations must be considered. One is when player i deviates at a time when he is being punished; the other is when player i deviates at a time when player j is being punished. The former poses no problem because player i cannot raise his single-period payoff by deviating from s^i unless $T_k < T_{k-1}$. Therefore, $T_k \geq T_{k-1}$ is required. Deviation from the path $\mu^{j,k-1}$ will lower the payoff of player i if

$$\bar{z}_i + T_k v_i < T_{k-1}\underline{z}_i + (T_k + 1 - T_{k-1})x_i \tag{4.20}$$

On the left is the payoff for the $T_k + 1$ periods before the path μ^{ik} reverts to s^*. The most player i can get is $\bar{z}_i$ for one period, followed by v_i for T_k periods. If he does not deviate, then over the same $T_k + 1$ period time span he may suffer as low a payoff as $\underline{z}_i$ for T_{k-1} periods followed by x_i for

$T_k + 1 - T_{k-1}$ periods. Equation (4.20) is equivalent to

$$T_k > \frac{\bar{z}_i - x_i + T_{k-1}(x_i - \underline{z}_i)}{x_i - v_i} \tag{4.21}$$

It is clear from equation (4.21) that $T_k \geqslant T_{k-1}$ because $\underline{z}_i \leqslant v_i$. Choose T_k to be the smallest integer satisfying equation (4.21) for all i. The strategy combination σ^* is now completely specified and is subgame perfect. To verify this consider any possible history h_t. Such a history determines a current path. The current path must be μ^0 or one of the μ^{jk} and the construction of σ^* ensures that no player could profitably deviate from that current path. QED

Note that all the punishment paths μ^{jk} have a two-phase character. In the early phase, the deviator is held to a payoff of v_i per period and, following this early punishment the path reverts to s^*, yielding the payoff vector x in each period. Abreu (1986) has aptly named these *stick-and-carrot punishments*. The high payoffs that follow the low payoffs of the first phase are needed not to deter the deviator from deviating, but rather to deter the punishers from defecting on the punishment. Without the carrot, the punishment to be meted out to the deviating player would not be a credible threat.

3.3 Extending the folk theorem to games with discounting

Fudenberg and Maskin (1986) provide a parallel theorem to Theorem 4.6 for repeated games with discounting. Interestingly, they prove their result by constructing a simple strategy combination that is subgame perfect. They do have an additional requirement, a version of which is stated here as Assumption 4.1.

ASSUMPTION 4.1 *If* $x \in H^*(N, S, P)$, $z \in R^n$, *and* $v \leqslant z \leqslant x$, *then* $z \in H^*(N, S, P)$.

Assumption 4.1 is stronger than is necessary. The role of this assumption can be explained along with a sketch of the equilibrium strategies. The equilibrium strategy combination is characterized by $n + 1$ paths, $\mu^0, \mu^1, \ldots, \mu^n$. The path $\mu^0 = (s^*, s^*, \ldots)$ satisfies $P(\mu_t^0) = x$ and each path μ^i is a punishment path for a player i. These punishments are of the stick-and-carrot sort. The first phase of μ^i holds the deviator, i, to a payoff of v_i by selecting s^i; however, the reversion phase does not return to s^*. Instead, μ^i reverts to a payoff level that is worse than $P(s^*)$ for everyone, but the reversion payoff to the deviator, player i, is lower than player i receives in the reversion phase of μ^j $(j \neq i)$. The reasons for this come out in the proof of Theorem 4.7. The purpose of Assumption 4.1 is to ensure that there are suitable payoffs below $P(s^*)$ to provide the various reversion points.

THEOREM 4.7 *Let* $\Gamma = (N, S, P, \alpha) = (N, \mathcal{S}^*, G)$ *be a repeated game with discounting and let* $x \in H^*(N, S, P)$. *Then, for some* $\alpha^* \in (0, 1)^n$, *if* $\alpha \in \times_{i \in N} (\alpha_i^*, 1)$, Γ *has a subgame perfect equilibrium point* σ *such that* $P(u_t(\sigma)) = x$ *for all* t.

Proof. A subgame perfect simple strategy combination $\sigma^\mu = (\mu^0, \mu^1, \ldots, \mu^n)$ will be constructed. Let $\mu^0 = (s^*, s^*, \ldots)$, where $P(s^*) = x$. μ^i is given by $\mu_t^i = s^i$ for $t = 1, \ldots, T_i$ and $\mu_t^i = w^i$ for $t > T_i$. The s^i are, of course, the actions that hold player i to a single-period payoff of v_i. It remains to determine the w^i and the T_i. It is permissible, and will simplify some later notation and derivations, if all players have the same T_i. The common value will be denoted T. From here the proof proceeds by selecting the w^i. Then T is jointly determined along with the α_i^*.

Choose $x' \in H^*(N, S, P)$ with $x' \ll x$. By Assumption 4.1 there is $\varepsilon > 0$, and a $w^i \in S$ for each i, such that $y^i = (x_1' + \varepsilon, x_2' + \varepsilon, \ldots, x_{i-1}' + \varepsilon, x_i', x_{i+1}' + \varepsilon, \ldots, x_n' + \varepsilon) \in H^*(N, S, P)$ and $P_j(w^i) = y_j^i$ for $j \in N$. Now choose an integer T so that

$$T > \frac{\bar{z}_i - v_i}{x_i' - v_i} \tag{4.22}$$

for all $i \in N$. If $0 \leq \alpha_i \leq 1$, then

$$\frac{\alpha_i - \alpha_i^{T+1}}{1 - \alpha_i} = \sum_{t=1}^{T} \alpha_i^t \leq T \tag{4.23}$$

First, consider deviation by player i from μ^0. Such a deviation is unprofitable if

$$\frac{x_i}{1 - \alpha_i} > \bar{z}_i + (\alpha_i + \cdots + \alpha_i^T) v_i + (\alpha_i^{T+1} + \cdots) x_i' \tag{4.24}$$

An upper bound on the gain from deviating is given by the right-hand side of equation (4.24) minus the left. It is

$$(\bar{z}_i - x_i) - (x_i - v_i) \frac{\alpha_i - \alpha_i^{T+1}}{1 - \alpha_i} - \frac{\alpha_i^{T+1}}{1 - \alpha_i} (x_i - x_i') \tag{4.25}$$

Equation (4.25) is bounded above by

$$\{\bar{z}_i - x_i) - (x_i - v_i) \frac{\alpha_i - \alpha_i^{T+1}}{1 - \alpha_i} \tag{4.26}$$

and equation (4.26) is bounded above by

$$(\bar{z}_i - x_i') - (x_i' - v_i) \frac{\alpha_i - \alpha_i^{T+1}}{1 - \alpha_i} \tag{4.27}$$

Equation (4.27), and consequently equation (4.25), is less than 0 if

$$\frac{\bar{z}_i - x_i'}{x_i' - x_i} < \frac{\alpha_i - \alpha_i^{T+1}}{1 - \alpha_i} \tag{4.28}$$

Therefore, if $\alpha_i > (\bar{z}_i - x_i')/(\bar{z}_i - v_i)$, there is a finite value, T^*, such that if $T > T^*$ then equation (4.25) is negative and deviation from μ^0 does not pay. Let I be the set of positive integers and $\Xi = \{(\alpha, T) \in (0, 1)^n \times I \mid \alpha_i > (\bar{z}_i - x_i')/(\bar{z}_i - v_i)$ for $i \in N\}$. Note that if $(\alpha, T) \in \Xi$ and $(\alpha', T') \geqslant (\alpha, T)$, then $(\alpha', T') \in \Xi$. Furthermore, Ξ is not empty.

Next examine deviation in a punishment phase. Suppose the current path is μ^i and i deviates. Then μ^i is restarted, which is worse for player i than continuing with the punishment. To see this in detail, recall that μ^i has two phases. If deviation occurs in the early phase, restarting μ^i neans extending the number of periods in which the lower payoff, v_i, is received. If deviation occurs during the later phase, the forgone payoff is $x_i'/(1 - \alpha_i)$ and the new payoff stream is bounded above by the right-hand side of equation (4.24). Therefore, deviation is unprofitable if

$$\frac{x_i}{1 - \alpha_i} > \bar{z}_i + (\alpha_i + \cdots + \alpha_i^T)v_i + (\alpha_i^{T+1} + \cdots)x_i' \tag{4.29}$$

Equation (4.29) is equivalent to equation (4.27), so the same conditions apply here as those for deviations from μ^0.

Finally, suppose player i deviates when the current path is μ^j $(j \neq i)$. Such a deviation is not profitable if

$$\frac{1 - \alpha_i^T}{1 - \alpha_i} \underline{z}_i + \frac{\alpha_i^T}{1 - \alpha_i}(x_i' + \varepsilon) > \bar{z}_i + (\alpha_i + \cdots + \alpha_i^T)v_i + (\alpha_i^{T+1} + \cdots)x_i' \tag{4.30}$$

The left-hand side is a lower bound on player i's payoff if she does not deviate. The path μ^j could give player i as little as $\underline{z}_i$ for T periods, after which she would receive $x_i' + \varepsilon$ in each period. The right-hand side is an upper bound on her payoff if she does deviate and is the same as shown in equation (4.24). The gain from deviating is bounded above by

$$(\bar{z}_i - \underline{z}_i) + \frac{\alpha_i - \alpha_i^T}{1 - \alpha_i}(v_i - \underline{z}_i) - \frac{\alpha_i^T}{1 - \alpha_i}\varepsilon \tag{4.31}$$

Equation (4.31) is less than 0 if

$$\frac{(1 - \alpha_i)(\bar{z}_i - \underline{z}_i)}{\varepsilon + v_i - \underline{z}_i} + \frac{\alpha_i(v_i - \underline{z}_i)}{\varepsilon + v_i - \underline{z}_i} < \alpha_i^T \tag{4.32}$$

As $\alpha_i \to 1$ the left-hand side of equation (4.32) converges to $(v_i - \underline{z}_i)/(\varepsilon + v_i - \underline{z}_i) < 1$ and the right-hand side converges to 1. As the left-hand side falls and the right-hand rises as α_i rises, for fixed T, there is a value of $\alpha_i \in (0, 1)$, $\psi_i(T)$, for which equality holds. Let $\Psi_i = \{(\alpha_i, T) \in (0, 1) \times I \mid \alpha_i > \psi_i(T)\}$, and $\Psi = \{(\alpha, T) \in (0, 1)^n \times I \mid (\alpha_i, T) \in \Psi_i$ for $i \in N\}$. Note that if $(\alpha, T) \in \Psi$ and $(\alpha', T') \geqslant (\alpha, T)$ then $(\alpha', T') \in \Psi$, and Ξ is not empty. Therefore, $\Xi \cap \Psi$ is not empty and $\sigma^\mu = (\mu^0, \mu^1, \ldots, \mu^n)$ is a subgame perfect equilibrium if $(\alpha, T) \in \Xi \cap \Psi$. QED

An equilibrium σ^{μ} has $2n + 1$ associated single-period payoff vectors that might be used. One is $P(s^*)$, the payoff that is achieved when no one ever deviates. The remaining $2n$ appear in pairs, one pair for each player. One member of each pair is $P(s^i)$, the payoff vector at which player i is held to v_i, and the other is $P(w^i) = y^i$, the reversion payoff vector that is part of the punishment path for player i. These are illustratated in Figure 4.5. Several aspects of equilibrium are shown. Note that v, the vector of individually rational payoffs, is not in the feasible set of payoffs. This may or may not happen; the point to note is that v need not be feasible in an n-person game. Notice that at s^2, when player 2 is being held to v_2, player 1 receives a payoff of $P_1(s^2)$ that is lower than v_1. In the illustration, the same fate does not befall player 2 when player 1 is held to v_1. In general, some players may receive payoffs beneath their individually rational levels when punishing another player. This is why the left-hand side of equation (4.30) uses $\underline{z}_i$ as the lower bound on the payoff of player i when player j is being held to v_j. Turning to the brighter side of the figure, H^* is the set of feasible payoffs that gives more to each player than v. Any point in this set can be supported by an appropriate σ^{μ} if the discount parameters are near enough to unity. Suppose the point marked x is supported by σ^{μ}. Below and to the left of x is x', the benchmark for the reversion payoffs under the punishment paths. If player 1 is the deviator, then, after spending T

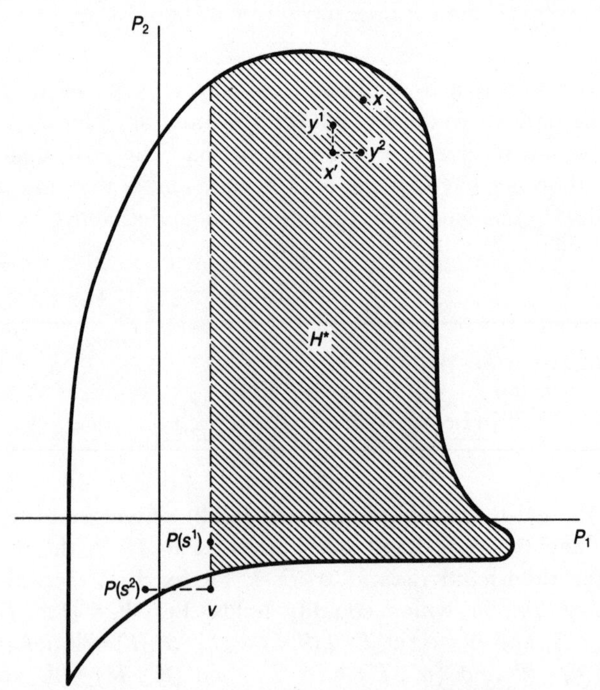

FIGURE 4.5 Illustration for the folk theorem generalized to games with discounting.

periods at $P(s^1)$, the path switches to y^1. At y^1 the deviator, player 1, receives a payoff of x_1', but the punisher, player 2, receives a payoff of $x_2' + \varepsilon$. Thus, in a reversion phase, any player would receive an extra ε per period as a punisher over and above what he would receive as a deviator. The reason for this extra ε can be seen in equation (4.32). It is necessary to take into account the possibility that player i as a punisher may, in the early phase, have to accept a payoff lower than v_i in order to hold a deviator to his individually rational payoff. That is, it is necessary to guarantee that the threats of punishment are credible.

3.4 Trigger strategy equilibria in supergames with discounting

Theorem 4.4 can easily be extended to general supergames without time dependence. Let $\{(N, S_t, P_t)\}_{t=0}^{\infty}$ be a sequence of single-shot games. Each game (N, S_t, P_t) has the same set of players, N; however, the single-shot payoff functions and the S_t can vary arbitrarily. Then the set of histories at time t, H_t, is $\times_{\tau=0}^{t-1} S_\tau$ and a decision rule for player i at time t is some ν_{it} whose domain is H_t and whose values are in S_{it}. Denote by V_{it} the set of such decision rules. Then the supergame strategy set of player i is $\mathscr{S}_i^* = \{\sigma_i = (\nu_{i0}, \nu_{i1}) \mid \nu_{it} \in V_{it},\ t = 0, 1, \ldots)$, and the joint strategy set is $\mathscr{S}^* = \times_{i \in N} \mathscr{S}_i^*$. Any $\sigma \in \mathscr{S}^*$ induces a path $u(\sigma) = (u_0(\sigma), u_1(\sigma), \ldots)$ and

$$G_i(\sigma) = \sum_{t=0}^{\infty} \alpha_{it} P_{it}(u_t(\sigma)) \tag{4.33}$$

is the payoff function of player i. The restrictions applying to equation (4.33) are (1) $\alpha_{i0} = 1$, (2) $0 < \alpha_{i,t+1} < \alpha_{it}$ for $i \in N$ and $t \geqslant 0$, and (3) $G_i(\sigma)$ is bounded on $\mathscr{S}^*$ for $i \in N$. The parameter α_{it} is the value of one unit of payoff received at time t and discounted to time 0. Therefore, the one-period discount parameter from period t to period $t-1$ is $\alpha_{it}/\alpha_{i,t-i}$. If $\alpha_{it}/\alpha_{i,t-1} = \alpha_i$ for all t, as in earlier sections, then $\alpha_{it} = \alpha_i^t$.

A trigger strategy is characterized by two paths (μ^*, μ^c). Any deviation from μ^* or μ^c results in μ^c being followed. $\mu^c = (s_0^c, s_1^c, \ldots)$, where s_t^c is an equilibrium point of (N, S_t, P_t). Thus, if any deviation has occurred prior to time t, then (μ^*, μ^c) specifies that s_t^c be chosen at time t. This choice holds whether there have been many past deviations or only one. Defining

$$\phi_{it}(s_t) = \max_{s_{it}' \in S_{it}} P_{it}(s_t \backslash s_{it}') \tag{4.34}$$

the following conditions ensure that (μ^*, μ^c) is a subgame perfect equilibrium point

$$\sum_{\tau=t}^{\infty} \alpha_{i\tau} P_{i\tau}(\mu_\tau^*) \geqslant \alpha_{it}\phi_{it}(\mu^*) + \sum_{\tau=t+1}^{\infty} \alpha_{i\tau} P_{i\tau}(\mu_\tau^c) \tag{4.35}$$

for $t = 0, 1, \ldots$ and $i \in N$. The conditions expressed by equation (4.35) are that player i at time t would obtain at least as high a discounted payoff along the path μ^* as would be obtained by receiving $\phi_{it}(\mu^*)$ at time t followed by single-shot noncooperative outcomes thereafter. This is true for all t, which means that defection from μ^* will not pay player i at any time. The same holds for all players in N. Because of the nonstationarity of the game, each time period must be separately checked.

4 Finite-horizon repeated games

It was long thought that the only subgame perfect equilibria in a finitely repeated game were strategy combinations calling for the selection of single-shot equilibria at each iteration of the game. This conjecture is false in general, but is true in finitely repeated games having a unique single-shot equilibrium, or in games in which all single-shot equilibria have the same payoffs. If several single-shot equilibria exist and if no player obtains precisely the same payoff at all of them, then a theorem establishing existence of subgame perfect trigger strategy equilibria can be proved, as well as a theorem establishing the Aumann-Shapley-Rubinstein theorem (Theorem 4.6) for finitely repeated games. Section 4.1 presents needed definitions and a theorem that states that a repeated game with a unique single-shot equilibrium payoff vector has no subgame perfect equilibria except for strategy combinations consisting of the play of a sequence of single-shot equilibria. Section 4.2 shows the existence of trigger strategy equilibria in finitely repeated games. This section follows Friedman (1985) with Rawthorn's (1988) correction. In Section 4.3 the fine paper of Benoit and Krishna (1985) that extends the Aumann-Shapley-Rubinstein theorem to finitely repeated games is discussed, but is not formally developed. Finally, a different approach to finitely repeated games due to Radner (1980) is discussed that admits an element of bounded rationality into the model.

4.1 The finitely repeated game model

A finitely repeated game is characterized by the single-shot game (N, S, P) that is played; the discount parameters of the players, $\alpha \in (0, 1]^n$; and the number of times, $T+1$, that (N, S, P) will be played. Thus, it may be represented as $\Gamma = (N, S, P, \alpha, T)$. A strategy $\sigma_i = (\nu_{i0}, \ldots, \nu_{iT})$ for a player is identical to a strategy in an infinitely repeated game except that it has only $T+1$ decision rules, one for each play of (N, S, P). For player i the repeated game strategy set is $\mathscr{S}^*_{iT} = \{\sigma_i = (\nu_{i0}, \ldots, \nu_{iT}) \mid \nu_{it}: S^t \to S_i, t = 0, \ldots, T\}$. Let $\mathscr{S}^*_T =$ the joint strategy set be $\times_{i \in N} \mathscr{S}^*_{iT}$. A strategy combination σ induces a path $u(\sigma) \in S^{T+1}$ in the obvious way, and the payoff function of player i is

$$G_i(\sigma) = \sum_{t=0}^{T} \alpha_i^t P_i(u_t(\sigma)) \tag{4.36}$$

In view of these definitions, a finitely repeated game can be denoted $\Gamma = (N, \mathcal{S}_T^*, G)$.

DEFINITION 4.17 *A* **finitely repeated game** $\Gamma = (N, \mathcal{S}_T^*, G)$ *is given by strategy spaces* $\mathcal{S}_{iT}^* = \{\sigma_i = (\nu_{i0}, \ldots, \nu_{iT}) \mid \nu_{it}: S^t \to S_i,\ t = 0, \ldots, T\}$ *for each player* $i \in N$, $\mathcal{S}_T^* = \times_{i \in N} \mathcal{S}_{iT}^*$, *payoff functions* G_i *specified in equation* (4.36), *and* $G = (G_1, \ldots, G_n)$.

Denote by S^c the equilibria of (N, S, P).

THEOREM 4.8 *Let* $\Gamma = (N, \mathcal{S}_T^*, G)$ *be a finitely repeated game satisfying rule 4.1. If* $P(s') = P(s'')$ *for all* $s', s'' \in S^c$, *then the only subgame perfect equilibria are* $\sigma \in \mathcal{S}_T^*$ *such that* $\nu_t(h_t) \in S^c$ *for all* $h_t \in S^t$ *and* $t = 0, \ldots, T$.

Proof. Suppose that $\sigma_i^* = (\nu_{i0}^*, \ldots, \nu_{iT}^*)$ is an equilibrium point of the game. Then $\nu_{iT}^*(h_T) \in S^c$. If the latter were not true, then for any h_T such that $\nu_{iT}^*(h_T) \notin S^c$ there would be at least one player who could improve her period T payoff without any effect on the payoffs of earlier periods. Therefore, $P_i(\nu_{iT}^*(h_T))$ is independent of h_T for all i. Looking now at period $T-1$, the choice of s_{T-1} will have no effect on $P(\nu_{iT}^*(h_T))$; the same argument may be applied to show that a subgame perfect equilibrium requires $\nu_{iT-1}^*(h_{T-1}) \in S^c$ for all $h_{T-1} \in S^{T-1}$. Proceeding recursively backward to $t = 0$ completes the proof. QED

COROLLARY *If* S^c *contains only one element,* s^c, *then* $\sigma = (s^c, \ldots, s^c)$ *is the only subgame perfect equilibrium of* Γ.

Proof of the corollary is left to the reader.

Keeping Theorem 4.8 in mind, a roughly opposite assumption is stated and used in Section 4.2.

ASSUMPTION 4.2 *There exist* $s^0, s^1, \ldots, s^n \in S^c$ *and that, for each* $i \in N$, $P_i(s^0) > P_i(s^i)$ *and* $P_i(s) \geqslant P_i(s^i)$ *for all* $s \in S^c$.

Assumption 4.2 could be weakened to require that, for each player i, there is some pair of strategy combinations s^i, $b^i \in S^c$ satisfying $P_i(b^i) > P_i(s^i)$. To require, as Assumption 4.2 does, that $b^i = s^0$ for all i merely simplifies exposition. That s^i gives player i the lowest payoff among all members of S^c imposes no restriction, as S^c must be compact.

4.2 Trigger strategy equilibria

A subgame perfect strategy combination must choose a single-shot equilibrium at time T, but if Assumption 4.2 holds, a meaningful choice of such equilibria is available, which means there is scope to punish a player i in time T for a defection at time $T-1$. The punishment is to choose s^i. Typically, if the current path is the same at T as at $t = 0$, then s^0 will be chosen at time T. The strategies adopted here, called *discriminating trigger strategies,* are simple strategies in the sense of Abreu (1988) and of Section 2.3 of this chapter and they can be described by means of $n + 1$ paths

$(\mu^0, \mu^1, \ldots, \mu^n)$. These strategies discriminate in the same way that the equilibrium strategies developed in Theorems 4.6 and 4.7 discriminate: the response to defection depends on which player defected.

DEFINITION 4.18 *A* **discriminating trigger strategy combination** *is* $(\mu^0, \mu^1, \ldots, \mu^n)$, *where* $\mu^i \in S^{T+1}$, $i = 0, \ldots, n$; $\mu_t^0 = s^*$ *for* $t = 0, \ldots, t^*$; $\mu_t^0 = s^0$ *for* $t = t^* + 1, \ldots, T$; *and* $\mu_t^i = s^i$ *for* $t = 0, \ldots, T$ *and* $i \in N$. *At* $t = 0$ *the current path is* μ^0. *If at any time t player i defects from the current path, then at time* $t + 1$ *the current path is* μ^i *and it is followed from* μ_{t+1}^i. *When two or more players defect simultaneously, the player with the smallest index is designated as the defector for purposes of choosing a new current path. If no defection occurs at time t, then the current path is unchanged.*

In view of Definition 4.18 a discriminating trigger strategy combination can be denoted $\sigma(s^0, \ldots, s^n, s^*, t^*)$.

THEOREM 4.9 *Let* $\Gamma = (N, \mathcal{S}_T^*, G)$ *be a finitely repeated game satisfying Assumption 4.2 and Rule 4.1, and let* $\sigma(s^0, \ldots, s^n, s^*, t^*)$ *be a discriminating trigger strategy combination with* $P_i(s^*) > P_i(s^i)$ *for all* $i \in N$. *Then there exists a finite positive integer* $\gamma(\alpha)$ *such that* $\sigma(s^0, \ldots, s^n, s^*, t^*)$ *is a subgame perfect equilibrium of* Γ *if* $T - \gamma(\alpha) \geqslant t^* \geqslant 0$,

$$\alpha_i > \frac{\phi_i(s^*) - P_i(s^*)}{\phi_i(s^*) - P_i(s^*) + P_i(s^0) - P_i(s^i)} \tag{4.37}$$

for $P_i(s^*) > P_i(s^0)$ *and*

$$\alpha_i > \frac{\phi_i(s^*) - P_i(s^*)}{\phi_i(s^*) - P_i(s^i)} \tag{4.38}$$

for $P_i(s^*) \leqslant P_i(s^0)$.

Proof. Suppose all players $j \neq i$ to be using $\sigma(s^0, s^1, \ldots, s^n, s^*, t^*)$ and consider the payoff to player i supposing each of the following: (i) player i uses $\sigma_i(s^0, s^1, \ldots, s^n, S^*, t^*)$; (ii) player i follows $\sigma_i(s^0, s^1, \ldots, s^n, s^*, t^*)$ for periods $0, \ldots, t_0 - 1$, for period $t_0 \leqslant t^*$, player i chooses $s_{it_0} \neq s_i^*$ and then he chooses s_i^i in periods $t_0 + 1, \ldots, T$. Under (i) the payoff is

$$\sum_{t=0}^{t^*} \alpha_i^t P_i(s^*) + \sum_{t=t^*+1}^{T} \alpha_i^t P_i(s^0) \tag{4.39}$$

Under (ii) the payoff is

$$\sum_{t=0}^{t_0-1} \alpha_i^t P_i(s^*) + \alpha_i^{t_0} \phi_i(s^*) + \sum_{t=t_0+1}^{T} \alpha_i^t P_i(s^i) \tag{4.40}$$

No alternative to $\sigma_i(s^0, s^1, \ldots, s^n, s^*, t^*)$ will deliver a higher payoff than the best alternative under (ii). It remains to compare equations (4.39) and (4.40) to show

$$\sum_{t=0}^{t^*} \alpha_i^t P_i(s^*) + \sum_{t=t^*+1}^{T} \alpha_i^t P_i(s^0) > \sum_{t=0}^{t_0-1} \alpha_i^t P_i(s^*) + \alpha_i^{t_0} \phi_i(s^*) + \sum_{t=t_0+1}^{T} \alpha_i^t P_i(s^i) \tag{4.41}$$

Equation (4.41) is equivalent to

$$\alpha_i > \frac{\phi_i(s^*) - P_i(s^*)}{\phi_i(s^*) - P_i(s') - \alpha_i^{t^*-t_0}(P_i(s^*) - P_i(s^0)) - \alpha_i^{T-t_0}(P_i(s^0) - P_i(s^i))} \quad \text{for } t_0 \leqslant t^* \qquad (4.42)$$

For arbitrary large T, the right-hand side of equation (4.42) converges from above to equation (4.37) for $P_i(s^*) > P_i(s^0)$ and 4.37 is bounded above by equation (4.38) for $P_i(s^*) \leqslant P_i(s^0)$, for any $t^* - t_0 \geqslant 0$. For given t_0 and t^*, let δ_i be the smallest integer value of $T - t^*$ that satisfies equation (4.42) and let $\gamma(\alpha) = \max_{i \in N} \delta_i$. Then $\sigma(s^0, s^1, \ldots, s^n, s^*t^*)$ is a non-cooperative trigger strategy equilibrium for Γ if $t^* \geqslant 0$ and $T \geqslant t^* + \gamma(\alpha)$.

To establish subgame perfection, choose $t \leqslant T$ and note that all past histories of the game at time t (i.e., actual choices $s_0, s_2, \ldots, s_{t-1}$) fall into two disjoint sets: The first are all histories such that the trigger strategy calls for choosing one of the s^i, and the second is the unique history that does not call for choosing any of the s^i (but rather for s^* if $t \leqslant t^*$ or s^0 if $t > t^*$). If $t = 0$, the first set is empty and the (null) history is taken to be in the second set. In either case, the trigger strategy used from t onward forms an equilibrium point for the subgame at time t with the given history. QED

From the proof of the theorem, it is clear that the equilibrium point $\sigma(s^0, s^1, \ldots, s^n, s^*, t^*)$ could easily yield efficient realizations during part of the game and the inefficient realizations could be limited to the last $\gamma(\alpha)$ periods. This and two additional points are established in the following corollary.

COROLLARY *If the conditions of Theorem* 4.9 *hold, then* (a) *for suitable T, trigger strategy equilibria exists with s^* on the payoff possibility frontier.* (b) *As $T \to \infty$, the fraction of periods allowing realizations on the payoff possibility frontier goes to one.* (c) *Restricting α so that $\alpha_i \in (0, 1)$, $i \in N$, and $T \geqslant 1$, a (possibly degenerate) trigger strategy equilibrium exists.*

Proof. For (a) there are two cases: $P(s^0)$ is on the payoff possibility frontier, or it is not. In the former case, let $s^* = s^0$ and apply the theorem. In the latter case, there must be s^* such that $P(s^*)$ dominates $P(s^0)$. Again, using this s^*, apply the theorem. Point (b) follows easily from the observation that γ does not change as T increases. Thus, $t^* = T - \gamma$ is permissible and the fraction of periods allowing a realization on the payoff possibility frontier is $1 - \gamma/(T + 1)$. Point (c) follows from the theorem by allowing $s^* = s^0$. (This is a degenerate trigger strategy equilibrium.) QED

Note that the key to Theorem 4.9 is that $s^0, s^1, \ldots, s^n$ are single-shot equilibrium points and $P_i(s^0) > P_i(s^i)$ for all $i \in N$. Nothing would change if $s^1 = s^2 = \cdots = s^n$. The following example may serve to illuminate the theorem. Let $n = 2$, $S_1 = [0, 20]$, $S_2 = [0, 25]$, $\alpha_1 = \alpha_2 = .9$, and let the

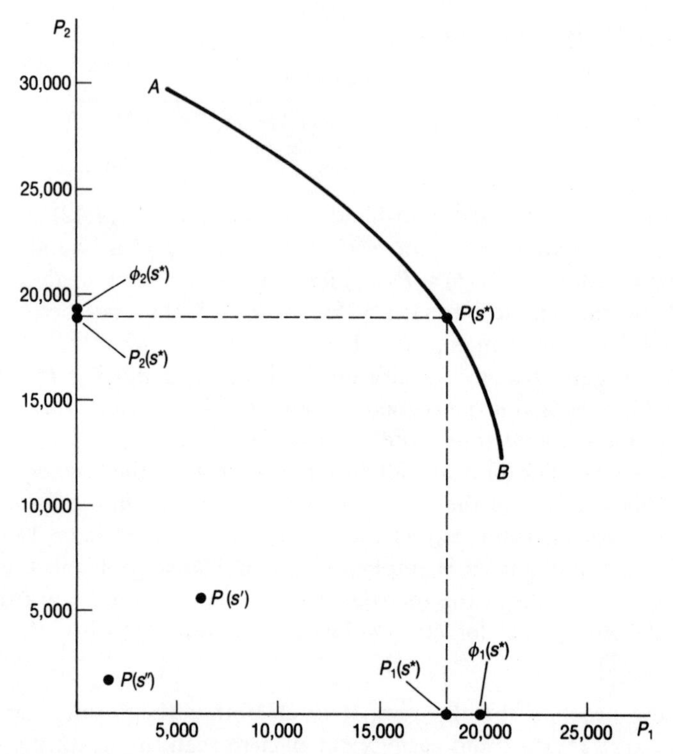

FIGURE 4.6 Trigger strategy equilibria in a finitely repeated game.

single-period payoff functions be

$$P_1(s) = -825s_1 - 15s_1^2 + 135s_1s_2 - 5s_1^3 \tag{4.43}$$

$$P_2(s) = -200s_2 - 5s_2^2 + 100s_1s_2 - \tfrac{2}{3}s_2^3 \tag{4.44}$$

The single-period payoff functions are illustrated in Figure 4.6. The payoffs associated with the two single-period equilibrium points are for $s'' = (5, 10)$ the payoffs are $(1{,}625, 1{,}833\frac{1}{3})$ and for $s' = (8, 15)$ they are $(6{,}080, 5{,}625)$.[4] The curve marked AB is the profit possibility frontier and a trigger strategy equilibrium can be based on the frontier point $s^* = (15, 25)$, whose associated payoffs are $(18{,}000, 18{,}598\frac{1}{3})$. The trigger strategy calls for letting $s_t^* = (15, 25)$ for $t = 0, \ldots, T-1$ and $s_T^* = s' = (8, 15)$. Then $s^1 = s^2 = s'' = (5, 10)$. The gains to short-run maximization for a player when the other player is using the trigger strategy are $\phi_1(15, 25) =$ 19801.15 for player 1, which is achieved at $s_1 = 12.0767$, and $\phi_2(15, 25) =$ 19144.37 for player 2, which is achieved at $s_2 = 23.1174$. If player 1 opts for the extra single-period gain of 1,801.15 (=19,801.15 − 18,000), he loses 16,375 (=18,000 − 1,625) in the next period alone if the time of defection is $t < T-1$, and 4,455 (=6080 − 1625) of the time of defection is $T-1$. With a discount parameter of .9, this is clearly not worthwhile. Similarly, for player 2, the short-term gain is 186.04 (=19,144.37 − 18,958.3) and the loss in the next period alone is 17,125 (=18,598.3 − 1,833.3) or 3,791.7 (=5,825 − 1,833.3), depending on which period the defection occurs.

4.3 Extending the folk theorem to finitely repeated games

In a remarkable paper Benoit and Krishna (1985) prove the Aumann-Shapley-Rubinstein result for finitely repeated games without discounting. The construction of their results is quite lengthy, so only a sketch is provided here. The model they use is close to the model used in Section 4.2. They require assumptions similar to Assumptions 4.1 and 4.2; Assumption 4.1 plays a role similar to its role in Theorem 4.7. The strategies they construct are based on *three-phase punishments* in which the first phase has the players holding a defector i to a payoff of v_i per period (Abreu's *stick*) and the second phase guarantees all players payoffs above their respective v_i (at least $x' \geqslant v$, Abreu's *carrot*); however, like Fudenberg and Maskin (1986), if x is a defector in the second phase, she receives x_i' per period, while if she were a punisher, she would receive $x_i' + \varepsilon$. In the third phase, the players select some single-shot noncooperative equilibrium actions in each period. The necessity for the third phase should be clear from Theorems 4.8 and 4.9; as the game nears the end, only single-shot equilibrium play can be subgame perfect.

Recall that the equilibrium strategies constructed for Theorem 4.6 (Aumann-Shapley-Rubinstein) keep count of the level of defection and require ever more severe punishments for successive levels. In the second phase of these punishments, the players revert to the same per period payoff vector that they enjoyed on the original path μ^0. In contrast, the proof of Theorem 4.7 (Fudenberg and Maskin) avoids this feature and uses the same punishment for a given player regardless of the level of defection. This is done by treating punishers better than defectors in the second phase, along with a second phase reversion that does not take any player to a higher payoff per period than was enjoyed along the path μ^0. Benoit and Krishna use features of both proofs. Like Aumann-Shapley-Rubinstein, they require ever more severe punishments at successive levels of defection and the second-phase reversion treats punishers and the defector differently, in the manner of Fudenberg and Maskin.

Benoit and Krishna prove a limit result that states an average per period payoff vector within ε of a given $z \in H^*(N, S, P)$ can be maintained by a subgame perfect equilibrium. That is, for an arbitrary positive ε there is a finite time T_0 such that, if $T > T_0$, then a subgame perfect equilibrium strategy combination can be found for which the average payoff per period will be within ε of z.

4.4 *Epsilon-equilibria in finite-horizon games using trigger strategies*

The bounded rationality approach to modeling human behavior is characterized by compromise between purely instinctive and/or traditional behavior on the one hand and full rationality on the other. Fully rational modeling in the social sciences usually presumes that decision makers possess large quantities of information and that they incur no costs or significant time delays in processing information. Obviously these information and processing cost conditions are not really met in practice. It remains an open question whether assuming they are met is the most fruitful method of modeling. Any of countless real-life decisions made by individual consumers or businesses is based on far less information than the total relevant information available. For example, a person purchasing an automatic clothes washer is unlikely to obtain and study carefully the specifications of any, much less all, available brands. Roughly speaking, the cost of doing so is unlikely to be less than the expected cost of making a less than optimal choice. Avoiding an extra hundred hours of studying and evaluating information will provide the resources needed to buy a large amount of repair service. Recognizing this, there are two ways to proceed formally. One is to insist that the cost of obtaining and processing information be incorporated explicitly into the model. This approach would preserve full optimization; however, the conditions of the model would be altered to recognize that information is not free, and the optimal amount would presumably be bought and used in the best possible way. This route complicates the model, making it likely to be analytically intractible. The second approach, that of bounded rationality, resorts to objective functions that are admittedly shortcuts, employing approximations and/or rules of thumb. Such models can preserve greater analytical manipulability. These models always look a bit unsatisfactory, because one can see ways that the decision makers in the model could use information they have in order to make better decisions. The rules of the model disallow the decision makers from doing this, however because the bounded rationality approach explicitly limits the avenues open to the decision makers for choosing actions. This can be defended on two grounds. First, if bounded rationality models were to yield better empirical predictions than other models, the approach would be vindicated. Second, it can be argued that decision makers do not make an exhaustive study of the optimal amount of information or of the optimal amount of information processing. Introspection and observation suggest that people do use intuition and rules of thumb. Of course the empirical superiority of bounded rationality models has not yet been established, and the apparent use of intuition may be best approximated by a model that explicitly allows for full optimization.

Radner (1980) approaches the finitely repeated game using bounded rationality. His players differ from fully rational players by being unable to distinguish between supergame strategies falling within a margin ε of being best replies. Such strategies may be called ε-best replies and the resulting equilibrium, an ε-equilibrium. That is, σ_i is an *ε-best reply* to σ' if the payoff to i from $\sigma' \backslash \sigma_i$ is within ε of the payoff associated with the best reply of i to σ'. Similarly, σ is an *ε-equilibrium* if, for each $i \in N$, σ_i is an ε-best reply to σ. Formal definitions are given in Definitions 4.19 and 4.20. The player's objective function is his average payoff over the remaining horizon: $G_i(\sigma) = \sum_{t=0}^{T} P_i(u_t(\sigma)/(T+1)$ as of time 0.[5]

DEFINITION 4.19 *In a game $\Gamma = (N, \mathscr{S}^*, G)$ the strategy $\sigma_i \in \mathscr{S}^*$ for player i is an ε-**best reply** to σ' if $G_i(\sigma' \backslash \sigma_i) \geqslant G_i(\sigma' \backslash \sigma_i'') - \varepsilon$ for all $\sigma_i'' \in \mathscr{S}^*$, and for $\varepsilon \geqslant 0$.*

DEFINITION 4.20 *$\sigma \in \mathscr{S}^*$ is an ε-**equilibrium** for the game $\Gamma = (N, \mathscr{S}^*, G)$ if, for $\varepsilon \geqslant 0$, σ_i is an ε-best reply to σ for each $i \in N$.*

Why a player will be as content with somewhat less than a best reply is not immediately clear, unless costs of calculation are assumed to lurk in the background. If such costs are presumed, then the bounded rationality approach can be taken as an approximation to optimal behavior in the presence of these costs. An ε-equilibrium σ^* is subgame perfect if σ^* induces an ε-equilibrium on each subgame.

The main result in this section is that trigger strategy equilibria remain viable even when the game is finite. Under these trigger strategy ε-equilibria, the collusive move is repeated for a while, then, for a duration of several periods at the end of the game, play reverts to a single-period equilibrium point move. The number of periods at the end depends on the payoff functions and the two single-period moves on which the trigger strategy is based; therefore, as the horizon of the game is allowed to grow, the span of time during which cooperation takes place grows at an equal rate.

DEFINITION 4.21 *$\sigma = (s^*, s^c, t^*)$ is a* **finite-duration trigger strategy combination**. *Player i chooses s_i^* in period 0 and on through period t^* except if some player j deviates from s_j^* at time $t' < t^*$; however, player i chooses s_i^c from period $t^* + 1$ onward. If some player j deviates from s_j^* in a period $t' < t^*$, then player i reverts to s_i^c in period $t' + 1$, and continues to choose s_i^c in all remaining periods.*

THEOREM 4.10 *Let $(N, \mathscr{S}^*, G)$ be a repeated game with finite T that satisfies Rule 4.1 and has a period t objective function of $\sum_{\tau=t}^{T} P_i(u_\tau(\sigma)/(T+1-t)$, $i \in N$, $t = 0, \ldots, T$. Let $\varepsilon > 0$, s^*, $s^c \in S$, with $P(s^*) \gg P(s^c)$ and with s^c being an equilibrium point of (N, S, P). Then the finite duration trigger strategy combination (s^*, s^c, t^*) is a subgame perfect ε-equilibrium if*

$$\frac{\phi_i(s^*) - P_i(s^*)}{T+1-t^*} \leqslant \varepsilon, \qquad i \in N \tag{4.45}$$

Proof. Suppose that the trigger strategy combination is being followed by all players except player i. Obviously, player i can do no better than to choose s_i^c from period $t^* + 1$ onward. But suppose that player i deviates from the trigger strategy in some period $t \leqslant t^*$. Then his payoff for periods t through T is, at best,

$$\frac{\phi_i(s^*) + (T-t)P_i(s^c)}{T+1-t} \tag{4.46}$$

but, by following the trigger strategy, it would be

$$\frac{(t^*+1-t)P_i(s^*) + (T-t^*)P_i(s^c)}{T+1-t} \tag{4.47}$$

The trigger strategy is an ε-best reply for player i if equation (4.46) minus equation (4.47) is less than ε. This difference is

$$\begin{aligned}\frac{[\phi_i(s^*) - P_i(s^*)] - [t^* - t][P_i(s^*) - P_i(s^c)]}{T+1-t} &\leqslant \frac{\phi_i(s^*) - P_i(s^*)}{T+1-t} \\ &\leqslant \frac{\phi_i(s^*) - P_i(s^*)}{T+1-t^*} \\ &\leqslant \varepsilon \end{aligned} \tag{4.48}$$

The same argument applies to all players $i \in N$; hence, $(s^*, s^c\ t^*)$ is an ε-equilibrium. Because it is an ε-equilibrium at each stage of the game, it is subgame perfect. QED

COROLLARY *Given s^* and s^c, the maximum value for t^* under which (s^*, s^c, t^*) is an ε-equilibrium is given by equation* (4.45) *and by the condition that*

$$\frac{\phi_i(s^*) - P_i(s^*)}{T-t^*} > \varepsilon \tag{4.49}$$

for at least one player $i \in N$.

Proof. The corollary follows from the proof of Theorem 4.10. QED

As an example, consider the single-period payoff functions in equation (4.10) depicting the five owners of rental cottages. Recall that, using $s_i^* = 90$ and $s_i^c = 30$, $i \in N$, $P_i(s^*) = 8{,}100$, $P_i(s_i^c) = 4{,}500$, and $\phi_i(s^*) = 14{,}580$. Letting $\varepsilon = 100$, it is possible to find $T - t^*$ by solving the equation

$$\frac{\phi_i(s^*) - P_i(s^*)}{T-t^*} = \varepsilon \tag{4.50}$$

Doing so yields $T - t^* = (14{,}580 - 8{,}100)/100 = 64.8$. Thus, the critical value of $T - t^*$ is 64. If, for example, $T = 1{,}000$, a trigger strategy based on $s_i^* = 90$ and $s_i^c = 30$ is viable with t^* equal to 936 or smaller. Note that ε can be any positive number, no matter how small, and the critical value of $T - t^*$ will still be finite; thus, if the horizon is sufficiently long, although finite, trigger strategies are possible. Furthermore, for fixed s^*, s^c, and ε, the minimal number of end-game periods when s^c must be chosen is constant—it does not grow as T grows. Thus, as T goes to infinity, the number of periods in which s^* can be chosen goes to infinity at the same rate.

Using average payoff over the remaining horizon rather than the discounted payoff as the player's objective function plays a crucial role in the proof of Theorem 4.10. This may be seen by supposing that player i faces finite duration trigger strategies on the part of the other players under which s^* is called for in periods 0 through t^*, with s^c for called for thereafter. Consider the position of player i in period t^*: If he adheres to the trigger strategy, his discounted payoff stream is

$$P_i(s^*) + P_i(s^c)(\alpha_i + \alpha_i^2 + \cdots + \alpha_i^{T-t^*}) \tag{4.51}$$

and if he maximizes against the trigger strategies of the others in period t^*, player i's discounted payoff stream is

$$\phi_i(s^*) + P_i(s^c)(\alpha_i + \alpha_i^2 + \cdots + \alpha_i^{T-t^*}) \tag{4.52}$$

Obviously, equation (4.52) is larger than equation (4.51) by $\phi_i(s^*) - P_i(s^*)$; therefore, the trigger strategy is not an ε-best reply if ε is smaller than this difference. For small ε, the trigger strategy must break down no matter what the length of the (finite) horizon. The average payoff per period criterion and a discounting criterion can be made commensurate. Choose a value for ε and, using a discounting criterion, apply the rule that an ε-best reply must come within $(T + 1 - t)/\varepsilon$ of being a best reply in time period t, when the end period of the game is period T. With $\alpha_i = 1$, this is equivalent to the formulation used in Theorem 4.10.

Another variant of bounded rationality is obtained by assuming that, in each period, the player looks ahead T periods. That is, as of period t, the objective function is

$$\sum_{\tau=t}^{T+t-1} \alpha_i^{\tau-t} P_i(s_i) \tag{4.53}$$

Bounded rationality is present because, although the game actually continues for infinitely many periods, the players consistently act as if the end point is finitely many periods ahead. Rather than presume the players do not recognize the truth, it makes more sense to assume that costs of calculating optimal behavior are reduced by looking only a finite distance into the future. The results of Theorem 4.8 to 4.10 can be applied to such a model. If a subgame perfect noncooperative equilibrium is sought, then there are two possibilities: First, if the single-period game has a unique equilibrium point, then a subgame perfect trigger strategy equilibrium cannot exist. The players will perform the backward induction argument at any period t starting from the period they regard as the last one, period $T + t - 1$. Second, if there are two single-shot equilibrium points satisfying the conditions of Theorem 4.9, then a trigger strategy equilibrium is possible; however, if T exceeds 2, then the actual path of observed moves will be stationary. That is, the players never reach a last period in which the better of the two single-shot equilibria is chosen. Similarly, if the

ε-equilibrium concept used in Theorem 4.10 is followed, it is used in a moving way. The trigger strategy will, in any period t, call for choosing s^* for the next t^* periods; that is, until period $t + t^* - 1$, and the time of the planned reversion to s^c, period $t + t^*$ never actually arrives.

5 A comparison between trigger strategy equilibria and folk theorem equilibria

The main comparisons I seek to make are between assumptions and equilibrium strategies under Theorems 4.4 (trigger strategy equilibrium), 4.6 (subgame perfect folk theorem without discounting), and 4.7 (subgame perfect folk theorem with discounting). Looking first at assumptions, quasiconcavity is central to Theorem 4.4 but plays no role in the other theorems. Its essential role is to help ensure existence of an equilibrium, s^c, for the single-shot game (N, S, P). Of course, the mode of punishment under Theorem 4.4 is reversion to s^c, consequently existence of s^c is crucial to existence of trigger strategy equilibrium. The equilibria under Theorems 4.6 and 4.7 entail punishing a player i by driving her payoff to v_i for a while. Existence of v_i is a consequence of Assumptions 3.1 and 3.2; quasiconcavity does not enter. Theorem 4.7 uses a dimensionality condition, Assumption 4.1, that is not used elsewhere. Intuitively this condition does not seem restrictive. With respect to assumptions, each theorem uses something the others do not, but the quasiconcavity condition used by the trigger strategy theorem seems, intuitively, the most restrictive.

All of these theorems were proved by construction, and the constructions have some marked differences. The trigger strategy combinations are, by far, the simplest. Punishment is always the same no matter who is to be punished. Theorem 4.6 is almost as simple; it differs by tailoring the punishment according to who deviated, and it calls for a reversion from punishment to the original actions after some finite time. The trigger strategy equilibria can use finite reversion, but they do not require it. Theorem 4.6 requires it. Theorem 4.7 calls for strategies of impressive complexity. Punishments are tailored according to who deviated and reversion is required, but the reversion cannot be to the original actions. Instead, the reversion is also tailored according to who deviated. Thus, the trigger strategy equilibria are much the simpler, particularly when compared to Theorem 4.7, which also includes discounting.

A comparison can also be made concerning the set of payoff outcomes that can be supported by the two species of equilibria. For trigger strategies, outcomes can be supported if all players get higher payoffs than at some single-shot equilibrium; for folk theorem equilibria, outcomes can be supported if all players get higher payoffs than at v, the minimax payoff vector. If s^c is any single-shot equilibrium, then necessarily, $v \leqslant P(s^c)$, since any player i can always be forced by the others to $P_i(s^c)$, which implies the inequality.

6 Applications of repeated games

A prime application of supergames is to oligopoly, and several applications shown here are variants of a particular oligopoly model that is introduced in Section 6.1. This model is of n firms in an infinitely repeated market under Rule 4.1. In Section 6.2, the model is retained, and n, the number of firms, is allowed to increase without bound. An interesting question is whether trigger strategy equilibria can be maintained both for large n and in the limit. In Section 6.3, the model of Section 6.1 is used again; the number of firms is fixed, but Rule 4.1 is relaxed. A firm never observes rival decisions; yet it is still possible to maintain trigger strategy equilibria, because a firm remains able to tell whether any firm has defected from the cooperative action. The last section discusses the concept of *altruism*, and it is shown that trigger strategy equilibria provide a means by which ordinary selfish behavior would look altruistic.

6.1 A model of differentiated products oligopoly

Suppose a market in which there are n producers of a differentiated product. Because of differentiation, the demand function facing one firm i is a continuous function of the prices of all n firms in each time period $t: q_{it} = f_i(p_t)$, where $p_t = (p_{1t}, \ldots, p_{nt}) \in R^n_+$ is the price vector and q_{it} is the output and demand of firm i in period t. The ith firm's total cost function is $C_i(q_{it}) = C_i\ (f_i(p_t))$, and the single-period profit function is

$$\pi_i(p_t) = p_{it} f_i(p_t) - C_i(f_i(p_t)) \tag{4.54}$$

The conditions imposed on the profit functions are stated below.

CONDITION 4.1 *The demand function $q_i = f_i(p)$ is nonnegative and continuous for $p \in R^n_+$. For $p \gg 0$ and $f_i(p) > 0$, $f_i(p)$ is twice continuously differentiable, $f^j_i(p) > 0, j \neq i$, and $\sum_{j=1}^m f^j_i(p) \leq \varepsilon < 0$. There is $p^+ \in R^n_+$ such that $f_i(p^+) = 0$, $i \in N$.*

CONDITION 4.2 *The cost function $C_i(q_i)$ is convex for $q_i \geq 0$. For $q_i > 0$, $C_i(q_i)$ is twice continuously differentiable and $C'_i(q_i) \geq 0$, $C_i(0) \geq 0$.*

CONDITION 4.3 *For any $p \gg 0$ at which $f_i(p) > 0$ and $p_i - C'_i(f_i(p)) \geq 0$, the profit function $\pi_i(p) = p_i f_i(p) - C_i(f_i(p))$ is concave in p_i.*

CONDITION 4.4 *There is $p^c \in R^n_{++}$ such that $\pi^i_i(p^c) = f_i(p^c) + [p_i - C'_i(f_i(p^c))] f^i_i(p^c) = 0$, $i \in N$.*

Condition 4.1 places conventional restrictions on the demand functions: As a firm's price rises, its sales decline; as the price of a rival firm rises, the firm's sales increase; the effect of changes in the firm's own price is larger in absolute value than the effects of all rival firms combined, and there are prices so high that firms sell nothing. Condition 4.2 requires nonnegative fixed and marginal costs, and stipulates that marginal cost is not falling as output rises. Condition 4.3, concavity of π_i with respect to p_{it} (for price vectors where price exceeds marginal cost), is a technical assumption in the sense that economic considerations do not call for it. It is a restriction that ensures the existence of a single-period Nash equilibrium. Condition 4.4 is merely a convenience in the exposition; it asserts existence of an interior Nash equilibrium (i.e., one at which all firms are active; their output levels are strictly positive). (For a more detailed exposition of these models see Friedman (1983).)

It is easily seen that single-period payoffs at p^c are inside the payoff possibility frontier, and it is possible to find $p^* \gg p^c$ such that $\pi_i(p^*) > \pi_i(p^c)$, $i \in N$. To see this, examine π_i^i and π_i^j evaluated at p^c:

$$\pi_i^i(p^c) = f_i(p^c) + [p_i^c - C_i'(f_i(p^c))]f_i^i(p^c) = 0 \tag{4.55}$$

$$\pi_i^j(p^c) = [p_i^c - C_i'(f_i(p^c))]f_i^j(p^c) \tag{4.56}$$

At p^c, $\pi_i^i = 0$ and $f_i(p^c) > 0$. In addition, $f_i^i < 0$ everywhere; hence, $p_i^c > C_i'(f_i(p^c))$. Using this information in equations (4.54), and recalling that $f_i^j > 0$ $(j \neq i)$, it is clear that $\pi_i^j > 0$ for all $i, j \in N$ $(i \neq j)$, evaluated at p^c. Thus, by continuity of the first derivatives, it is possible to find $p^* \gg p^c$ for which all firms have larger profits $(\pi_i(p^*) > \pi_i(p^c),\ i \in N)$.

Suppose now that the discount parameter of firm i is α_i, making its objective function

$$\sum_{t=0}^{\infty} \alpha_i^t \pi_i(p_t) \tag{4.57}$$

Characterizing a trigger strategy equilibrium is very easy. Let $\phi_i(p^*) = \max_{p_i \in [0, p_i^+]} \pi_i(p^* \backslash p_i)$. Then, for p^* such that $\pi_i(p^*) > \pi_i(p^c)$, $i \in N$, a trigger strategy combination, σ, based on p^* and p^c would be defined. That is, player i would let $p_{i0} = p_i^*$ and would choose $p_{it} = p_i^*$ if $p_\tau = p^*$ for all $\tau < t$. Otherwise, $p_{it} = p_i^c$ would be selected. This combination σ is a trigger strategy equilibrium if

$$\alpha_i \geq \frac{\phi_i(p^*) - \pi_i(p^*)}{\phi_i(p^*) - \pi_i(p^c)} \tag{4.58}$$

which is the condition given in Theorem 4.4.

Turning to a numerical example, suppose $n = 3$, costs are nil, and demand functions for the firms are

$$q_i = 100 - 3p_i + \sum_{j \neq i} p_j \tag{4.59}$$

Then the single-period profit function of each firm i is

$$\pi_i(p) = 100p_i - 3p_i^2 + p_i \sum_{j \neq i} p_j \tag{4.60}$$

Because the model is symmetric, all firms have the same prices, output levels, and profits at the single-period Nash equilibrium. These are $p_i = 25$, $q_i = 75$, and $\pi_i = 1875$. A reasonable trigger strategy outcome to examine is the joint profit maximum, at which $p_i = 50$, $q_i = 50$, and $\pi_i = 2{,}500$.

Finally, $\phi_i(50, 50, 50)$, the profit that one firm can achieve in just one period by maximizing against joint maximum prices for the others, is $3{,}333\frac{1}{3}$. This profit is achieved at a price of $33\frac{1}{3}$ and an output level of 100. It is easily seen that the right-hand side of equation (4.58), is, in this case, 4/7, which indicates that the trigger strategy combination is an equilibrium point if all firms have discount parameters of $\alpha_i = 4/7$ or more.

6.2 Oligopoly for large and increasing n

As Green (1980) has shown, an infinite number of firms is compatible with trigger strategy equilibria. Sufficient conditions are that (1) equation (4.58) remains satisfied and (2) each firm remains aware in each period t of the price choices of all other firms in all past periods. The former means that following the trigger strategy is superior to defecting from it for each firm, and the latter means that each firm can see the defection of any other firm in the period following the defection. While the information conditions of the model guarantee that defection will be promptly spotted by all players, there is no guarantee that equation (4.58) will remain satisfied as n increases. It is easy to make up examples of either sort. The examples included here show that the same three-firm model can be a special case of two different n-firm models that differ in the way that they change as n grows. These examples stem from the work of Lambson (1984), who has studied extensively the conditions under which trigger strategy equilibria remain viable as the number of players in a game increases.

For the first example, let the demand and single-period profit functions be

$$q_i = 100 - 3p_i + \frac{2}{n-1}\sum_{j\neq i} p_j \tag{4.61}$$

$$\pi_i = 100p_i - 3p_i^2 + \frac{2p_i}{n-1}\sum_{j\neq i} p_j \tag{4.62}$$

If $n = 3$, the model reduces to the model of Section 6.1. For any value of n, price, output, and single-period profit for the single-period Nash equilibrium are $p_i = 25$, $q_i = 75$, and $\pi_i = 1{,}875$; for the joint profit maximum, $p_i = 50$, $q_i = 50$, and $\pi_i = 2{,}500$; and for the single period of defection, $p_i = 33\frac{1}{3}$, $q_i = 100$, and $\pi_i = 3{,}333\frac{1}{3}$. Thus, nothing is changed as n grows, and there are trigger strategy equilibria for all values of n.

For the second example, demand and single-period profits are

$$q_i = \frac{400 - 4np_i + 4\sum_{j\neq i} p_j}{n+1} \tag{4.63}$$

$$\pi_i = \frac{400p_i - 4np_i^2 + 4p_i\sum_{j\neq i} p_j}{n+1} \tag{4.64}$$

The demand system in equation (4.63) is based on the inverse demand functions

$$p_i = 100 - \tfrac{1}{2}q_i - \tfrac{1}{4}\sum_{j \neq i} q_j, \qquad i \in N \tag{4.65}$$

If equation (4.65) is taken to be the correct structure for all n, and these inverse demand functions are inverted to express output as a function of prices, equation (4.63) is obtained. The single-period noncooperative equilibrium data in this model are $p_i = 100/(n+1)$, $q_i = 400n/(n+1)^2$, and $\pi_i = 40{,}000n/(n+1)^3$; the joint maximum data are $p_i = 50$, $q_i = 200/(n+1)$, and $\pi_i = 10{,}000/(n+1)$; and the period of defection data are $p_i = 25(n+1)/n$, $q_i = 100$, and $\pi_i = 2{,}500(n+1)/n$. The right-hand side of equation (4.58) is

$$\frac{\dfrac{2{,}500(n+1)}{n} - \dfrac{10{,}000}{n+1}}{\dfrac{2{,}500(n+1)}{n} - \dfrac{40{,}000n}{(n+1)^3}} = \frac{1 - 4n/(n+1)^2}{1 - 16n^2/(n+1)^4} \tag{4.66}$$

Obviously, equation (4.66) converges to 1 as n goes to infinity; therefore, for any fixed value of $\alpha_i < 1$, as n grows a value of n will be reached, beyond which the inequality in equation (4.58) will cease to hold; it will be better to defect than to adhere to a trigger strategy. Thus, trigger strategy equilibria will not be possible for fixed α_i and arbitrarily large n. Of course, equation (4.66) is calculated for a particular trigger strategy; however, a similar expression and result would obtain in this example for any point on which one chose to have a trigger strategy.

6.3 Trigger strategy equilibria when firms cannot observe prices

Suppose that the market is characterized by the model in Section 6.1, but that Rule 4.1 does not hold. In particular, a firm never observes the actual price choices made by the other firms. It is clear in this case that trigger strategy equilibria are still possible because the firms can still discern whether their rivals are sticking to the trigger strategy prices or defecting. The means for this observation is the firm's own output level. In each period, firms simultaneously select prices, following which each firm observes its own output level. If any firm defects from its trigger strategy price, other firms will find their demand and output differ from what they expected. No firm can tell which rival defected, but that knowledge is not necessary for it to get the signal that it should switch to its single-period Nash equilibrium price.

Porter (1983) analyzes a model in which firms do not observe the choices made by rivals, as above, and in which randomness enters the demand functions. The distribution of the random variable is known; however, its presence means that all firms can be following trigger strategies and, at the same time, a firm will have lower sales than expected. Again, it remains possible to have trigger strategy equilibria, but two alterations are called for in those strategies. Assume that the random variable in demand, u_t, is additive and the mean of its distribution is 0. Then the firm's demand function is

$$q_{it} = 100 - 3p_{it} + \sum_{j \neq i} p_{jt} + u_t \tag{4.67}$$

This is the demand function from the example in Section 6.1, with the random variable u_t added. All firms are subject to the same realization of the random variable.

The first alteration in the trigger strategy is to lower the trigger output level to something below the expected output level. Suppose that the firms had determined a price of 50 for each to achieve the joint profit maximum. The associated expected output levels are 50 each. To state the trigger strategy so that a firm reverts to a price of 25 in the period after observing an output level below 50 is unduly stringent, because it makes no allowance for the random variable ever being negative. Assuming a symmetric distribution, there is a probability of .5 that the random variable is negative in period 0 and a probability of $1 - .5^{t+1}$ that it will have been negative at least once by the end of period t. This standard is clearly too severe; the firms are better off choosing a trigger output level of something less than 50. They face a dilemma: A trigger of 50 leaves little room for a firm to choose a price below 50 and go undetected, but output levels below the trigger will appear very quickly from a negative realization of the random variable. A trigger very far below 50 will ensure that a random variable realization is unlikely to send output below the trigger, but it gives lots of room for firms to cheat without detection. These considerations call for a carefully calculated intermediate value for the trigger that balances off these two elements. A second approach is to do a statistical analysis of the supposed values of the sequence of random variables. If firms are adhering to the trigger strategies, then, in each period a firm can infer the value of the random variable. The value inferred by firm i is $\bar{u}_{it} = q_{it} - 50$. Over time, firm i observes $(\bar{u}_{i0}, \bar{u}_{i1}, \ldots, \bar{u}_{it})$, and at each time t it can test the hypothesis that the observed sequence is drawn from a population with mean 0 and the known variance of u_t. If the standards for that test are agreed on in advance by all firms, then all firms adhering to the trigger strategy will come to identical conclusions in each period.

The second alteration in the trigger strategy is that the firms should, when reverting to single-period equilibrium point prices, revert only for a finite time. Clearly this length of reversion, T, must be the same for all firms and known to all. The reason for using a finite length is that, no matter what positive trigger value is chosen for output, if all firms always adhere to choosing prices of 50, there will still be a time when the trigger level is reached and the firms revert to choosing prices of 25. The only way to have the higher prices continue into the distant future is to use finite period reversions. Doing this, of course, increases the value of defection, but the choice of T, the duration of the reversion, and of the trigger value of output would be made jointly.

6.4 Altruism and repeated games

Altruism is defined as "devotion to the welfare of others, regard for others, as a principle of action; opposed to egoism or selfishness." In a world of rational people who choose their actions according to a plan to maximize the value of an objective function, devising an operational definition of altruism can be difficult. For example, suppose there are two people in a world of m commodities and no production. Suppose that person 1 has an endowment of $w^1 \in R^m$, and person 2 has an endowment of $w^2 \in R^m$. Now assume that $w^1 \gg 0$ and $w^2 = 0$; hence, person 2 will be miserable unless person 1 gives him part of his endowment. Say person 1 divides his endowment into $y^1 \gg 0$ and $y^2 \gg 0$, where $y^1 + y^2 = w^1$, and he reserves y^1 for his personal consumption while giving y^2 to person 2. To complete the picture, let this whole situation be a one-shot thing and assume that, of all the choices open to person 1, the division into y^1 and y^2 is what he decides is best. This is easily explained with conventional consumer choice theory: Person 1 has preferences over commodity allocations $(x^1, x^2) \in R^{2m}$, where x^1 will be consumed by person 1 and x^2 by person 2. Under suitable assumptions (see Green (1976) or Hildenbrand and Kirman (1976)), a continuous utility function $u_1(x^1, x^2)$ can represent the preferences of person 1.

The preceding illustration leads to a definition of altruism under which a person is altruistic if increases in the consumption bundles of (some) other people will increase her own satisfaction. It is natural to state a companion definition of a selfish person as one whose satisfaction depends only on her own personal consumption bundle. One may wish to argue that a person cannot be altruistic when choosing actions that maximize her own utility, and the fact that x^2 enters the utility functions of person 1 merely means that person 1 is being selfish when giving something to person 2. Her motive is to suit herself.

A second definition of altruism, proposed by Kurz (1976), is that a person is altruistic if her satisfaction does not depend on the consumption of others and she gives valuable commodities to another person without having a binding agreement under which she will receive something of value in return. Kurz shows that altruistic behavior can be characteristic of trigger strategy equilibria.

My purpose here is not to settle the question of how best to define altruism within the context of economics; rather, I wish only to suggest briefly two possible definitions and to draw on Kurz (1976) to show the connection between trigger strategy equilibria and the second definition. Suppose there are two people in a trading situation as depicted in Figure 4.7. Their initial endowment of goods is, in each period, at the point E.

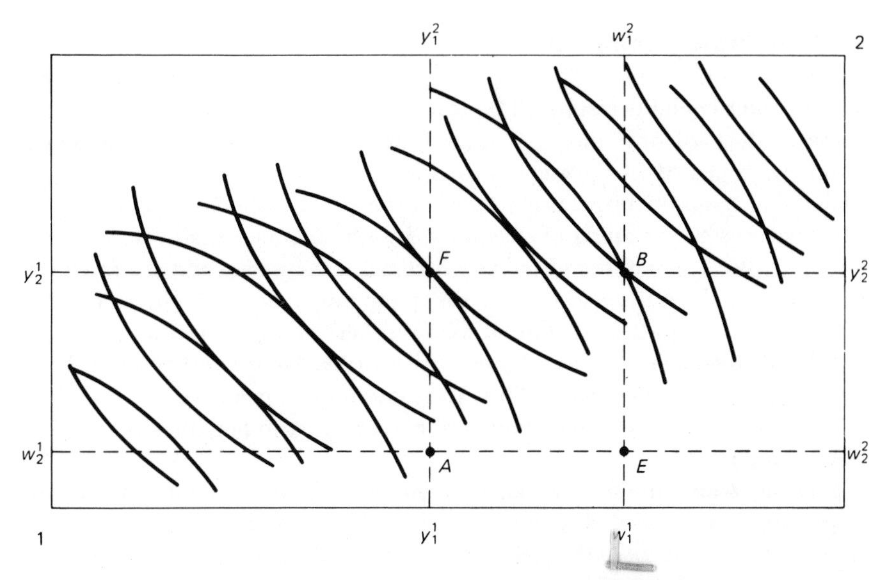

FIGURE 4.7 Trigger strategy equilibria and altruism.

Player 1 receives $w^1 = (w_1^1, w_2^1)$ and player 2 receives $w^2 = (w_1^2, w_2^2)$. Now suppose a trigger strategy σ_1 for player 1 under which he gives player 2 $\delta_1 = w_1^1 - y_1^1$ units of good 1 in the initial period. In each subsequent period, he gives the same to player 2 if player 2 has, in all past periods, given player 1 $\delta_2 = w_2^2 - y_2^2$ units of good 2. Otherwise, player 1 gives player 2 nothing. Let σ_2 be defined in a parallel way for player 2. Each player is engaging in voluntary gifts—there is no binding contract between them. To say that no strings are attached goes too far, but certainly either player in any period is free to give nothing or to give any gift falling within his endowment. If, for example, player 1 gave nothing in a particular period, the two players' consumption bundles would be at B in Figure 4.7. Following σ, forever after this act neither player would give any gifts and they would consume at E. On the other hand, if they follow their trigger strategies $\sigma = (\sigma_1, \sigma_2)$ indefinitely, they will consume F in Figure 4.7. Letting $u_i(x^i)$ be the single-period utility (payoff) function of player i and α_i his discount parameter, σ is a trigger strategy equilibrium if

$$\alpha_1 > \frac{u_1(w_1^1, y_2^1) - u_1(y^1)}{u_1(w_1^1, y_2^1) - u_1(w^1)} \tag{4.68}$$

$$\alpha_2 > \frac{u_2(y_1^2, w_2^2) - u_2(y^2)}{u_2(y_1^2, w_2^2) - u_2(w^2)}. \tag{4.69}$$

both hold. This is an equilibrium with behavior that is altruistic under the Kurz definition.

7 Concluding comments

This chapter contains results on noncooperative equilibrium in multiperiod games. Most of the models studied are repeated games, sometimes infinitely repeated and sometimes finitely. Several points about the underlying noncooperative orientation of the chapter should be made. The first concerns the meaning of *cooperative*. In a game, one could consider the structure of the game, the observed outcome of the game, and the mode of play. Any of these three aspects of the game might be cooperative. By the usual definition, and the definition followed here, *a cooperative game is a game that has a cooperative structure. A game has a cooperative structure if and only if binding agreements are permitted.* This designation of cooperative game makes no reference whatever to the choices of players (i.e., to how they determine their moves).

Looking back on the folk theorem and the theorems related to it, an important implication of these results is that *cooperative outcomes may be achieved in some noncooperative games.* I am implicitly regarding outcomes on the payoff possibility frontier of a game (N, S, P) as being *cooperative,* because they are not, in general, noncooperative outcomes in (N, S, P) itself and because much of cooperative game theory is built on the premise that outcomes will surely be selected on the frontier when binding agreements are permitted. To my knowledge, the term *cooperative outcome* has not received a formal definition. For present purposes it is defined as *a Pareto optimal outcome.* An implication of some theorems in the present chapter is that *cooperative outcomes may result from noncooperative equilibria in repeated games.* In this context, *cooperative behavior* would naturally be behavior that leads to cooperative outcomes. So we have here noncooperative games in which some noncooperative equilibria entail cooperative outcomes and cooperative behavior.

Many existence-of-equilibrium theorems were proved by the construction of equilibrium strategy combinations that consisted of paths and rules concerning when to change from one path to another. Simultaneous deviations were treated almost cavalierly in the sense that there could be, at most, one official deviator at a time and the official deviator was selected in an arbitrary manner when simultaneous deviations occurred. This is justified by the noncooperative equilibrium concept. Recall that σ is an equilibrium if $G_i(\sigma) \geqslant G_i(\sigma \backslash \sigma_i')$ for any σ_i' and i. That is, if σ is an equilibrium, then no player would obtain a larger payoff by using a different strategy while holding fixed the strategies of the other players. This definition says that *no single player can profitably deviate from a noncooperative equilibrium.* Simultaneous deviations are thus not relevant for determining if a strategy combination is an equilibrium.

Reconsider the notion of *credible threat.* Throughout this chapter credible threat has been defined implicitly as a threat embodied within a subgame perfect equilibrium strategy combination. Such a definition may be clear, but may not be acceptable to everyone. To see this, consider a grim trigger strategy equilibrium. If someone defects from s^*, the players switch to s^c and stay there forever. This is costly; they could forgive the defector, give him a second chance, and return to choosing s^*. This has appeal but, it destroys the deterrent effect of the threat to revert to s^c and, in so doing, removes the enforcement mechanism that allowed the choice of s^*. That is, following deviation from s^* the players have some incentive to renegotiate the way they play the remainder of the game, and so grim trigger strategy combinations fail to be *renegotiation-proof.* If an equilibrium were renegotiation-proof, then, following any history at any time, the equilibrium strategy combination would prescribe behavior from which renegotiation was effectively impossible. This sounds like a requirement that the equilibrium associated with each subgame be Pareto optimal.

Exercises

1. A finite two-person bimatrix game is given by

5, 9	5, 7	−3, 0	20, 5
3, 10	2, 20	4, 5	15, 17
−4, 1	10, 3	2, 2	0, −5
0, 1	8, −2	6, 4	10, 0

   a. What are the pure strategy equilibrium points of this game?
   b. Based on pure strategies, is there a finitely repeated game equilibrium point that is subgame perfect if the game is repeated T times? Can such a strategy have the pure strategies $(2, 4)$, with payoff realization $(15, 17)$ for part of the game?
   c. For the trigger strategy in part (b), what is the minimum number of periods during which pure strategy $(2, 4)$ cannot be selected?
   d. What are the lower bounds for the discount parameters in part (c) such that the trigger strategy combination you have identified is an equilibrium point?
2. Assume that the finite game in problem 1 is to be repeated infinitely. Identify a subgame perfect trigger strategy combination and give the associated minimum values of the discount parameters for which this trigger strategy combination is an equilibrium point.
3. Suppose a game (N, S, P) given by $N = \{1, 2\}$, $S_1 = [0, 50]$, $S_2 = [0, 50]$, $P_1(s) = 100s_1 - 10s_1^2 + 10s_1s_2$, $P_2(s) = 200s_2 - 15s_2^2 + 10s_1s_2$.
   a. What is an equilibrium point of this game? Is it uniuqe? What are the equilibrium payoffs?
   b. What payoffs would the players obtain if they chose s^* in order to maximize $P_1(s) + P_2(s)$? Can these payoffs be supported as a subgame perfect trigger strategy noncooperative equilibrium point? If the answer is yes, with what range of values for the discount parameters?
   c. What are the minimax payoffs of the players? When player 1 forces player 2 to her minmax payoff. v_2, does player 1 receive a payoff at, above, or below v_1?

d. Suppose the game in part (a) will be finitely repeated with $\alpha_1 = \alpha_2 = 1$. For $\varepsilon = 50$ what is the minimum value of $T - t^*$ that allows the joint profit maximum found in *b* to be attained? How are ε and $T - t^*$ related?

4. For the model illustrated in Figure 4.6 with payoff functions given in equations (4.43) and (4.44), find the single-shot payoff associated with $s = (6, 12)$. Suppose $\alpha_1 = \alpha_2 = 1$ and that the game will be finitely repeated. Show that an equilibrium path μ^0 can be supported under which $\mu_t^0 = P(6, 12)$ for $t = 0, \ldots, T-1$ and $\mu_T^0 = P(8, 15)$.
5. Let $p = \max\{0, 100 - \sum_{i=1}^{n} q_i\}$ be the inverse demand function for a Cournot oligopoly and suppose each firm has a total cost function $C_i = 10q_i$. Find the single-shot noncooperative equilibrium payoff and the minimax payoff for each player as a function of n. For the case of $n = 8$ and $\alpha_i = .95$ for all firms, can you construct a subgame perfect strategy combination that will realize a payoff of 22 per period?

Notes

1. From an analytical standpoint, it does not matter whether the action taken in a single time period is a *move* in the sense of game theory or a *strategy* that involves several moves. Either way, the actions and payoffs within one time period constitute a game, and the sequence of actions and payoff functions constitute another game—the supergame. The latter is built out of the former. It does happen that the single-period actions are moves (single-move strategies) in many applications of these models.
2. When the concept of *strategy* was first introduced, it was emphasized that the move of a player at a particular node would generally depend on the information the player would have at that point in the game. To allow time dependent strategies is merely to allow this same property to apply.
3. The earliest source I know of for a written statement along the lines of the folk theorem is Luce and Raiffa (1957) in their discussion of the repeated prisoners' dilemma.
4. There is one other equilibrium: $(0, 0)$ with associated payoffs of 0 for each player. For my example, I use the two interior equilibria; however, any two of the three equilibria are suitable for the illustration.
5. Although Radner mentions the objective function used here (average payoff over the remainder of the horizon; $\sum_{\tau=t}^{T} P_i(s_\tau)/(T - t + 1)$ at time t), he actually uses average payoff from the beginning; that is $\sum_{\tau=0}^{T} P_i(s_\tau)/(T + 1)$ at time t. His formulation is easier to work with but is not as reasonable. It is intuitively more appealing that players would, from any time t onward, seek to maximize their average payoffs from that time on; however, both formulations lead to qualitatively similar results.

5

Time-dependent supergames, limited information, and bounded rationality

The models studied in this chapter range over several topics that come under the heading of games in semiextensive or extensive form. The first of these, supergames with time-dependent structures, covered in Sections 1 and 2 is a straightforward generalization of the infinite horizon models of Chapter 4. The single-period payoff function for a player in time t depends on the actions of all players in the previous period (s_{t-1}), as well as on their current actions (s_t). Most of Section 1 is devoted to models with a stationary structure; that is, the single-period payoff function, $P_i(s_{t-1}, s_t)$, does not change over time. Existence of an equilibrium point and conditions under which any path of equilibrium actions must converge to a unique steady state are given. These models and the associated convergence results can be generalized to nonstationary payoff structures. Section 2 contains an extension of the Fudenberg–Maskin theorem (Theorem 4.7) to time-dependent supergames.

Section 3 takes up another sort of generalization of the infinite horizon models of Chapter 4—in this case to stochastic games. The setup involves a fixed set of players (N) who play an ordinary single-shot game in each time period; however, the game they play at time t is randomly selected from a finite set of games. When the game of time t is to be played, all players know which game has been chosen. Time dependence enters in the specification of the random process by which the next game is chosen. The probability of playing game Γ_t in time t depends on which game Γ_{t-1} was played in time $t-1$ and on the actions, s_{t-1}, through the stochastic mechanism that selects which game is played. The results in this section draw heavily on techniques of stochastic dynamic programming, as the game is essentially made up of n interconnected stochastic dynamic programming problems.

Section 4 is concerned with a bounded rationality approach to an infinite-horizon game of incomplete information; in this case, a large number of players are divided into pairs with each pair playing a two-person noncooperative game. This process is repeated indefinitely, often with each player knowing very little about the past play of his present opponent and having limited information about the payoff structure of the current game. Each player can be assumed to undertake a full Bayesian analysis of the game. Then, at each play, a player attempts to identify his current opponent by the available information and guesses how the player will act on the basis of this individual evaluation. This approach requires an extremely elaborate set of computations by each player in each period. Rather than the full rationality Bayesian approach, a bounded rationality method is used that has each player treating the possible opposing players in a statistical fashion. The player knows some characteristics of the opposing set of players, and treats each opponent actually faced as being a random draw from this population. The opponent's play can depend on some current, specific information.

Section 5 looks at some basic challenges to the noncooperative equilibrium. These challenges are fundamentally different from the refinements literature (see Chapter 2, Section 5). The latter looks to eliminate some noncooperative equilibria by imposing appealing additional conditions; it thus accepts the concept of the equilibrium point and claims it does not go sufficiently far. In contrast, Section 5 of this chapter contains two games, each having a unique noncooperative equilibrium, but the equilibrium is unconvincing: Intuition suggests that "reasonable players" would behave differently. Section 6 examines some additional questions raised by noncooperative equilibrium and discusses the minimum requirements of rational behavior. Section 7 examines some applications and Section 8 includes concluding comments.

1 Time-dependent supergames

In practice, it is easy to cite examples in which the payoffs received by the players in a game for time period t depend not only on the actions taken in that period, but on actions taken in one or more past periods as well. For example, imagine an election campaign in which the candidates' must state their positions on various issues of public interest. Supposing this election to be one period of an ongoing political game, the credibility of a candidate's announcement of her position would naturally depend on the positions she has taken in the past. Consequently, the relationship between the payoff received in the current period and the positions announced in the current period is affected by the past announcements of positions by the players. To announce extreme departures from past positions is likely to deprive a candidate of votes. Short- and long-run tradeoffs occur when, for example, a candidate who has been, say, far to the left in the political spectrum and who sees the whole country moving to the right, starts a slow, orderly change of her own position toward the right. This may deprive her of votes in the short run by disappointing her traditional constituency, but it may greatly strengthen her position several years hence by keeping her nearer the national mainstream. A member of the House of Representatives who has been winning by a wide margin and who looks to win higher office may make such a trade.

A second example is a market for a durable good. The demand facing any supplier in a given time period will depend on the size and age profile of the stock of the good held by consumers. Households that purchased new refrigerators in 1980 can pretty well be ruled out as potential buyers in 1981 or 1982.

The remainder of this section is divided into four sections, the first of which describes the stationary model and has some basic results that help to characterize equilibria. Section 1.2 examines the existence and stability of steady-state equilibria in the stationary model. In Section 1.3, the nonstationary model is described, and it is shown how the results of Sections 1.1 and 1.2 generalize. Section 1.4 explores the relationship between the stationary model and conventional state variable formulations.

1.1 The stationary model

In Section 1.1.1, the assumptions used for the stationary time-dependence supergame model are given and the class of games is defined. These games have open-loop equilibria. This is proved as a corollary to Theorem 3.1. Section 1.1.2 is concerned with the interpretation of a time-dependent supergame as a sequence of single-shot games. There may be many ways to make such an interpretation, but not all of them will be fruitful; however, a particular interpretation is suggested that is intimately connected to the characterization of equilibrium points in the supergame, and that connects them to a sequence of equilibrium points in the (appropriately defined) single-shot games.

1.1.1 Definition of stationary time-dependent supergames and existence of open-loop equilibria

Formulating a stationary infinite-horizon game model is straightforward. The single-period payoff is written as $P_i(s_{t-1}, s_t)$. More than one past period could be incorporated, but the basic principles are illustrated in the simpler situation. Assumption 3.1 is retained and Assumptions 3.2 and 3.3C are suitably modified:

ASSUMPTION 5.1 *$P_i(s_{t-1}, s_t) \in R$ is bounded and continuous for all $(s_{t-1}, s_t) \in S \times S$.*

ASSUMPTION 5.2 *$P_i(s_{t-1}, s_t)$ is a concave function of $(s_{i,t-1}, s_{it}) \in S_i \times S_i$.*

DEFINITION 5.1 *A* **stationary time-dependent supergame,** $\Gamma = (N, S, s_0, P, \alpha)$, *satisfies Assumptions* 3.1 *and* 5.1. *For all $i \in N$, $\alpha_i \in [0, 1)$. If Assumption* 5.2 *holds, the game is a* **concave stationary time-dependent supergame.** *The set of open-loop strategies for player i consists of $\Sigma_i = s_{i0} \times S_i \times S_i \times \ldots$, where $s_{i0} \in S_i$ is a fixed initial condition for period* 0 *and, for each $t > 0$, $s_{it} \in S_i$.*

The payoff function of player i, in terms of open-loop strategies, is

$$\sum_{t=1}^{\infty} \alpha_i^{t-1} P_i(s_{t-1}, s_t) \tag{5.1}$$

The condition that s_{i0} has a single possible value means merely that the action in the game begins in period 1 and that s_0 is an initial condition at that time. Although this initial condition is given from the players' standpoint, we shall sometimes consider various possible values for it. That a concave stationary time-dependent supergame has an equilibrium point follows almost immediately from Theorem 3.1. Restricting attention to open-loop strategies, equation (5.1) is a concave function of σ_i and existence of equilibrium follows from a proof parallel to the proof of Theorem 3.1 but using the fixed-point theorem of Fan (1952).

COROLLARY *A concave stationary time-dependent supergame has an equilibrium point in open-loop strategies.*

The open-loop strategy set of player i (denoted Σ_i, as in Chapter 4) is $s_{i0} \times S_i \times S_i \times \ldots$ and the closed-loop strategy set (denoted Σ_i^*, as in Chapter 4) is defined by Definition 4.4. As noted in Chapter 4, a strategy combination $\sigma \in \Sigma^*$ induces a path $\{u_t(\sigma)\}_{t=0}^{\infty}$ and the supergame payoff of player i is naturally given as $G_i(\sigma) = \sum_{t=1}^{\infty} \alpha_i^{t-1} P_i(u_{t-1}(\sigma), u_t(\sigma))$. From time to time strategies and strategy combinations will be written using notation such as $\sigma^* = (\sigma_1^*, \ldots, \sigma_n^*)$ or $\sigma^i = (\sigma_1^i, \ldots, \sigma_n^i)$, and so forth. It will always be understood that $u_0(\sigma^*) = s_0^* = s_0$ and $u_0(\sigma^i) = s_0^i = s_0$, where s_0 is the fixed and unchanging initial condition.

1.1.2 *On the interpretation of a time-dependent supergame as a sequence of one-shot games*

Three interesting questions relating to stationary supergame equilibrium points are (a) whether the single-period actions, are, in some sense, equilibrium points of single-shot games, (b) whether the supergame has a steady-state equilibrium, and (c) whether a steady-state equilibrium is stable. To be more precise, let $\sigma^* = (s_0^*, s_1^*, s_2^*, \ldots)$ be an equilibrium point of the supergame. Question (a) asks whether there is a natural, reasonable way to form payoff functions for each individual period t so that s_t^* is an equilibrium point for the individual period t game. Question (b) asks whether there is some special $s^* \in S$ such that, if the initial condition is $s_0 = s^*$, then $s_t^* = s^*$, $t = 0, 1, 2, \ldots$ is an equilibrium point of the supergame. Question (c) can be posed this way: Suppose that s_0^*, the initial condition, is an arbitrary member of S. Then, for the equilibrium sequence of actions $\{s_t^*\}$, must s_t^* converge independently of s_0^* to a steady-state s^* as t goes to infinity? Question (a) is answered in the remainder of this section. Questions (b) and (c) are taken up in Section 1.2.

$s_t = s^*$ (the supergame steady-state equilibrium value) is not an equilibrium point of the single-shot game $(N, S, P(s_{t-1}^*, s_t))$. To see this suppose that

$$P_1(s_{t-1}, s_t) = s_{1,t-1}^2 - s_{1,t-1}s_{2t} + 3s_{1t} - s_{1,t-1}s_{1t}^2 \tag{5.2}$$

$$P_2(s_{t-1}, s_t) = s_{2,t-1}^2 - s_{2,t-1}s_{1t} + 3s_{2t} - s_{2,t-1}s_{2t}^2 \tag{5.3}$$

and $\alpha_1 = \alpha_2 = .5$. Then, for player 1, and open loop σ

$$G_1(\sigma) = \sum_{t=1}^{\infty} .5^{t-1} P_1(s_{t-1}, s_t) \tag{5.4}$$

and an equilibrium strategy for player 1 will satisfy

$$.5^{t-1}\left[\frac{\partial P_1(s_{t-1}, s_t)}{\partial s_{1t}} + .5\frac{\partial P_1(s_t, s_{t+1})}{\partial s_{1t}}\right]$$
$$= .5^{t-1}[3 - 2s_{1,t-1}s_{1t} + .5(2s_{1t} - s_{2,t+1} - s_{1,t+1}^2)] = 0, \qquad t = 1, 2, \ldots \tag{5.5}$$

If $s_{10} = s_{20} = 1.2$ and $s_{2t} = 1.2$ for $t = 1, 2, \ldots$ then the optimal strategy for player 1 is $\sigma_1^* = (1.2, 1.2, 1.2, \ldots)$. Due to the symmetry of the model, if player 1 uses σ_1^*,the best reply of player 2 is $\sigma_2^* = (1.2, 1.2, 1.2, \ldots)$. Now suppose that $s_\tau = (1.2, 1.2)$, $\tau = 0, \ldots, t-1$ and that player 1 believes that $s_{2t} = 1.2$ will be chosen by player 2. What would player 1 select if he wished to maximize $P_1(s_{t-1}, s_t)$? Clearly, this is determined by

$$\frac{\partial P_1(s_{t-1}, s_t)}{\partial s_{1t}} = 3 - 2s_{1,t-1}s_{1t} = 3 - 2.4s_{1t} = 0 \tag{5.6}$$

or $s_{1t} = 1.25$. Thus, the equilibrium supergame strategy combination $\sigma^* = (s_0^*, s_1^*, \ldots)$ is not composed of actions s_t^* that are equilibrium points for the myopic games $(N, S, P(s_{t-1}^*, s_t))$.

In other words, equilibrium behavior does not, in general, mean that a player maximizes the payoff in each period by his choice for that period. The reason, of course, is that the choice in period t will affect the payoff in period $t+1$ as well as that of period t, which suggests that s_t^* is an equilibrium point of some specially formed myopic game that takes into account the payoffs of periods t and $t+1$. This is stated as Lemma 5.1, where, as the foregoing numerical example indicates, the relevant payoff functions are seen to be

$$P_i(s_{t-1}^*, s_t) + \alpha_i P_i(s_t, s_{t+1}^*) = P_i^*(s_{t-1}^*, s_t, s_{t+1}^*).$$

LEMMA 5.1 $\sigma^* = (s_0^*, s_1^*, s_2^*, \ldots)$ *is an open-loop equilibrium point of the concave stationary time-dependent supergame* (N, S, s_0^*, P, α) *if and only if* s_t^* *is an equilibrium point of the game* $(N, S, P^*(s_{t-1}^*, s_t, s_{t+1}^*))$ for $t = 1, 2, \ldots$.

Proof. It is first shown that if σ^* is an equilibrium point of (N, S, s_0^*, P, α), then s_t^* is an equilibrium point of $(N, S, P^*(s_{t-1}^*, s_t, s_{t+1}^*))$. If, for some t, s_t^* is not an equilibrium point, then there is at least one player i who can choose $s_{it}' \neq s_{it}^*$ and increase his payoff. Look at the supergame payoff function in detail, evaluated at $\sigma^* \backslash s_{it}'$:

$$\begin{aligned} G_i(\sigma^* \backslash s_{it}') &= P_i(s_0^*, s_1^*) + \cdots + \alpha_i^{t-2} P_i(s_{t-2}^*, s_{t-1}^*) + \alpha_i^{t-1} P_i(s_{t-1}^*, s_t^* \backslash s_{it}') \\ &\quad + \alpha_i^t P_i(s_t^* \backslash s_{it}', s_{t+1}^*) + \alpha_i^{t+1} P_i(s_{t+1}^*, s_{t+2}^*) + \cdots \\ &= P_i(s_0^*, s_1^*) + \cdots + \alpha_i^{t-2} P_i(s_{t-2}^*, s_{t-1}^*) + \alpha_i^{t-1} P_i^*(s_{t-1}^*, s_t^* \backslash s_{t-1}', s_t^* \backslash s_{it}', s_{t+1}^*) \\ &\quad + \alpha_i^{t+1} P_i(s_{t+1}^*, s_{t+2}^*) + \cdots \end{aligned} \tag{5.7}$$

By selecting s'_{it}, player i increases his payoff for the combined periods t and $t+1$, but payoffs are unchanged for all remaining periods; thus, if s_t^* is not an equilibrium point for $(N, S, P^*(s_{t-1}^*, s_t, s_{t+1}^*))$, then σ^* is not an equilibrium point for (N, S, s_0^*, P, α).

Thus, if σ^* is an equilibrium point of $(N, S, \sigma_0^*, P, \alpha)$, s_t^* is an equilibrium point of $(N, S, P^*(s_{t-1}^*, s_{t+1}^*))$ for all $t > 1$. If s_t^* is an equilibrium point of $(N, S, P^*(s_{t-1}^*, s_t, s_{t+1}^*))$ for $t = 1, 2, \ldots$, then σ_i^* is a local maximum of G_i; however, a local maximum of a concave function is a global maximum (see Roberts and Varberg (1973, page 123)). Therefore, σ^* is an equilibrium point of (N, S, P, α) and the theorem is proved. QED

Lemma 5.1 allows a simple way of describing the open-loop equilibrium points of (N, S, s_0, P, α) by means of the equilibrium points of the games $(N, S, P^*(s_{t-1}, s_t, s_{t+1}))$. For any $s_{t-1}, s_{t+1} \in S \times S$, there is a nonempty set of equilibrium points of $(N, S, P^*(s_{t-1}, s_t, s_{t+1}))$. This set of equilibrium points is denoted $\psi(s_{t-1}, s_{t+1})$ and ψ is called the *equilibrium correspondence*. More formally:

DEFINITION 5.2 *The* **equilibrium correspondence** *of* (N, S, s_0, P, α) *is denoted* $\psi(s', s'')$ *and is defined by the conditions that* $\psi(s', s'') \subset S$ *and,* $s^* \in \psi(s', s'')$ *if and only if* s^* *is an equilibrium point of* $(N, S, P^*(s', s, s''))$.

In view of Definition 5.2, the content of Lemma 5.1 can be stated as follows: $\sigma^* = (s_0^*, s_1^*, \ldots)$ is an open-loop equilibrium point of the time-dependent supergame if and only if $s_t^* \in \psi(s_{t-1}^*, s_{t+1}^*)$ for $t = 1, 2, \ldots$.

1.2 Existence and stability of a steady-state equilibrium

The assumptions made thus far, Assumptions 3.1, 5.1, and 5.2, are not sufficient to ensure existence of a steady-state equilibrium. In Section 1.2.1, where existence of equilibrium points is proved, an additional assumption is made that places restrictions on the equilibrium correspondence. It is shown in Section 1.2.2 that these same conditions guarantee stability of the open-loop equilibria of time-dependent supergames.

1.2.1 Existence of a unique steady-state equilibrium

The existence of a steady-state equilibrium, that is, an equilibrium $\sigma^* = (s_0^*, s_1^*, \ldots)$ where $s_t^* = s_{t'}^*$ for all t and t', can be stated in terms of the equilibrium correspondence: There is a steady-state equilibrium point if and only if there is some s^* such that $s^* \in \psi(s^*, s^*)$. This, however, is a fixed-point problem; there is a steady-state equilibrium if and only if the equilibrium correspondence has a fixed point. The following assumption is one that ensures that there is a fixed point and, in addition, that starting from any initial condition s_0, an equilibrium is both unique and converges to the steady state.[1]

ASSUMPTION 5.3 *The equilibrium correspondence* $\psi(s', s'')$ *is a single-valued function that obeys the following Lipschitz condition*:

$$\|\psi(s_{t-1}, s_{t+1}) - \psi(s'_{t-1}, s'_{t+1})\| \leqslant k_1 \|s_{t-1} - s'_{t-1}\| + k_2 \|s_{t+1} - s'_{t+1}\| \tag{5.8}$$

where $k_1 + k_2 \leqslant k < 1$ *for all* $s_{t-1}, s'_{t-1}, s_{t+1}, s'_{t+1} \in S$.

This assumption restricts the equilibrium correspondence in two ways. First, the condition that ψ is a function satisfying a Lipschitz condition k, with $k < 1$, means that the game $(N, S, P^*(s^*_{t-1}, s_t, s^*_{t+1}))$ has a unique equilibrium point. Second, this equilibrium point, which changes as s^*_{t-1} and s^*_{t+1} change, cannot change very fast. In particular, suppose S is a subset of R^m and suppose that the equilibrium correspondence is differentiable. Then Assumption 5.3 would require

$$\sum_{j=1}^{m} \sum_{i=1}^{n} \left| \frac{\partial \psi}{\partial s_{ij,t-1}} \right| \leqslant k_1 \qquad \sum_{j=1}^{m} \sum_{i=1}^{n} \left| \frac{\partial \psi}{\partial s_{ij,t+1}} \right| \leqslant k_2 \tag{5.9}$$

A function that satisfies the inequalities in equations (5.8) or (5.9) is called a *contraction.* A useful fact about contractions is that a contraction never has more than one fixed point, and, if the contraction maps a set into itself, it has a fixed point (see Bartle (1964, page 170)).[2]

LEMMA 5.2 *If a concave stationary time-dependent supergame satisfies Assumption* 5.3, *then there is a unique steady-state equilibrium. That is, there is a unique* $s' \in S$ *such that, if* $s_0 = s'$, *then* $\sigma' = (s', s', \ldots)$ *is an equilibrium point.*

Proof. If s_{t-1} is required to equal s_{t+1}, then $\psi(s, s)$ can be treated as a function from S to S. A fixed point of this mapping is a steady-state equilibrium. Such a fixed point exists and is unique because ψ is a contraction. QED

1.2.2 Stability of open-loop equilibria

Showing stability of steady-state equilibria is more difficult than proving existence of a unique steady state. Suppose that $\sigma^* = (s^*_0, s^*_1, \ldots)$ is an equilibrium point of a supergame and that s' is the unique steady state. To prove stability is to prove that s^*_t converges to s' as t goes to infinity. In other words, no matter what the initial condition s^*_0, eventually any (open-loop) equilibrium behavior will have the players choosing actions arbitrarily close to s'.

THEOREM 5.1 *Let* (N, S, s^*_0, P, α) *be a concave stationary time-dependent supergame satisfying Assumption* 5.3, *let* $\sigma^* = (s^*_0, s^*_1, \ldots)$ *be an open-loop equilibrium of that game, and let* s' *be the* (*unique*) *steady-state equilibrium action. Then* $\lim_{t\to\infty} s^*_t = s'$.

Proof. The basic fact exploited in the proof is that ψ obeys the contraction conditions in Assumption 5.3. A sequence of functions, $g_t(s_{t+1}, s_0)$, $t =$

$1, 2, \ldots$, is defined using ψ and having the property that $s_t^* = \psi(s_{t-1}^*, s_{t+1}^*)$ implies $s_t^* = g_t(s_{t+1}^*, s_0^*)$. The functions g_t are all shown to be contractions, which, in turn, allows the theorem to be proved.

Define g_1 by $s_1 = \psi(s_0, s_2) = g_1(s_2, s_0)$. To define g_t for $t > 1$ when g_{t-1} is known, let

$$s_t = \psi(s_{t-1}, s_{t+1}) = \psi(g_{t-1}(s_t', s_0), s_{t+1}) \tag{5.10}$$

Holding fixed the values of s_0, and s_{t+1}, the function $s_t = \psi(g_{t-1}(s_t', s_0), s_{t+1})$ is a mapping of s_t' into s_t. It is the fixed point of this mapping, shown presently to be unique, that defines $g_t(s_{t+1}, s_0)$. From Assumption 5.3, ψ is a contraction, and this implies both that g_t exists and is a contraction. Suppose that g_{t-1} exists and is a contraction obeying the Lipschitz condition λ_{t-1} with respect to s_t and μ_{t-1} with respect to s_0. Then, by way of deriving g_t, let $s_t' = \psi(g_{t-1}(s_t^1, s_0^1), s_{t+1}^1)$ and $s_t'' = \psi(g_{t-1}(s_t^2, s_0^2), s_{t+1}^2)$ and note that

$$\begin{aligned} \|s_t' - s_t''\| &= \|\psi(g_{t-1}(s_t^1, s_0^1), s_{t+1}^1) - \psi(g_{t-1}(s_t^2, s_0^2), s_{t+1}^2)\| \\ &\leqslant k_1 \|g_{t-1}(s_t^1, s_0^1) - g_{t-1}(s_t^2, s_0^2)\| + k_2 \|s_{t+1}^1 - s_{t+1}^2\| \\ &\leqslant k_1(\lambda_{t-1} \|s_t^1 - s_t^2\| + \mu_{t-1} \|s_0^1 - s_0^2\|) + k_2 \|s_{t+1}^1 - s_{t+1}^2\| \end{aligned} \tag{5.11}$$

If $\lambda_{t-1} < 1$, then g_t is well defined. That is, equation (5.10) as a function of s_t' into s_t is a contraction; hence it has a fixed point. Suppose this contraction condition holds for $t-1$ (it surely holds for $t-1=1$), and let us see if it holds for t. Using the fact that g_t is well defined, s_t' may be set equal to s_t^1 and s_t'' to s_t^2. Doing this and rearranging equation (5.11) gives

$$\|s_t^1 - s_t^2\| \leqslant \frac{\|s_0^1 - s_0^2\| \, k_1\mu_{t-1} + \|s_{t+1}^1 - s_{t+1}^2\| \, k_2}{1 - k_1\lambda_{t-1}} \tag{5.12}$$

If $\lambda_{t-1} + \mu_{t-1} \leqslant k$ (which holds for $t-1=1$) then, because $\lambda_t = k_2/(1 - k_2\lambda_{t-1})$ and $\mu_t = k_1\mu_{t-1}/(1 - k_1\lambda_{t-1})$, it follows that

$$\lambda_t + \mu_t = \frac{k_2 + k_1\mu_{t-1}}{1 - k_1\lambda_{t-1}} \leqslant k_1 + k_2 \leqslant k < 1 \tag{5.13}$$

Thus, g_t is a contraction, and by induction, $g_t(s_{t+1}, s_0)$ is well defined and is a contraction obeying the Lipschitz conditions λ_t, μ_t, where $\lambda_t + \mu_t \leqslant k$ for all $t \geqslant 1$.

Thus, the contractions g_t also have the properties, inherited from ψ, that $s_t^* = g_t(s_{t+1}^*, s_0^*)$ for all $t \geqslant 1$ if $\sigma^* = (s_0^*, s_1^*, \ldots)$ is an equilibrium point, and that $s' = g_t(s', s')$ for all $t \geqslant 1$, where $s' = \psi(s', s')$ is the steady-state equilibrium action. It remains to use the sequence of functions $\{g_t\}$ to show that s_t^* converges to s'. Define $\delta_t = \|s_t^* - s'\|$, $t = 0, 1, 2, \ldots$, and use

$$\begin{aligned} \|s_t^* - s'\| &= \|g_t(s_{t+1}^*, s_0^*) - g_t(s', s')\| \leqslant \lambda_t \|s_{t+1}^* - s'\| + \mu_t \|s_0^* - s'\| \\ &= \lambda_t\delta_{t+1} + \mu_t\delta_0 \end{aligned} \tag{5.14}$$

to see that

$$\delta_1 \leqslant \lambda_1\delta_2 + \mu_1\delta_0 \leqslant \lambda_1(\lambda_2\delta_3 + \mu_2\delta_0) + \mu_1\delta_0 \leqslant \delta_{T+1}\prod_{t=1}^{T}\lambda_t + \delta_0\sum_{t=1}^{T}\mu_t\prod_{\tau=0}^{t-1}\lambda_\tau \tag{5.15}$$

where λ_0 is defined to be unity. Clearly, the first term in equation (5.15) goes to 0 as T goes to infinity, because the δ_t are bounded due to S being compact, and the λ_t are less than $k<1$. Therefore;

$$\delta_t \leqslant \delta_0\sum_{t=1}^{\infty}\mu_t\prod_{\tau=0}^{t-1}\lambda_\tau \leqslant k\delta_0 \tag{5.16}$$

Thus, in general, $\delta_t \leqslant k\delta_{t-1}$ and by induction, $\delta_t \leqslant k^t\delta_0$, which goes to 0 as t goes to infinity, which completes the proof. QED

While the assumption restricting ψ to be a function satisfying a Lipschitz condition is rather strong, it does permit an interesting result in Theorem 5.1. Furthermore, as an example in Section 6 illustrates, this result can be useful in applications to economics. The open-loop equilibria of Theorem 5.1 are clearly not perfect equilibria. To see this, suppose that $\sigma^* = (s_0^*, s_1^*, \ldots)$ is an equilibrium point and assume that the history of the game at time t is $h_t = (s_0', s_1', \ldots, s_{t-1}')$, where $s_{t-1}' \neq s_{t-1}^*$. In general, the subgame strategy combination $\sigma_t^* = (s_{t-1}', s_t^*, s_{t+1}^*, \ldots)$ for the subgame starting an period t with an initial condition s_{t-1}' is not an equilibrium point, because $s_t^* \in \psi(s_{t-1}', s_{t+1}^*)$ would not usually hold. Although these equilibria are not perfect, they are not faulty or intuitively unacceptable in the sense of incorporating noncredible threats. Insofar as one is dealing with completely rational players who possess complete information and who are capable of carrying out the actions they plan, these equilibria are not faulty.

1.3 Nonstationary supergames and trigger strategy equilibria

Theorem 5.1 can be generalized to nonstationary time-dependent supergames. The payoff function of player i in period t is written $P_{it}(s_{t-1}, s_t)$, and the equilibrium correspondence is no longer stationary. Letting $P_{it}^*(s_{t-1}, s_t, s_{t+1}) = P_{it}(s_{t-1}, s_t) + \alpha_i P_{i,t+1}(s_t, s_{t+1})$, $\psi_t(s_{t-1}, s_{t+1})$ is the set of equilibrium points for (N, S, P_t^*). In allowing a different single-period payoff function for each time period, it is necessary to ensure that the supergame payoffs are bounded. Lemma 5.1 takes an appropriately different form: σ^* is an open-loop equilibrium for the supergame, if and only if $s_t^* \in \psi_t(s_{t-1}^*, s_{t+1}^*)$ for all $t \geqslant 1$. Finally, Assumption 5.3 must be restated so that each ψ_t satisfies Lipschitz conditions k_{1t} and k_{2t} with $k_{1t} + k_{2t} \leqslant k < 1$. Because the payoff functions can change arbitrarily from period to period, there is no hope of finding a steady state to which any equilibrium converges; however, a turnpike result is possible. That is, all equilibria converge to the same ultimate path: if σ' and σ'' are any two equilibria, then $||s_t' - s_t''||$ goes to 0 as t goes to infinity. Details can be found in Friedman (1981).

It is natural to inquire about trigger strategy equilibria in time-dependent supergames. As long as an open-loop equilibrium for the supergame has the property that s_t^* is not on the payoff possibility frontier of P_t^*, and the difference between P_t^* at equilibrium and the frontier is appropriately bounded away from 0, there is clearly room to show the existence of such equilibria. Even without such a condition, criteria that are sufficient for trigger strategy equilibria are easy to state in a way that is obviously analogous to their statement in Chapter 4; however, it is difficult to state conditions that ensure that a trigger strategy equilibrium is on the global payoff possiblity frontier of the supergame, as opposed to its merely being on the frontiers of the P_t^* (see Friedman (1974)). These trigger strategy equilibria will not be subgame perfect for the same reason that the open-loop equilibria are not subgame perfect: although trigger strategies are closed loop by definition, under certain conditions the trigger strategies can give rise to subgames in which the players follow open-loop equilibrium strategies. Because these open-loop equilibria are not subgame perfect, the trigger strategy equilibria of which they are part cannot be subgame perfect. Again, despite the absence of perfectness, these equilibria contain no threats that are noncredible.

1.4 State variables and the partitioned states model

A common way of modeling processes in which the situation at time t depends upon the history of the system to that time is by means of *state variables*. To see the appeal of state variables, consider first an even more general formulation, in which the situation at time t depends upon the history directly. Let a_{it} be the *action of player i at time t* and let π_i denote the single-period payoff to player i. then a_t is the action combination and π is the single-period payoff vector. The history of the game at time t will then be $h_t = (a_0, \ldots, a_{t-1})$, the observed actions taken in the past. One could specify that π_i is a function of (a_t, h_t), by stating that the payoff received in time t by the player will depend on the actions of all players and the history of play at that time.

Such a model, in its full generality, would be hard to work with. Stationarity of structure appears to be inherently absent, as each period's payoff functions depends upon something different. Suppose now that the effects of the past upon π_{it} are summarized in some unchanging number of variables, $x_t \in R^l$. The x_t are state variables and evolve over time according to $x_t = \phi(x_{t-1}, a_{t-1})$. Thus, the state variable has a fixed dimension, l, and its value at time t depends upon the actions taken at time $t-1$ and the actual state variable value at time $t-1$. The period t payoff of player i is then $\pi_i(x_t, a_t)$. Stationarity of structure is present as the model is developed here. A nonstationary structure could be obtained by allowing the single-period payoff functions to vary over time (a time subscript is added to π_i) or by allowing the state transition to vary over time (a time subscript is added to ϕ_i).

In a model of an economy, capital stocks are typical state variables. The production possibilities within a time period depend on the amount of an input, say labor, that firms buy; their respective capital stocks, however, are fixed by past decisions. The capital stocks for the next time period will depend on current investment decisions and the current sizes of capital stocks. Another example of a model with state variables is an oligopoly in which firms have productive capital and in which advertising has longlasting effects, building up a goodwill stock for the firm.

The state variable model developed here covers many interesting economic applications. The model is called a *partitioned states model* and satisfies Assumptions 5.4–5.6, set out in this chapter. The model, which is shown to satisfy Assumptions 3.1, 5.1, and 5.2, is a special case of the state variable models sketched earlier. It differs by requiring that the state variables can be partitioned into n subsets, one for each player. The particular state variables associated with player i have their values at time t determined by the actions of only player i at time $t-1$ and only their own values at time $t-1$. Think of the capital stock of a firm as an example. Its value at time t depends only on its own value at $t-1$ and the actions of the firm at that time.

Suppose that the payoff function for each player i in each time t takes the form $\pi_i(x_t, a_t)$, where $x_t = (x_{1t}, \ldots, x_{nt}) \in R^{nk}$ is the state variable vector at time t and $a_t = (a_{1t}, \ldots, a_{nt}) \in R^{nm}$ is the action combination at time t, with a_{it} being the action of player i in time t. The transition mechanism has a decomposition property so that $x_{i,t+1} = \phi_i(x_{it}, a_{it}) \in R^k$, $i \in N$, and $\phi = (\phi_1, \ldots, \phi_n)$. Thus, the state variables x_{it} are identified with player i in the sense that it is the actions of player i alone, along with the previous values of only these state variables, that determine the evolution of these variables over time.

ASSUMPTION 5.4 *The individual action spaces, A_i, are compact and convex subsets of R^m for all $i \in N$.*

ASSUMPTION 5.5 *The state transition function is $\phi = (\phi_1, \ldots, \phi_n)$, where $\phi_i : X_i \times A_i \rightarrow X_i$ and X_i is a compact and convex subset of R^k, $i \in N$. Each ϕ_i is continuously invertible with respect to a_{it}. The inverse is written $a_{it} = \phi_i^{-1}(x_{it}, x_{i,t+1})$. $\phi^{-1} = (\phi_1^{-1}, \ldots, \phi_n^{-1})$. Each ϕ_i is concave in (x_{it}, a_{it}) for all $i \in N$.*

ASSUMPTION 5.6 *The single-period payoff functions, $\pi_i(x_t, a_t)$, are continuous in (x_t, a_t), increasing in (x_{it}, a_{it}) and concave in (x_{it}, a_{it}) for all $i \in N$.*

From Assumption 5.5, the single-period payoff function can be written $\pi_i(x_t, a_t) = \pi_i[x_t, \phi^{-1}(x_t, x_{t+1})]$. The nature of the state variable partition means that, given x_{it}, choosing a_{it} is equivalent to choosing $x_{i,t+1}$; therefore, it is legitimate to think of the period t action of player i as being the choice of $x_{i,t+1}$. This allows s_{it} from the model of Section 1 to be identified with $x_{i,t+1}$. Making this identification for all i, gives $s_{it} \equiv x_{i,t+1}$ and $S_i \equiv X_i$ for all

$i \in N$ and $t \geqslant 0$. Thus,

$$\pi_i[x_t \phi^{-1}(x_t, x_{t+1})] = \pi_i[s_{t-1}, \phi^{-1}(s_{t-1}, s_t)]$$
$$\equiv P_i(s_{t-1}, s_t) \tag{5.17}$$

Consequently the following lemma holds:

LEMMA 5.3 *A model satisfying Assumptions* 5.4 *and* 5.5 *satisfies Assumptions* 3.1 *and* 5.1. *If Assumption* 5.6 *is satisfied, then Assumption* 5.2 *is satisfied as well.*

Proof. The argument preceding the statement of the lemma establishes that Assumptions 3.1 and 5.1 are implied by Assumptions 5.4 and 5.5. That Assumption 5.6 implies Assumption 5.2 follows from ϕ_i being concave in (x_{it}, a_{it}) and π_i being both increasing and concave in (x_{it}, a_{it}) by means of Berge (1963, Theorem 4, page 191). QED

2 A modification of the folk theorem for time-dependent supergames

The results stated in this section come from Friedman (1990). The time dependence in the model makes it difficult to specify the payoffs to which a player can be held; however, once that is resolved, a proof can be constructed along the lines of the proof of Theorem 4.7. The minimax payoffs for a player at time t depend on both the actions of the preceding period s_{t-1}, and the discount parameter, α_i. It is easy to determine the lowest period t payoff, $P_i(s_{t-1}, s_t)$, to which player i can be held, but this is not really the relevant consideration. If player i is to be held to low payoffs for two or more periods, then the effect of s_t on the payoffs of both period t and $t+1$ must be considered. Thus, the discounted payoff to which a player can be held over the relevant horizon of time must be determined.

The lowest payoff stream to which a player can be held by means of open-loop strategies over the whole of the game, starting from an arbitrary s_0, is found in Section 2.1. The paths that accomplish this are characterized as well. Then, in Section 2.2 the level of the average payoff per period to which a player is held along the minimax paths developed in Section 2.1 is studied. Section 2.3 contains an additional assumption needed to ensure existence of equilibrium. Finally, the results from these sections are used in Section 2.4 to establish the extension of the folk theorem to time-dependent supergames.

2.1 The structure of minimax paths

For any player i, a *best-reply open-loop supergame payoff* is defined as $\Phi_i(\sigma) = \max_{\sigma_i \in \Sigma_i} G_i(\sigma \backslash \sigma_i')$, and the *open-loop minimax payoff* is $\gamma_i(s_0) = \min_{\sigma \in \Sigma} \Phi_i(\sigma)$. Denote by $\sigma^i = (s_0^i, s_1^i, s_2^i, \ldots) \in \Sigma$ a path (open-loop strategy combination) that achieves $\gamma_i(s_0)$ optimally. (It is understood that $s_0^i \equiv s_0$.) That is, $G_i(\sigma^i) = \Phi_i(\sigma^i) = \gamma_i(s_0)$, so player i cannot do better by choosing other than σ_i^i, nor can she be held to a lower payoff if the other players choose differently. In the remainder of this section it is proved that $\gamma_i(s_0)$ is the minimax payoff of player i even when closed-loop strategies are considered.

LEMMA 5.4 *Given Assumptions* 3.1 *and* 5.1, Φ_i, $i \in N$, *is both defined and continuous for all* $\sigma \in \Sigma$.

Proof. Φ_i is defined for each $\sigma \in \Sigma$ because a continuous function (G_i) achieves a maximum on a compact set (Σ). That each Φ_i is continuous follows from the continuity of G_i and the nature of the max operator. QED

From Lemma 5.4 it is immediately clear that $\gamma_i(s_0)$ exists, because it is the minimum of a continuous function defined on a compact set. A characteristic of σ^i is that s_t^i minimaxes player i in the constituent game $(N, S, P^*(s_{t-1}^i, \cdot, s_{t+1}^i))$, $t > 0$. This is proved as Lemma 5.5.

LEMMA 5.5 *Given Assumptions* 3.1 *and* 5.1, $\sigma^i = (s_0^i, s_1^i, s_2^i, \ldots) \in \Sigma$ *satisfies the conditions that* s_t^i *minimaxes player* i *in the games* $(N, S, P^*(s_{t-1}^i, \cdot, s_{t+1}^i))$, $t = 1, 2, \ldots$.

Proof. The proof is virtually identical to the proof of Lemma 4.1, and is thus omitted. QED

Due to Lemma 5.5 the minimax paths for a player i can be characterized by a *minimax correspondence for player* i: Ψ_i defined on $S \times S$, with $\Psi_i(s_{t-1}, s_{t+1}) \subset S$ is the set of action combinations that minimax player i for the game $(N, S, P^*(s_{t+1}, \cdot, s_{t+1}))$. Thus, an open-loop strategy combination $\sigma = (s_0^i, s_1^i, s_2^i, \ldots)$ minimaxes player i if and only if $s_t^i \in \Psi_i(s_{t-1}^i, s_{t+1}^i)$ for all $t > 0$. As $\Psi_i(s_{t-1}, s_{t+1})$ is nonempty for any $(s_{t-1}^i, s_{t+1}^i) \in S \times S$, it is possible to commence a minimax path for a player i from any $s_{t-1} \in S$. Such paths will be needed later; they are denoted

$$\mu^i(s_{t-1}) = (m_{t-1}^i(s_{t-1}), m_t^i(s_{t-1}), \ldots) \tag{5.18}$$

where $m_{t-1}^i(s_{t-1}) \equiv s_{t-1}$.

A little can be said about the relationship between the open-loop minimax payoff to player i and the minimax payoff that would obtain using closed-loop strategies. Denote by $\delta_i(s_0)$ the open-loop payoff that player i can guarantee himself. This is called the *open-loop maximin payoff* and is found by reversing the order of maximizing and minimizing that is used to obtain $\gamma_i(s_0)$. Let $\gamma_i^*(s_0)$ denote the closed-loop minimax payoff to player i and let $\delta_i^*(s_0)$ be the closed-loop maximin payoff. The known facts are that a player can guarantee himself a payoff that is less than or equal to the payoff to which he can be held. Thus, $\delta_i(s_0) \leq \gamma_i(s_0)$ and $\delta_i^*(s_0) \leq \gamma_i^*(s_0)$. More can be said. First, $\gamma_i^*(s_0) \leq \gamma_i(s_0)$, because, the members of $N \backslash \{i\}$ are free to select the open-loop minimax strategies that will guarantee them that player i will be kept down to $\gamma_i(s_0)$. Therefore, the availability of closed-loop strategies will not help player i but might help the others. On the other hand, $\delta_i(s_0) \leq \delta_i^*(s_0)$ on similar reasoning; namely, player i can choose his open-loop maximin strategy and assure himself at least $\delta_i(s_0)$. Having closed-loop strategies available may help him, and cannot hurt him. Thus, $\delta_i(s_0) \leq \delta_i^*(s_0) \leq \gamma_i^*(s_0) \leq \gamma_i(s_0)$, which informs us of two things: First, $\gamma_i^*(s_0)$, the open-loop maximin payoff, is bounded below by $\delta_i(s_0)$ and above by $\gamma_i(s_0)$. These bounds will generally be easier to find than will $\gamma_i^*(s_0)$ itself. Second, if $P_i(s_{t-1}, \cdot, s_{t+1})$ is quasiconcave in s_i and quasiconvex in $(s_{1t}, \ldots, s_{i-1,t}, s_{i+1,t}, \ldots, s_{nt})$, then $\delta_i(s_0) = \gamma_i(s_0)$, which implies $\delta_i(s_0) = \delta_i^*(s_0) = \gamma_i^*(s_0) = \gamma_i(s_0)$.

As in Chapter 4, the largest possible payoff that player i could receive in a single period ($\bar{v}_i$) and the smallest possible payoff ($\underline{v}_i$) are needed. They are

$$\bar{v}_i = \max_{(s',s'')\in S^2} P_i(s', s'') \tag{5.19}$$

$$\underline{v}_i = \min_{(s',s'')\in S^2} P_i(s', s'') \tag{5.20}$$

Without loss of generality the payoff functions can be normalized so that $\underline{v}_i = 0$ for all i.

2.2 The level of the per period avergage minimax payoff

The minimax payoffs worked out in the preceding section are too severe; for, as in Theorem 4.7, an equilibrium strategy combination will have punishment provisions that push a deviating player's payoff to a very low level for only a finite number of periods. That phase will be followed by payoffs that are better for all players than their minimax payoffs, so it is necessary to know the payoffs to which a player can be held for finite stretches of time. As a practical matter, this can be reduced to determining the average payoff per period to which a player can be held for finite intervals of time. The problem is approached in several steps, the first of which is to develop $\gamma_i^T(s_t, \alpha_i)$, the discounted payoff to player i over periods $t+1$ through $t+T$ when player i is being subjected to minimax play for these periods, the discount parameter is α_i, and the period t action combination is s_t.

Let $\Sigma_i(s_t) = \{\sigma_i' = (s_{i0}', s_{i1}', \ldots) \in \Sigma_i \mid s_{it}' = s_{it}\}$ and let $\Sigma(s_t) = \times_{i\in N} \Sigma_i(s_t)$. Then $\Sigma(s_t)$ is the set of paths for which $s_t' = s_t$. For $\sigma' \in \Sigma(s_t)$ let

$$G_i^T(\sigma', s_t, \alpha_i) = \sum_{\tau=t+1}^{t+T} \alpha_i^{\tau-t-1} P_i(s_\tau') \tag{5.21}$$

$G_i^T(\sigma', s_t, \alpha_i)$ is the payoff to player i over the periods $t+1$ through $t+T$ discounted back to time $t+1$ under an arbitrary path $\sigma' \in \Sigma$ that is constrained only by having its tth element, s_t', equal to s_t. Obviously, the first t move combinations of σ' (i.e., $s_0', s_1', \ldots, s_{t-1}'$) are irrelevant to the value of $G_i^T(\sigma', s_t, \alpha_i)$. To develop the minimax payoff for this time interval let the best-reply payoff of player i be

$$\Phi_i^T(\sigma', s_t, \alpha_i) \max_{\sigma_i \in \Sigma_i(s_t)} G_i^T(\sigma \backslash \sigma_i', s_t, \alpha_i) \tag{5.22}$$

and let

$$\gamma_i^T(s_t, \alpha_i) = \min_{\sigma_i' \in \Sigma_i(s_t)} \Phi_i^T(\sigma', s_t, \alpha_i) \tag{5.23}$$

$\gamma_i^T(s_t, \alpha_i)$ is the lowest payoff to which player i can be held by the other players over the time interval $t+1$ through $t+T$, given that the players $N\backslash\{i\}$ start to play to hold down his payoff at time $t+1$, quit doing so after time $t+T$ is finished, and s_t is the move combination that was actually chosen at time t. Of course, s_t is an *initial condition relative to the punishing of player* i over this time interval.

LEMMA 5.6 *$\gamma_i^T(s_t, \alpha_i)$ is continuous in $\alpha_i \in [0, 1]$ for each $T > 0$ and $s_t \in S_t$.*

The proof is obvious and is omitted.

From here, $\gamma_i^T(s_t, \alpha_i)$ is used to establish bounds on minimax payoffs when the discount parameter of the player is close to 1. Let

$$v_i^m(T) = \frac{1}{T} \sup_{s_t \in S} \gamma_i(s_t, 1) \tag{5.24}$$

$v_i^m(T)$ is the least upper bound on the average payoff per period to which player i can be held, irrespective of the initial condition s_t.

It is shown below that the initial condition can be made virtually irrelevant to the value of $v_i^m(T)$ if the time span, T, is long enough. This is done by looking at average punishment payoffs to which player i could be held if there were no initial condition to consider, and then showing that such average punishment payoffs are attainable, at least in the limit. Consider first the payoff to which player i can be held for T periods, say, periods 2 through $T+1$, given that s_1 through s_T can be freely chosen, but $s_{T+1} = s_1$ is required. Denote this payoff $v_i^M(T)$. Over a horizon of $T+1$ periods, the punishers could hold player i to a total payoff of $\bar{v}^i + Tv_i^M(T)$. Using the punishment policy in a cyclically repetitive way for, say, k times the punishers could hold player i to a total payoff of $\bar{v}^i + Tv_i^M(T)$. As k grows the average payoff per period is approaching $v_i^M(T)$. $v_i^M(T)$ is defined in a way analogous to all the other minimax payoffs that have been defined in this chapter and Chapter 4. Let $\eta_i^T(s_1, \ldots, s_T)$ be the maximum over $(s_{i1}, \ldots, s_{iT}) \in S_i^T$ of $P_i(s_T, s_1) + \sum_{t=1}^{T-1} P_i(s_t, s_{t+1})$; then $v_i^M(T)$ is the minimum of $\eta_i^T(s_1, \ldots, s_T)$ over $(s_1, \ldots, s_T) \in S^T$. It is shown in Lemma 5.7 that $v_i^M(T)$ has a limit, v_i, as $T \to \infty$, and that this limit is necessarily a lower bound on the values attained by members of the sequence $\{v_i^M(T)\}$. In Lemma 5.8 it is shown that $\gamma_i^T(s_t, \alpha_i)$ can be held arbitrarily close to Tv_i if the discount parameter is close enough to 1 and the time horizon, T, is sufficiently long.

LEMMA 5.7 $v_i = \lim_{T\to\infty} v_i^M(T)$ *exists.*

Proof. For each $T > 1$ let $v'_{iT} = \min\{v_i^M(1), \ldots, v_i^M(T)\}$. Clearly, for any given T, $v_i \leq v'_{iT}$, because for any very large $t > T$, $v_i^M(T)$ is bounded above by a number arbitrarily close to v'_{iT}. Consider an horizon of $KT + 1$ periods starting at time t with arbitrary s_t. In the first period of this horizon player i might receive as much as $\bar{v}_i$, but over each of the K remaining cycles of length T he can be held to an average of v'_{iT} per period. Thus, the total payoff is bounded above by $\bar{v}_i + KTv'_{iT}$ and the average payoff per period is bounded above by

$$v'_{iT} + \frac{\bar{v}_i - v'_{iT}}{KT + 1} \tag{5.25}$$

Consequently, for an arbitrary $\varepsilon > 0$, $T > 0$, and $t > 0$ there is a finite K' such that for $K > K'$, $v_i^M(KT + 1) \leq v'_{it} + \varepsilon$. Meanwhile, the sequence $\{v'_{it}\}$ is non-increasing and bounded below, so it has a limit. This limit is the v_i that is sought. QED

Let $v = (v_1, \ldots, v_n)$.

LEMMA 5.8 *For an arbitrary* $\varepsilon > 0$, *there is a finite* T' *and* $\alpha'_i \in (0, 1)$ *such that, for* $T > T'$ *and* $\alpha_i > \alpha'_i$,

$$\gamma_i^T(s_t, \alpha_i) \leq T(v_i + \varepsilon) \tag{5.26}$$

Proof. $\gamma_i^T(s_t, \alpha_i)/T$ converges to $v_i^m(T)$ as $\alpha_i \to 1$ and, from Lemma 5.7, $v_i^M(T)$ converges to v_i as $T \to \infty$. QED

To summarize, then, v_i is the average payoff per period to which a player can nearly be held for long lengths of time and v is the vector of such payoffs. The lowest discounted payoff to which a player can be held for a T period time span, $\gamma_i^T(s_t, \alpha_i)$, is close to Tv_i when T is sufficiently large and α_i is close to 1.

2.3 An additional assumption on the model

A further assumption is needed relating the payoffs on punishment paths to the payoffs available otherwise. This assumption, Assumption 5.7, plays the same role in Theorem 5.2 as does Assumption 4.1 in Theorem 4.7. For each $s \in S$ let

$$V(s) = \{P(s, s') \in R^n \mid s' \in S \text{ and } P(s, s') \gg v\} \tag{5.27}$$

$V(s)$ is the set of attainable single-period payoff vectors at a time t, given that $s_{t-1} = s$. $V(S)$, defined below, is the set of period t payoff vectors that are attainable no matter what s_{t-1} may be.

$$V(S) = \bigcap_{s \in S} V(s) \tag{5.28}$$

Let $\mathring{V}$ be the interior of $V(S)$, and let

$$V^* = \{x \in V(S) \mid \text{there exists } x' \in \mathring{V},\ x' \ll x\} \tag{5.29}$$

An element, $x \in V^*$ is a single-period payoff vector that dominates v and that can be attained period after period, indefinitely, and that strictly dominates some other such payoff vector, $x' \in \overset{o}{V}$ (i.e., $x \gg x' \gg v$). Furthermore, x' has an open ε-neighbourhood around it within which each point is attainable.

ASSUMPTION 5.7 *V^* is not empty.*

It is natural to wonder whether Assumption 5.7 is unduly restrictive. Consider the steady-state single-period payoffs $P(s, s)$ and the set $\{P(s, s) \mid s \in S\}$. If this set has a nonempty interior containing a point that strictly dominates v, then Assumption 5.7 is satisfied. This condition will easily hold in models in which the state variables are capital stocks.

Finally, it will prove helpful to have notation for the set of open-loop strategy combinations whose associated payoffs are at least as high in each period as something in V^*. To that end let

$$\Sigma'(V^*) = \{\sigma \in \Sigma \mid P(s_{t-1}, s_t) \geqslant x \text{ for some } x \in V^* \text{ and all } t > 0\} \quad (5.30)$$

The paths in $\Sigma'(V^*)$ are all supportable outcomes of a subgame perfect noncooperative equilibrium (for suitable values of the players' discount parameters), as will be seen in the following section.

2.4 The folk theorem modified for time-dependent games

The folk theorem presented here differs from the theorems for repeated games by asserting that payoffs that are never below payoffs in V^* in each period can be sustained as the result of a subgame perfect noncooperative equilibrium. That is, let v^* be an arbitrary element of V^* and let $\sigma^* \in \Sigma(V^*)$. Then there is a subgame perfect equilibrium for which the realized payoffs in each period will be those associated with σ^*. In the special case where $P_i(s_{t-1}, s_t) = P_i(s'_{t-1}, s_t)$ for all $s_{t-1}, s'_{t-1}, s_t \in S$ (i.e., where time dependence is eliminated), $v_i = \gamma_i^T(s, \alpha_i)$ for all $s \in S$, $\alpha_i \in (0, 1)$, and $T \geqslant 1$; the minimax payoffs are constant. In this case, Theorem 5.2 becomes very similar to Theorem 4.7. Because of the time-dependent structure, the attainable payoffs, as well as the minimax payoffs, discounted to the current period, may vary over time because s_t may vary with t. To avoid immensely more complicated statements that allow only slightly more general results, the constant payoffs in V^* are used to delineate a set of attainable outcomes. The equilibria are constructed along the lines of those in the proof of Theorem 4.7, and the proof is a suitably modified variation of that proof.

THEOREM 5.2 *Let the game $\Gamma = (N, S, s_0, P, \alpha)$ satisfy Assumptions* 3.1, 5.1, *and* 5.7. *For any $v^* \in V^*$ and $\sigma^* = (s_0^*, s_1^*, s_2^*, \ldots) \in \Sigma$ with $P(s_{t-1}^*, s_t^*) \geqslant v^*$ for all $t \geqslant 1$ there exists $\alpha_i^* \in (0, 1)$, $i \in N$, such that for discount parameters $\alpha_i \in (\alpha_i^*, 1)$, $i \in N$, there exists a subgame perfect noncooperative equilibrium under which the payoff realizations in every period $t \geqslant 1$* are $P(s_{t-1}^*, s_t^*)$.

Proof. Let $v^* \in V^*$ and let $\sigma^* \in \Sigma$ satisfy $P(s^*_{t-1}, s^*_t) \geq v^*$ for all $t \geq 1$. The proof proceeds in four parts. (1) A strategy combination is specified that describes (A) the behavior to follow if no one defects from the path σ^*, known as the *initial path*: (B) the behavior to follow if a player defects from σ^*; and (C) the behavior to follow if a player defects from the behavior described in (B) or (C). Proceeding along the initial path (A) is sometimes called *cooperative behavior*; stage (B) and (C) behavior describe *punishment paths*. (2) It is proved that for some $\alpha^1_i < 1$, if $\alpha_i \in (\alpha^1_i, 1)$, $i \in N$, no player would gain by defecting from the initial path. (3) It is proved that for some $\alpha^2_i < 1$, if $\alpha_i \in (\alpha^2_i, 1)$, $i \in N$, no player would gain by defecting from stage (B) behavior when he is the player being punished. (4) It is proved that for some $\alpha^3_i < 1$, if $\alpha_i \in (\alpha^3_i, 1)$, $i \in N$, no player would gain by defecting from the punishment of another player. Parts (1)–(4) imply that the strategy combination is a subgame perfect noncooperative equilibrium.

(1) There exists $\xi > 0$ and $x \in \mathring{V}$ such that

$$x_i + \xi < v^*_i \leq P_i(s^*_{t-1}, s^*_t) \tag{5.31}$$

for all $i \in N$ and $t \geq 1$, and

$$(x_1 + \delta_1, \ldots, x_n + \delta_n) \in \mathring{V} \tag{5.32}$$

for all $\delta \in [0, \xi]^n$. Therefore, defining $w^i_t(s_t) \equiv s_t$, for each $i \in N$ and $s_t \in S$, there exists a $\omega^i(s_t) = (w^i_t(s_t), w^i_{t+1}(s_t), w^i_{t+2}(s_t), \ldots) \in \Sigma$ satisfying

$$P_j(w^i_{t+\tau}(s_t), w^i_{t+\tau+1}(s_t)) = x_j + \xi \tag{5.33}$$

for all $j \neq i$ and $\tau \geq 0$ and

$$P_i(w^i_{t+\tau}(s_t), w^i_{t+\tau+1}(s_t)) = x_i \tag{5.34}$$

for all $\tau \geq 0$. The strategy of player i is as follows:

(A) In period 1, choose s^*_{i1}. In period $t > 1$ choose s^*_{it} if the observed actions of the previous periods, $\tau = 1, \ldots, t-1$, were s^*_t.

(B) (1) If the player j (including player i) defects from σ^* at time t and chooses $s_{jt} \neq s^*_{jt}$, then, from $t+1$ to $t+T^*$, choose $m^j_{i\tau}(s^*_t \backslash s_{jt})$, $\tau = t+1, \ldots, t+T^*$. (2) From $t+T^*+1$ onward choose $w^j_{i\tau}[m_{t+T^*}(s^*_t \backslash s_{jt})]$. The preceding actions are selected by player i as long as no player defects from these actions.

(C) If any player k (including player i) defects during the phases described under (B), then restart phase (B1) and continue with (B2) with player k substituted for player j.

(2) To see how T^* is chosen, it is necessary to find a separate time horizon for each type of defection, and then let T^* be the longest of these. We begin by finding a T^1 for which defection from stage (A) behavior will be unprofitable. To this end, look first at the payoffs that player i receives if no player ever defects. These payoffs can be forced to average at least $v_i + \xi$ per period. Suppose player i defects and, following this defection, stage (B) is carried out with stage (B1) lasting T periods and with no player ever

defecting thereafter. If $\varepsilon_i(\alpha_i, T)$ is the tolerance around v_i within which the average payoffs of player i can be held, given a time span of T and a discount parameter of α_i, then an upper bound on player i's payoff stream for the periods starting with the period of defection and lasting to the end of stage (B1) is

$$\bar{v}_i + T[v_i + \varepsilon_i(\alpha_i, T)] \tag{5.35}$$

Now choose finite T_i^0 and $\alpha_i^0 < 1$ so that $\varepsilon_i(\alpha_i, T) \leq \xi/3$ for all $\alpha_i \geq \alpha_i^0$ and $T \geq T_i^0$. Then for sufficiently large $\alpha_i \geq \alpha_i^0$ and $T \geq T_i^0$ equation (5.35) is bounded above by $\bar{v}_i + T(v_i + \xi/3)$ and the payoff associated with remaining on the (A) path over the same span of time is bounded below by $(v_i + 2\xi/3)(T+1)$. Denote these bounds by α_i^1 (<1) and T_i^1 $(<\infty)$. It is easily verified that

$$\bar{v}_i + T\left(v_i + \frac{\xi}{3}\right) < \left(v_i + \frac{2\xi}{3}\right)(T+1) \tag{5.36}$$

if

$$T > \frac{(\bar{v}_i - v_i)}{\xi} \tag{5.37}$$

Therefore, T^1, is chosen to be at least as large as all of the T_i^1 and also large enough to satisfy equation (5.37) for all i. Given T^1 and the α_i^1, if $\alpha_i^1 < \alpha_i < 1$, it will not be in the interest of any player i to defect during stage (A), because his payoff will be less than it is in stage (A).

(3) Suppose a player i has defected from the phase (A) behavior. He cannot gain by defecting from either phase (B1) or (B2). Phase (B2) yields a payoff that is larger than phase (B1). Defection restarts (B1), then returns to phase (B2), so defection from (B2) is not profitable. Defection from (B1) causes (B1) to restart, which delays the return to the more profitable (B2); hence, defection from (B1) is also not profitable. Consequently, a player who cannot gain by defecting from phase (A) cannot gain by defecting from phase (B1) or (B2) when he is the current defector.

(4) It remains to see whether player i should defect from phase (B) when the prior defection that got the game into phase (B) was due to some other player j. Defection in phase (B2) clearly cannot be worthwhile; the player gives up a larger payoff than he would if he were the defector and then he receives the defectors punishment. Defection in phase (B1) gives the player, at most, the left-hand side of equation (5.38) and nondefection gives him, at least, the right-hand side.

$$\bar{v}_i + \left(v_i + \frac{\xi}{3}\right)T^1 + \frac{\alpha_i^{T^1+1}}{1-\alpha_i}x_i < \underline{v}_i(T^1+1) + \frac{\alpha_i^{T^1+1}}{1-\alpha_i}(x_i + \xi) \tag{5.38}$$

Clearly there exists $\alpha_i^2 \in (\alpha_i^1, 1)$ such that if $\alpha_i \in (\alpha_i^2, 1)$, it will not be in the interest of player i to defect during stage (B1) (i.e., equation (5.38) will hold).

Now, to tie everything together, $\alpha_i^* = \alpha_i^2$, and $T^* = T^1$. There is no time at which it is in the interest of any player to defect. Any defection at any time by any player in any stage of play will start phase (B) with that player singled out. The ensuing payoffs will be inferior to the payoffs enjoyed by that player. One last detail must be specified. Suppose that two or more players defect simultaneously. To handle such a situation designate one of the players *the defector*. If $i, j, k, \ldots$ defect at once, let that player with the smallest index ($\min\{i, j, k, \ldots\}$) be designated as the defector. Now consider an arbitrary time period, t, and an arbitrary history of play up to that time. What should player i do at time t? It is easy to verify what stage the game is in and whether any player defected in the previous period. No matter what the history of play has been, it corresponds to some period of some stage that has been defined and for which behavior is specified by the strategies outlined above. Furthermore, it is proved that defection is never in the interest of any player from any period of any stage; therefore, for any subgame of the original game, the strategies induced by the original strategies are Nash equilibria, which means that the original strategies themselves are a subgame perfect equilibrium. QED

3 Stochastic games

The basic feature distinguishing stochastic games from other games is that, in each time period, the particular payoff functions to be faced by the players are chosen randomly, and the exact probability distribution depends on the actions of the players in the previous period as well as on the particular payoff functions that were drawn at that time. The treatment here follows the line of development started by Shapley (1953a) and carried further by Rogers (1969) and Sobel (1971). The specific model developed is found in Friedman (1977, Chapter 10) and is a slight variant of Sobel's model. Essentially, there is a fixed group of n players and a fixed collection of K ordinary one-shot games. At the start of each time period, one of the K games is selected at random. The selection is made known to the players' then they make their choices for the period.

3.1 The transition mechanism and the conditions defining a stochastic game

The probability distribution governing the choice among the K games to be played in period $t+1$ depends on the actual game played in period t and on the actions selected by the players in period t. This probability distribution, called the *transition mechanism*, is defined in Definition 5.3. An individual single-shot game is also called a *state*; thus, to say that game k was encountered (or played) at time t is exactly the same as to say the state at time t was state k.

ASSUMPTION 5.8 *The set of states* $\Omega = \{1, \ldots, K\}$ *is finite.*

DEFINITION 5.3 *The* **transition mechanism** *is* $q(k' \mid k, s)$, *which is the probability that the next state will be state* k' *when the current state is* k *and the current action is* s.

ASSUMPTION 5.9 *The transition mechanism satisfies the following conditions*: (a) $q(k' \mid k, s) \geqslant 0$ *for all* $k, k' \in \Omega$ *and all* $s \in S_k$; (b) $\sum_{k' \in \Omega} q(k' \mid k, s) = 1$; *and* (c) *for any* $s', s'' \in S_k$, $k', k \in \Omega$, *and* $\lambda \in [0, 1]$,

$$q(k' \mid k, \lambda s' + (1 - \lambda)s'') = \lambda q(k' \mid k, s') + (1 - \lambda)q(k' \mid k, s'') \quad (5.39)$$

Conditions (a) and (b) in Assumption 5.9 are usual conditions requiring probabilities to be nonnegative and sum to 1. Condition (c) bears closer examination. It states that the probabilities are linear in s, the actions of the players. Although this condition is rather strong and is not explicitly stated in the earlier literature (Sobel (1971) and earlier work), it is implicitly assumed there as well. For example, Sobel (1971) deals with finite single-period games and, of course, allows mixed strategies. As a consequence, his framework implies condition (c).

DEFINITION 5.4 *A* **stochastic game** *is denoted* $\Gamma = (N, \Omega, \{S_k\}, \{P_k\}, q, \alpha)$, *where* (N, S_k, P_k) *is, for each* $k \in \Omega$, *a game satisfying Assumptions* 3.1, 3.2, *and* 3.3Q. Γ *has* q *as its transition mechanism,* Γ *satisfies Assumptions* 5.8 *and* 5.9, *and, for each player* $i \in N$, *the objective function is expected discounted payoff using the discount parameter* $\alpha_i \in [0, 1)$.

3.2 *Existence of equilibrium*

The existence of equilibrium points is shown in several stages. The first of these, carried out in Section 3.2.1, is to define *policies*, which are essentially closed-loop strategies under which the move of a player in each period is a function of the current state. It is known from stochastic dynamic programming that the set of optimal rules for a decision maker in a stationary decision problem includes one or more policies. This result can be applied to the stochastic game model to allow attention to be restricted to strategies that are policies. Then, in Section 3.2.2, the transition probabilities and payoff functions are formulated for the case in which all players use policies. Following that, in Section 3.2.3, the decision problem facing a single player is examined, and finally, in Section 3.2.4, existence of equilibrium is proved.

3.2.1 Policies, strategies, and Blackwell's lemma

In a stochastic game, the history as of period t consists of the actual state and action for every past period plus the state for period t. It might be thought that the whole history would, at any time, be relevant for a decision, however, that need not be so. Under reasonable and interesting conditions, it is possible to restrict attention to strategies that are *policies*. These are defined, after which the needed conditions are stated.

DEFINITION 5.5 *A* **policy for player** *i, θ_i, is a function from Ω to $\times_{k\in\Omega} S_{ik}$. Θ_i is the set of policies for player i. $\theta = (\theta_1, \ldots, \theta_n)$, and $\Theta = \times_{i\in N}\Theta_i$. The distance between two policies, $d(\theta, \theta')$, is defined by $d(\theta_i, \theta_i') = \max_{k\in\Omega} d(\theta_i(k), \theta_i'(k))$ and $d(\theta, \theta') = \max_{i\in N} d(\theta_i, \theta_i')$.*

Thus, a policy for player i is a rule that picks out a particular action (single-shot game strategy), $s_{ik} \in S_{ik}$ for each state $k \in \Omega$. There is no time dependence in the sense that the value of t does not help determine the chosen action. Only the current state matters, and any time that a particular state k is encountered, the same action s_{ik} is taken. Now consider the situation of player i when all other players are known to be choosing policies. It turns out that among the best replies of player i, when any history dependent strategy is available, is a policy. In other words, if attention is restricted to policies and an equilibrium point in policies is found, then this strategy combination is still an equilibrium when more general strategies are allowed. This result is due to Blackwell (1965) and is stated, but not proved, as follows.[3]

BLACKWELL'S LEMMA *Let $\Gamma = (N, \Omega, \{S_k\}, \{P_k\}, q, \alpha)$ be a stochastic game and let $\theta \in \Theta$. If player i has a best reply to θ when she is allowed to choose from the set of all history dependent strategies, then one of the best replies is an element of Θ_i.*

3.2.2 Transition probabilities and the payoff functions when players use policies

Given a policy combination $\theta \in \Theta$, the transition probabilities are exactly determined. q_θ denotes the $K \times K$ transition matrix that governs the stochastic game when the players choose θ.[4]

$$q_\theta = \begin{bmatrix} q(1 \mid 1, \theta(1)) \cdots q(K \mid 1, \theta(1)) \\ \vdots \qquad\qquad \vdots \\ q(1 \mid K, \theta(K)) \cdots q(K \mid K, \theta(K)) \end{bmatrix} = \begin{bmatrix} q(1, \theta(1)) \\ \vdots \\ q(K, \theta(K)) \end{bmatrix} \tag{5.40}$$

To formulate the expected payoff of a player, let

$$F_{i\theta} = [P_{i1}(\theta(1)), \ldots, P_{iK}(\theta(K))] \tag{5.41}$$

$F_{i\theta}$ is a single-period payoff vector for player i in which the first coordinate, $P_{i1}(\theta(1))$, is the payoff that player i receives when the game is in state 1 and the policy (strategy) combination θ is being followed. Suppose now that the state in time 0 is k. Then the expected discounted payoff of player i, given the policy combination θ, is

$$P_{ik}(\theta(k)) + \alpha_i q(k, \theta(k))F_{i\theta} + \alpha_i^2 q(k, \theta(k))q_\theta F_{i\theta} + \cdots + \alpha_i^t q(k, \theta(k))q_\theta^{t-1}F_\theta + \cdots \tag{5.42}$$

and

$$G_{i\theta} = F_{i\theta} + \alpha_i q_\theta F_{i\theta} + \alpha_i^2 q_\theta^2 F_{i\theta} + \cdots + \alpha_i^t q_\theta^t F_{i\theta} + \cdots = [I - \alpha_i q_\theta]^{-1} F_{i\theta} \tag{5.43}$$

is the vector of expected discounted payoffs for player i. $[I - \alpha_i q_\theta]$ has an inverse because it is a strictly dominant diagonal matrix. The kth component of equation (5.43) is the expected discounted payoff for player i, given that the initial state is state k.

3.2.3 *The decision problem of a single player*

The decision problem faced by one player when the policies of the other players are given, or, to put it another way, finding a best reply for a player to a policy combination θ, is an exercise in stochastic dynamic programming. A maximum of

$$G_{i\theta\backslash\theta_i} = [I - \alpha_i q_{\theta\backslash\theta_i}]^{-1} F_{i\theta\backslash\theta_i} \tag{5.44}$$

over $\theta_i \in \Theta_i$ is sought; however, this maximum is of a different character from the best replies seen earlier. Ideally, we seek a best reply θ_i' to θ such that θ_i' leads to a maximum expected payoff *irrespective of the initial state*. On the face of it, there should be no surprise if the optimal policy differs according to what the initial state happened to be, but a result from stochastic dynamic programming establishes that a best reply exists that is best no matter what the initial state. This is proved in Denardo (1967) and can also be seen in Friedman (1977, Chapter 10).

DENARDO'S LEMMA *In a stochastic game Γ, the set of policies for player i that are best replies to a policy combination θ, denoted $r_i(\theta)$, is nonempty. If $\theta_i' \in r_i(\theta)$, then $G_{i\theta\backslash\theta_i'} \geq G_{i\theta\backslash\theta_i''}$ for all $\theta_i'' \in \Theta_i$.*

3.2.4 *Equilibrium in the stochastic game*

The lemmas of Blackwell and Denardo ensure that $r(\theta) = \times_{i \in N} r_i(\theta)$ is nonempty for all $\theta \in \Theta$ and that r maps Θ into Θ. Proving that the stochastic game has an equilibrium point is, of course, done if $r(\theta)$ can be shown to have a fixed point. This, in turn, can be based on the Kakutani fixed-point theorem if (a) Θ is compact and convex, (b) $r(\theta)$ is upper semicontinuous, and (c) the image sets, $r(\theta)$, are nonempty and convex. These properties are true and are established in the following three lemmas.

LEMMA 5.9 *In a stochastic game Γ, the policy space Θ is compact and convex.*

Proof. A policy can be thought of as a point in a Euclidean space of finite dimension. The dimension is $K \times m$, where K is the number of states and m is the dimensionality of S_{ik} for all i and k.[5] Thus, Θ is compact if it is closed and bounded. Boundedness follows from $K \times m$ being finite and from the boundedness of each S_{ik}. Similarly, closedness follows from the closedness of each S_{ik}. Finally, convexity of Θ follows from convexity of S_{ik}. That is, if $\theta_i'(k)$ and $\theta_i''(k)$ are both in S_{ik}, then $\lambda\theta_i'(k) + (1 - \lambda)\theta_i''(k)$ is also in S_{ik} for all i and k. QED

As to upper semicontinuity of $r(\theta)$, this follows directly if $G_{i\theta}$ is continuous with respect to θ for each i.

LEMMA 5.10 *$G_{i\theta}$ is continuous with respect to θ.*

Proof. Continuity of $G_{i\theta}$ is implied by continuity of $[I - \alpha_i q_\theta]^{-1}$ with respect to $(s_{11}, \ldots, s_{1K}, \ldots, s_{n1}, \ldots, s_{nK})$ and continuity of $P_{ik}(s_{1k}, \ldots, s_{nk})$ with respect to $(s_{1k}, \ldots, s_{nk})$ for all k. Continuity of the P_{ik} is Assumption 3.2, while continuity of $[I - \alpha_i q_\theta]^{-1}$ follows from $[I - \alpha_i q_\theta]$ being continuous in θ and nonsingular for all $\theta \in \Theta$. QED

DEFINITION 5.6 *Let $f(x)$ be a function from $A \subset R^n$ to R^m. Then $f(x)$ is* **quasiconcave** *if, for any $x \in A$, the set $\{y \in A \mid f(y) \geqslant f(x)\}$ is convex.*

LEMMA 5.11 *$G_{i\theta}$ is quasiconcave in θ_i, and $r(\theta)$ is convex for all $\theta \in \Theta$.*

Proof. If $G_{i\theta}$ is quasiconcave in θ_i, then $r_i(\theta)$ is convex; hence $r(\theta)$ is also convex. Thus, all that remains to show is that $G_{i\theta}$ is quasiconcave in θ_i. Suppose that $\theta \backslash \theta_i'$ and $\theta \backslash \theta_i''$ are any two policy combinations such that $G_{i\theta\backslash\theta'_i} = G_{i\theta\backslash\theta''_i}$, and let $\theta_i^* = \lambda\theta_i' + (1-\lambda)\theta_i''$ for $\lambda \in [0, 1]$. Then, $G_{i\theta}$ is quasiconcave if

$$G_{i\theta\backslash\theta^*_i} \geqslant \lambda G_{i\theta\backslash\theta'_i} + (1-\lambda) G_{i\theta\backslash\theta''_i} \tag{5.45}$$

for any such choice of θ, θ_i', θ_i'', and λ. To simplify notation, let $q' = q_{\theta\backslash\theta_i'}$, $q'' = q_{\theta\backslash\theta_i''}$, $q^* = q_{\theta\backslash\theta_i^*}$, $F_i' = F_{i\theta\backslash\theta_i'}$, $F_i'' = F_{i\theta\backslash\theta''_i}$, and $F_i^* = F_{i\theta\backslash\theta^*_i}$. Recalling that $[I - \alpha_i q']^{-1} F_i' = G_{i\theta\backslash\theta'_i} = G_{i\theta\backslash\theta'_i} = [I - \alpha_i q'']^{-1} F''_i$, it follows that

$$F''_i = [I - \alpha_i q''][I - \alpha_i q']^{-1} F_i' \tag{5.46}$$

and defining F_i^ε by the relation

$$F_i^* = \lambda F_i' + (1-\lambda) F_i'' + F_i^\varepsilon \tag{5.47}$$

it follows from Assumption 3.3Q that $F_i^\varepsilon \geqslant 0$. Now, using equation (5.47)

$$G_{i\theta\backslash\theta^*_i} = [I - \alpha_i q^*]^{-1} F_i^* = [I - \alpha_i q^*]^{-1} [\lambda F_i' + (1-\lambda) F_i'' + F_i^\varepsilon] \tag{5.48}$$

and using equation (5.46)

$$\begin{aligned} G_{i\theta\backslash\theta^*_i} &= [I - \alpha_i q^*]^{-1} \{\lambda F_i' + (1-\lambda)[I - \alpha_i q''][I - \alpha_i q']^{-1} F_i'\} \\ &\quad + [I - \alpha_i q^*]^{-1} F_i^\varepsilon = [I - \alpha_i q^*]^{-1} \{\lambda [I - \alpha_i q'] \\ &\quad + (1-\lambda)[I - \alpha_i q'']\} [I - \alpha_i q']^{-1} F_i' + [I - \alpha_i q^*]^{-1} F_i^\varepsilon \\ &= [I - \alpha_i q^*]^{-1} \{\lambda [I - \alpha_i q'] + (1-\lambda)[I - \alpha_i q'']\} G_{i\theta\backslash\theta'_i} + [I - \alpha_i q^*]^{-1} F_i^\varepsilon \end{aligned} \tag{5.49}$$

From condition (c) of Assumption 5.9;

$$\lambda [I - \alpha_i q'] + (1-\lambda)[I - \alpha_i q''] = [I - \alpha_i q^*] \tag{5.50}$$

therefore, equation (5.49) becomes

$$G_{i\theta\backslash\theta^*_i} = G_{i\theta\backslash\theta'_i} + [I - \alpha_i q^*]^{-1} F_i^\varepsilon \tag{5.51}$$

$G_{i\theta}$ is quasiconcave in θ_i if $[I - \alpha_i q^*]^{-1} F_i^\varepsilon \geqslant 0$. Because $F_i^\varepsilon \geqslant 0$, it remains to point out that $[I - \alpha_i q^*]^{-1} \geqslant 0$. The latter holds because $[I - \alpha_i q^*]$ is a dominant diagonal matrix whose principal diagonal elements are positive and whose off-diagonal elements are nonpositive. From McKenzie (1960), it is known that the inverse of such a matrix has all nonnegative elements; therefore, the lemma is proved. QED

THEOREM 5.3 *A stochastic game has an equilibrium point.*

Proof. The proof follows from Lemmas 5.9 to 5.11. QED

4 Supergames with incomplete information

One model due to Rosenthal (1979), is examined in this section. The model adopts an equilibrium concept that is computationally simpler than Harsanyi's (1967, 1968a, 1968b) equilibrium, with the players' choices based on relatively little information. An equilibrium concept is proposed that differs from the noncooperative equilibrium but that retains some resemblance to it. Section 4.1 describes the structure of the game, Section 4.2 contains the equilibrium notion and a proof of existence of equilibrium, and Section 4.3 comments on the model.

4.1 Description of the game

Suppose a collection of players, divided into two subsets, is to play a sequence of two-person games. Each such game is finite—that is, each player possesses a finite number of pure strategies. In each time period, the players are randomly matched into pairs, one member of the pair coming from subset I, the other from subset II. Each player knows her own payoff function, but is ignorant of the payoff function of any other player. Unlike in the Harsanyi (1967, 1968a, 1968b) modeling of incomplete information games, the player does not possess a subjective probability distribution over a family of payoff functions for the other players.

The player does know, at each time t, what pure strategy was played in period $t-1$ by her current opponent, although the opponent is not identified. If the opponent used a mixed strategy in period $t-1$, the player knows the pure strategy realization that was selected by use of the mixed strategy. She does not know the mixed strategy itself. Although the single-period payoff function of a player is the same in all periods, the behavior to be expected from opponents need not be the same in all periods, because different opponents will have different payoff functions. The player is assumed to approach the game as a problem in stochastic dynamic programming under which the behavior expected by her in period t depends on the information commonly available to her and her opponent—the pair of actions which the two of them realized, respectively, in their period $t-1$ games. An equilibrium, called a *Markovian equilibrium* point, is characterized by strategies fulfilling two kinds of conditions. First, the players all have empirically testable beliefs concerning how they will be paired over time and what behavior they will see by opponents. The evidence they observe must be consistent with these beliefs. Second, given those beliefs, it must be impossible to increase a player's payoff by means of another strategy.

4.1.1 *Assumptions of the model*

There are two pools of players, with the same (finite) number in each pool. In each time period, the players from the first pool are randomly paired into two-player games with players of the second pool. The games played in each period are finite (i.e., they have a finite number of pure strategies available to each player).

ASSUMPTION 5.10 *There are two types of players, denoted I and II; the set of players of each type is* $V = \{1, \ldots, v\}$. *Type j players have pure strategy sets* $M_j = \{1, \ldots, m_j\}$, *where* m_j *is finite and* $j = I, II$. *The payoff matrix for player* (j, i), $j \in \{I, II\}$, $i \in V$, *is* A_{ji}, *with* m_I *rows and* m_{II} *columns. The matrix entry* $a_{ji}(k, l)$ *is the payoff to player* (j, i) *when the type I player chooses pure strategy k and the type II player chooses pure strategy l. The only payoff matrix known to player* (j, i) *is* A_{ji}.

ASSUMPTION 5.11 *In each period t, a permutation of the type II players,* $(\phi(1), \ldots, \phi(v))$, *is chosen. Each of the* $v!$ *permutations has a probability of* $1/v!$ *of being chosen, and the pairing into two-person games is* $\{i, \phi(i)\}$, $i \in V$.

Assumptions 5.10 and 5.11 set up the basic structure concerning the single-period games and the pairing of players. A player of type I has equal probability of being paired with any type II player in each period. Each period the players engage in finite two-person noncooperative games of incomplete information. The players may use mixed strategies; however, it is convenient to defer the formal introduction of such strategies until the players' beliefs are sketched.

4.1.2 *Information conditions and the players' beliefs*

The players' beliefs must, of course, be described in relation to the information that they possess and that will come to them in the course of the play of the game. Denote by u_{jit} the actual pure strategy played by player (j, i), $(j \in \{\text{I}, \text{II}\}$, $i \in V)$ at time t. Thus, if a mixed strategy were used, $u_{jit} \in M_j$ would be the realization of the random process that the mixed strategy describes.

DEFINITION 5.7 *The* **immediate history** *for two players* (I, i) *and* (II, i') *who are paired together at time t in a single-period game is* $(u_{Ii,t-1}, u_{IIi',t-1}) = \mu = (\mu_I, \mu_{II})$.

From the vantage point of the paired players, μ is, in effect, a state variable. It is denoted below in the simple form $\mu = (\mu_I, \mu_{II})$ to avoid unnecessary notational baggage.

A player of one type will have beliefs about two aspects of the game that concern the players of the other type. First, at time t, anticipating the state that will prevail, a player i of type I will, of course, know $u_{\mathrm{I}i,t-1} = \mu_{\mathrm{I}}$, but, prior to being told μ_{II}, he will have beliefs given by the probability distribution $\pi_{\mathrm{I}} = (\pi_{\mathrm{I}}(1), \ldots, \pi_{\mathrm{I}}(m_{\mathrm{II}}))$. That is, $\pi_{\mathrm{I}}(k)$ is the probability that the next opponent will have $\mu_{\mathrm{II}} = k$. The distribution π_{I} is identical for all type I players and is a belief of those players. Similarly, there is a distribution π_{II} describing the common beliefs of the type II players. The reason for all players of one type having the same beliefs is that their beliefs must be consistent with the available evidence, and, as they all have the same evidence, they are led to the same beliefs.

DEFINITION 5.8 *π_j is the* **player selection belief** *of all players of type j. $\pi_j(k) \geqslant 0$ for $j \in \{I, II\}$ and all k, and $\sum_{k=1}^{m_{II}} \pi_I(k) = \sum_{k=1}^{m_I} \pi_{II}(k) = 1$.*

The second element of the players' beliefs is the pair of probability distributions $p_{\mathrm{I}} = (p_{\mathrm{I}}(1 \mid \mu), \ldots, p_{\mathrm{I}}(m_{\mathrm{II}} \mid \mu))$ and $p_{\mathrm{II}} = (p_{\mathrm{II}}(1 \mid \mu), \ldots, p_{\mathrm{II}}(m_{\mathrm{I}} \mid \mu))$. $p_{\mathrm{I}}(k \mid \mu)$ is the probability belief of a type I player that his current opponent will select $k \in M_{\mathrm{II}}$ given that the two players' immediate history is μ.

DEFINITION 5.9 *p_j is the* **player action belief** *of all players of type j. $p_j(k \mid \mu) \geqslant 0$ for $j \in \{I, II\}$ and all k, and $\sum_{k=1}^{m_{II}} p_I(k \mid \mu) = \sum_{k=1}^{m_I} p_{II}(k \mid \mu) = 1$ for all $\mu \in M_I \times M_{II}$.*

DEFINITION 5.10 *Letting $\pi = (\pi_I, \pi_{II})$ and $p = (p_I, p_{II})$, (π, p) is the* **belief structure of the game** *and (π_j, p_j) is the* **belief structure of the type** *j* **players.**

4.1.3 Players' strategies and the payoff function

Based on Blackwell's lemma and Denardo's lemma, if all players other than player (j, i) use policies that choose mixed strategies for each time period, contingent on only the immediate history, then among the discounted payoff maximizing strategies open to player (j, i) will be a policy of the same simple sort. Therefore, attention is restricted to these policies. In this interpretation of the game, it is assumed that each player has forgotten her own past history further than one period back.

DEFINITION 5.11 *A* **policy for player** *(j, i), θ_{ji}, is an array having $m_j \times m_I \times m_{II}$ elements. An individual element is denoted $\theta_{ji}(k \mid \mu)$ where $k \in M_j$ and $\mu \in M_I \times M_{II}$.*

ASSUMPTION 5.12 *θ_{ji} satisfies the following conditions: $\theta_{ji}(k \mid \mu) \geqslant 0$, $k \in M_j$, $\mu \in M_I \times M_{II}$; $\sum_{k=1}^{m_j} \theta_{ji}(k \mid \mu) = 1$, $\mu \in M_I \times M_{II}$; and θ_{ji} is defined for all $j = I$, II and $i \in V$.*

The policy combination for all type j players is $\theta_j = (\theta_{j1}, \ldots, \theta_{jv})$; the policy combination for all players is $\theta = (\theta_{\rm I}, \theta_{\rm II})$; the set of policies for player (j, i) is denoted Θ_{ji}; the set of policy combinations for type j players is $\Theta_j = \times_{i \in V} \Theta_{ji}$; and the set of policy combinations for all players is $\Theta = \Theta_{\rm I} \times \Theta_{\rm II}$.

The single-period payoff functions for the players are given by

$$P_{{\rm I}i}(\theta_{{\rm I}i}, p_{\rm I} \mid \mu) = \sum_{k \in M_{\rm I}} \sum_{l \in M_{\rm II}} \theta_{{\rm I}i}(k \mid \mu) p_{\rm I}(l \mid \mu) a_{{\rm I}i}(k, l) \tag{5.52}$$

$$p_{{\rm II}i}(\theta_{{\rm II}i}, p_{\rm II} \mid \mu) = \sum_{k \in M_{\rm I}} \sum_{l \in M_{\rm II}} \theta_{{\rm II}i}(l \mid \mu) p_{\rm II}(k \mid \mu) a_{{\rm II}i}(k, l) \tag{5.53}$$

Let $P_{ji}(\theta_{ji}, p_j) = (P_{ji}(\theta_{ji}, p_j \mid (1, 1)), \ldots, P_{ji}(\theta_{ji}, p_j \mid (m_{\rm I}, m_{\rm II})))$ be the single-period payoff vector for player (j, i). That is, it shows the single-period expected payoff associated with every state. The decision problem of a single player, say $({\rm I}, i)$, with $(\pi_{\rm I}, p_{\rm I})$ specified, looks like a straightforward stochastic dynamic programming exercise. Note that the single-period payoff, conditional on the state μ, depends on the player's policy $\theta_{{\rm I}i}$ and her beliefs about opponents' actions $p_{\rm I}$. The transition mechanism for this player is defined by

$$q_{{\rm I}i}(\mu' \mid \mu, \theta_{{\rm I}i}) = \theta_{{\rm I}i}(\mu'_i \mid \mu) \pi_{\rm I}(\mu'_{\rm II}) \tag{5.54}$$

which depends on the current state, μ; the player's own policy, $\theta_{{\rm I}i}$; and the player's beliefs about how her opponents are selected, $\pi_{\rm I}$. $q_{{\rm I}i}(\mu' \mid \mu, \theta_{{\rm I}i})$ is the probability that today's state for player $({\rm I}, i)$ will be μ', given that yesterday's state was μ and the player's policy is $\theta_{{\rm I}i}$. The transition mechanism itself is a square array of $m_{\rm I} \times m_{\rm II}$ rows and columns. Each row is a probability distribution over today's states, contingent on yesterday's state and the policy. The mechanism is denoted

$$q_{{\rm I}i\theta} = \begin{bmatrix} q_{{\rm I}i}((1,1) \mid (1,1), \theta_{{\rm I}i}), \ldots, q_{{\rm I}i}((m_{\rm I}, m_{\rm II}) \mid (1,1), \theta_{{\rm I}i}) \\ \vdots \qquad\qquad\qquad \vdots \\ q_{{\rm I}i}((1,1) \mid (m_{\rm I}, m_{\rm II}), \theta_{{\rm I}i}), \ldots, q_{{\rm I}i}((m_{\rm I}, m_{\rm II}) \mid (m_{\rm I}, m_{\rm II}), \theta_{{\rm I}i}) \end{bmatrix} \tag{5.55}$$

Denoting the discount parameters by α_{ji}, the (discounted, expected) supergame payoff for a type I player is

$$\sum_{t=0}^{\infty} \alpha_{{\rm I}i}^t q_{{\rm I}i\theta}^t P_{{\rm I}i}(\theta_{{\rm I}i}, p_{\rm I}) = G_{{\rm I}i}(\theta_{{\rm I}i}, p_{\rm I}, \pi_{\rm I}, \alpha_{{\rm I}i}) \tag{5.56}$$

Analogous expressions to equations (5.54) to (5.56) can be written for type II players.

4.1.4 *Defining the game*

With the material developed in Sections 4.1.1 to 4.1.3, it is possible to complete the description of the game. To that end, *a stationary two-person incomplete information supergame* is defined in Definition 5.12. Note that, in one sense, this game has $2v$ players, but, in each time period, a player is in a two-person, single-shot game.

DEFINITION 5.12 *A* **stationary two-person, incomplete information supergame**, $\Gamma = (M_I, M_{II}, V, \{A_{ji}\}, \{\alpha_{ji}\})$ *is defined by Assumptions* 5.10 *to* 5.12, Definitions 5.7 to 5.11, *and the information conditions*: (1) *all players know* M_I, M_{II}, *and* V, (2) *only player* (j, i) *knows* A_{ji}, *and* α_{ji}, *and* (3) *a player knows she is always paired with a player of the other type, but is never informed which player.*

4.2 Existence of equilibrium

In order to prove existence of equilibrium, the equilibrium concept itself must first be carefully specified. This is done in Section 4.3.1. Existence is proved in Section 4.2.2.

4.2.1 Markovian beliefs and the definition of a Markovian equilibrium

Two sorts of conditions, outlined in the following two definitions, are required. The first set of conditions defines consistency of beliefs with observed phenomena; the second is a best-reply condition.

DEFINITION 5.13 *(p, π) is a* **rational Markovian belief relative to** *θ if*

$$w_{Ii}(k) = \sum_{\mu_I \in M_I} w_{Ii}(\mu_I) \sum_{\mu_{II} \in M_{II}} \theta_{Ii}(k \mid \mu)\pi_I(\mu_{II}), \qquad i \in V \tag{5.57}$$

$$w_{IIi}(k) = \sum_{\mu_{II} \in M_{II}} w_{IIi}(\mu_{II}) \sum_{\mu_I \in M_I} \theta_{IIi}(k \mid \mu)\pi_{II}(\mu_I), \qquad i \in V \tag{5.58}$$

$$\pi_{II} = \frac{1}{v} \sum_{i \in V} w_{Ii} \tag{5.59}$$

$$\pi_I = \frac{1}{v} \sum_{i \in V} w_{IIi} \tag{5.60}$$

$$p_{II}(k \mid \mu) \sum_{i \in V} w_{Ii}(\mu_I) = \sum_{i \in V} w_{Ii}(\mu_I)\theta_{Ii}(k \mid \mu), \qquad \mu \in M_I \times M_{II} \tag{5.61}$$

$$p_I(k \mid \mu) \sum_{i \in V} w_{IIi}(\mu_{II}) = \sum_{i \in V} w_{IIi}(\mu_{II})\theta_{IIi}(k \mid \mu), \qquad \mu \in M_I \times M_{II} \tag{5.62}$$

Equations (5.57) to (5.62) define the steady-state distributions of the observed pure strategy plays of each player. The equilibrium concept is defined only with respect to long-run steady-state conditions. Equation (5.59) requires that the beliefs of type II players about the distribution of player selections conform to the actual distribution of the choices of type I players. Any discrepancy could, in fact, be detected. Equation (5.60) is a parallel requirement concerning the beliefs of type I players. Equation (5.61) is a consistency requirement that the player action beliefs of type II players be the same as the frequencies that are actually observed, given the policies and steady-state distributions of the type I players. Again, a discrepancy could be detected if a game were moving over time in a steady state. Equation (5.62) is a parallel statement concerning the beliefs of the type I players.

DEFINITION 5.14 *(θ, p, π) is a* **Markovian equilibrium point** *if (p, π) is a rational Markovian belief relative to θ, $\theta' \in \Theta$, and*

$$G_{ji}(\theta_{ji}, p_j, \pi_j, \alpha_{ji}) = \max_{\theta'_{ji} \in \Theta_{ji}} G_{ji}(\theta'_{ji}, p_j, \pi_j, \alpha_{ji}), \quad i \in V, \quad j \in \{\mathrm{I}, \mathrm{II}\} \tag{5.63}$$

Clearly Definition 5.14 is a best-reply condition. No single player could alter his behavior and increase his payoff.

4.2.2 The existence of a Markovian equilibrium

Existence of a Markovian equilibrium is proved in the following theorem.

THEOREM 5.4 *A stationary two-person, incomplete information supergame $\Gamma = (M_I, M_{II}, V, \{A_{ji}\}, \{\alpha_{ji}\})$ has a Markovian equilibrium point.*

Proof. The proof proceeds by means of a standard fixed-point argument using the Kakutani theorem. A correspondence is described presently whose fixed points are the equilibria whose existence is to be shown. Let $w(\mu) = (w_{\mathrm{I}}(\mu_{\mathrm{I}}), w_{\mathrm{II}}(\mu_{\mathrm{II}}))$, where $w_j(\mu_j) = (w_{j1}(\mu_j), \ldots, w_{jv}(\mu_j))$. Equations (5.57) to (5.63) define a correspondence, $f(w, p, \pi, \theta)$, that maps a compact, convex domain into itself. Letting an image of $f(w, p, \pi, \theta)$ be denoted by (w', p', π', θ'), the image elements are obtained as follows: w' is defined by equations (5.57) and (5.58). Clearly, w'_{ji} is always a probability distribution with nonnegative entries that sum to 1, and $f_w(w, \pi, \theta) = w'$, given by equations (5.57) and (5.68), is a continuous function. $p' \in f_p(w,)u$ is defined using equations (5.61) and (5.62) as follows:

$$p'_{\mathrm{II}}(k \mid \mu) = \frac{\sum\limits_{i \in V} w_{\mathrm{I}i}(\mu_{\mathrm{I}}) \theta_{\mathrm{I}i}(k \mid \mu)}{\sum\limits_{i \in V} w_{\mathrm{I}i}(\mu_{\mathrm{I}})}, \qquad i \in V, \qquad \mu \in M_{\mathrm{I}} \times M_{\mathrm{II}}, \tag{5.64}$$

when $\sum_{i \in V} w_{\mathrm{I}i}(\mu_{\mathrm{I}}) > 0$; otherwise, $(p'_{\mathrm{II}}(1 \mid \mu), \ldots, p'_{\mathrm{II}}(m_{\mathrm{I}} \mid \mu))$ takes on any values satisfying the conditions $p'_{\mathrm{II}}(k \mid \mu) \geqslant 0$ and $\sum_{k \in M_{\mathrm{I}}} p'_{\mathrm{II}}(k \mid \mu) = 1$. That is, when $\sum_{i \in V} w_{\mathrm{I}i}(\mu_{\mathrm{I}}) = 0$, then any probability distribution is in the image set. $p'_{\mathrm{I}}(k \mid \mu)$ is defined analogously. The correspondence f_p is clearly upper semicontinuous and has convex image sets. Finally, from Lemma 5.11, it is clear that $f_\theta(p, \pi)$, defined by equation (5.63), is upper semicontinuous and has convex image sets. Here, f corresponds to the best-reply mapping r, in Lemma 5.10. The mapping f is

$$f(w, p, \pi, \theta) = f_w(w, \pi, \theta) \times f_p(w, \theta) \times f_\pi(w) \times f_\theta(p, \pi) \qquad (5.65)$$

and it clearly satisfies the Kakutani fixed-point theorem; therefore, the game has an equilibrium point. QED

4.3 Comments on the model

4.3.1 *Extensions*

The model can easily be generalized to allow for n different types of player rather than only two. Under this arrangement, there would be v players of each type. They would be randomly grouped into v-different n-person games in each period, with each game having one player of each type. It is a straightforward matter to reformulate equations (5.57) to (5.63) to define rational Markovian beliefs for such a game, to define the Markovian equilibrium point, and to use essentially the same existence proof.

Straying from the finite game formulation to allow the single-period games the same structure as that allowed by Assumptions 3.1, 3.2, and 3.3Q may not be so easy. The difficulty lies in moving from discrete to continuous probability distributions. Doing so might prove analytically difficult. Also, it becomes less plausible to argue that the players, in a steady state, could obtain enough evidence to see that their conjectures are correct. Although the accumulating evidence need not contradict the conjectures, it would be insufficient to give strong confirmation to them.

4.3.2 *Comparison to Harsanyi's incomplete information model*

It is interesting to compare this game to Harsanyi's incomplete information games and to compare the Markovian equilibrium point with a Nash noncooperative equilibrium point. Doing the latter first, recall that the game with v players of type I and type II is, in fact, a noncooperative incomplete information game with $2v$ players. The Markovian equilibrium strategy of each player is an approximate best reply for the player, given the level of his information; however, if the information conditions were changed so that a player in each time period knew the identity of his opponent, he could have a strategy conditioned on this. Such a more finely tuned strategy would, in general, be different from the Markovian strategy, because it could take advantage of differences in the strategies of the various rival players.

The model differs from Harsanyi's incomplete information games in two respects. First, Harsanyi's game is formulated to be one shot; players thus have no opportunity to gather information over time and compare it with their beliefs. This frees beliefs in Harsanyi's model from having to conform to experience. Harsanyi's players are assumed to know the list of possible payoff functions with which their rival players may be endowed, while, in the Rosenthal model, no such information is assumed. Harsanyi's game is one in which there are truly n players, but the players act as if there are more, while in the Rosenthal model there are, in fact, as many players as the players presume. It is an advantage of the Rosenthal formulation that the players do not make presumptions about rival's payoff functions that may be arbitrary. Instead, they do not concern themselves with rivals' payoff functions; they follow rules of behavior that are optimal relative to assumptions about how other players will actually behave, and in equilibrium their beliefs about rivals' behavior are consistent with observations.

5 Reputations, rationality, and Nash equilibrium

There are games in which a noncooperative equilibrium outcome appears intuitively unreasonable. This section is devoted to examining two of them. The first is called the *chain store paradox,* presented and discussed by Selten (1978). The second, called the *centipede game,* is due to Rosenthal (1981). Rosenthal's paper was apparently inspired by Selten's work and the related work of others. The common thread running through both is that "irrational" equilibrium outcomes appear superior to strictly "rational" ones. The so-called rational outcome is the Nash noncooperative equilibrium; thus, these papers point to deficiencies in the Nash concept.

5.1 Selten's chain store paradox

The Selten (1978) example is presented in Section 5.1.1, following which Section 5.1.2 describes a way out of the paradox, suggested by Kreps and Wilson (1982a), that is based on altering the information conditions originally used by Selten.

5.1.1 Formulation of the chain store example

There are 21 players in this game. Player 0 is called the chain store and has branches in 20 different cities, numbered 1, . . . , 20. Player $i \in \{1, \ldots, 20\}$ is an entrepreneur who could open a competing store in city i. The game goes on for exactly 20 periods. In period i, player i decides whether or not to enter the market in his town. After player i announces his decision, player 0 decides to play either *soft* or *tough.* The single-period payoffs are shown in Table 5.1. In period 2, player 2 makes a similar decision, and the game proceeds in this way through period 20. A single period is also shown in Figure 5.1.

TABLE 5.1 One stage of the chain store paradox

		Player i	
		In	Out
Player 0	Soft	2, 2	5, 1
	Tough	0, 0	5, 1

Working backward from the final period, it is clear that player 20 should enter and player 0 should play *soft*. Given that player 20 decides *in*, player 0 is better off playing *soft*, because her payoff is higher. In period 19; the situation is the same. Player 0 cannot affect the later period through her current decision, and player 19 is best off to choose *in*. The backward induction argument holds, and the only perfect equilibrium point of the game involves each player $i \in \{1, \ldots, 20\}$ choosing *in*, with player 0 choosing *soft* in each period. The payoff matrix in Table 5.1 is known to all players, all know the structure of the game, and all know the past choices of all players.

The source of the paradox is that this (unique) perfect equilibrium point seems intuitively unreasonable compared with player 0, the chain store, playing *tough* in the early periods of the game to scare off later firms from entering the market. Very likely, by playing *tough* for the first several

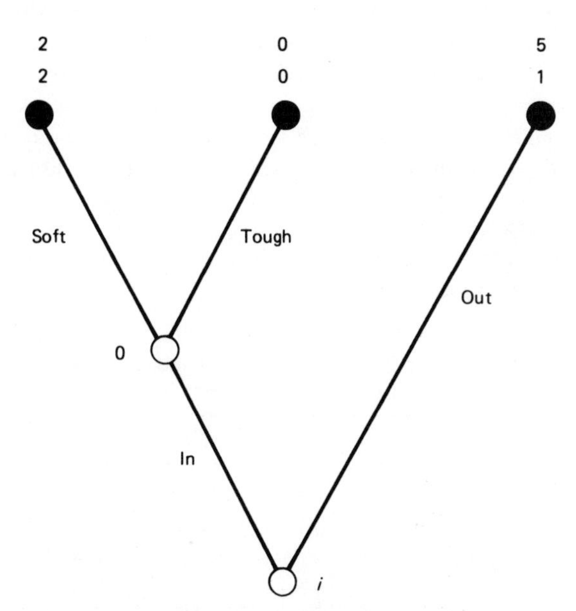

FIGURE 5.1 One round of the chain store paradox.

periods, many players will decide it is not worthwhile to come into the market. Players who act near the end of the game (say, $i = 18$, 19, and 20) may be undeterred from entering the market because they may believe that the remaining time is not sufficient to make it worthwhile for the chain store to attempt to deter entry. The chain store would likely take this view as well. Thus entrance in the last few periods, coupled with player 0 choosing *soft,* seems reasonable. To summarize, the chain store would play *tough* in any period a player chose *in,* from the start to the last few periods. In these final periods, the chain store would play *soft* in the face of entry. The other players would either never enter until the last few periods, when a *soft* response could be expected, or, perhaps the first few firms would try *in,* be greeted with *tough,* and convince all but the last few firms to play *out.* The problem is how to rationalize formally the intuitively plausible behavior just described. The method suggested below is appealing, although it changes the structure of the game.

5.1.2 *Escaping the paradox via incomplete information*

Kreps and Wilson (1982a) suggest a way out that requires a change in the information conditions of the game. Suppose the chain store's true payoff structure is not known with certainty to the other players, but instead they know it is as shown in either Table 5.1 or Table 5.2. The original payoff structure is called a *weak chain store* and the payoff structure in Table 5.2 is called *strong chain store.* A strong chain store will always play *tough* in the face of entry, because doing so maximizes its single-period payoff and has no adverse long-run consequences. Therefore, a chain store that plays *tough* in the face of entry could be (a) a strong chain store using its dominant strategy or (b) a weak chain store trying to discourage entry by mimicking the behavior expected of a strong chain store. Suppose the players i $(\neq 0)$ believe the probability the chain store is weak to be $p < \frac{1}{2}$, and that player 0 is, in fact, weak. Then the following strategies are an equilibrium point:

Player 1 chooses *out.*
Player i chooses *in* if player 0 has chosen *soft* at any past time and chooses *out* otherwise ($i = 2, \ldots, 20$).

TABLE 5.2 A single stage of an alternate version of the chain store paradox

		Player i	
		In	Out
Player 0	Soft	0, 0	5, 1
	Tough	2, −1	5, 1

Player 0 chooses *tough* in any period 1, . . . , 17 during which another player selects *in* and *soft* in any period 18, 19, 20 during which another player selects *in*.

Under these strategies, the equilibrium path will be that all players $i = 1, \ldots, 20$ choose *out*. None of them can increase expected payoff by selecting differently. Of course, this version of the game lacks the complete information assumed in Selten's original specifications; therefore, it does not resolve the paradox on Selten's terms.

5.2 *Another view of beneficial "irrational" behavior*

The second game, due to Rosenthal (1981), involves a version of the finitely repeated prisoners' dilemma with sequential decision. Suppose a single-shot payoff matrix as shown in Table 5.3. The rules of the game are that player 1 chooses, then his choice is revealed to player 2, who chooses. If one or both choose *s* (stop), the game ends with the completion of the first round of play. Otherwise, a second round is played in the same manner as the first. If both choose *c* (continue), another round is played until a total of five rounds has been played and the game ends. An abbreviated extensive form of the five-round game is shown in Figure 5.2. It is assumed that player 2 would never choose *c* after learning that player 1 has chosen *s*; this sequence is thus omitted. If player 1 chooses *s*, then player 2 also chooses *s*, but if player 1 chooses *c*, player 2 might choose either *s* or *c*. Now suppose that each player believes that the other player will choose to continue at each decision point with probability $\min\{1, .5 + .4D\}$, where D is the difference between the payoff to the player if he continues and stops. Then, at his last decision point, player 2 would choose *stop* with probability .9 and *continue* with probability .1. At the next to last decision node, it is the turn of player 1. He chooses to continue with probability $.5 + .4(.9 \times 7 + .1 \times 10 - 8) = .22$, and his strategy is (.78, .22). The game is displayed again in Figure 5.3, which shows at each branch the probability assigned to it, and at each node the expected payoff to the two players conditional on reaching that node using the strategies just outlined. An asterisk appears next to the payoff of the player who moves at the node. As with the chain store game, the only equilibrium point is for player 1 to choose to stop at the first node.

TABLE 5.3 One stage of a sequential game with uncertainty

		Player 2	
		s	*c*
Player 1	*s*	0, 0	3, −1
	c	−1, 3	2, 2

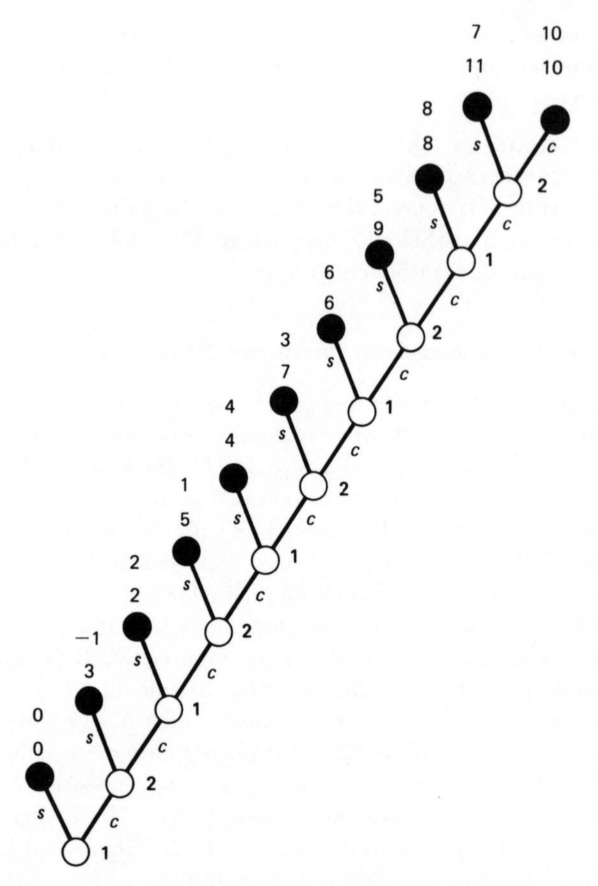

FIGURE 5.2 The centipede game, a game similar to the prisoners' dilemma with sequential decision.

The reasonable solution proposed by Rosenthal rests on a small irrationality or act of faith by the players: Each supposes the other will, with some positive probability, choose to continue the game at each node. I hesitate to use the word *irrational* to describe this behavior, because it prejudges that only best-reply behavior is rational. In the present game, at each node the player who moves takes on a risk: If he continues, the other player can terminate the game immediately, which lowers his payoff one unit, or he can continue, which gives him a payoff, two units higher. If the game continues to his own next decision node, then no matter how much further it goes, he must be better off as compared with immediate stopping. To continue and trust to the good sense of the other player also to continue seems a reasonable risk, because both players stand to gain.

6 Assorted topics relating to noncooperative games

This section touches lightly on several topics. Section 6.1 briefly examines three requirements that can be used to select among noncooperative

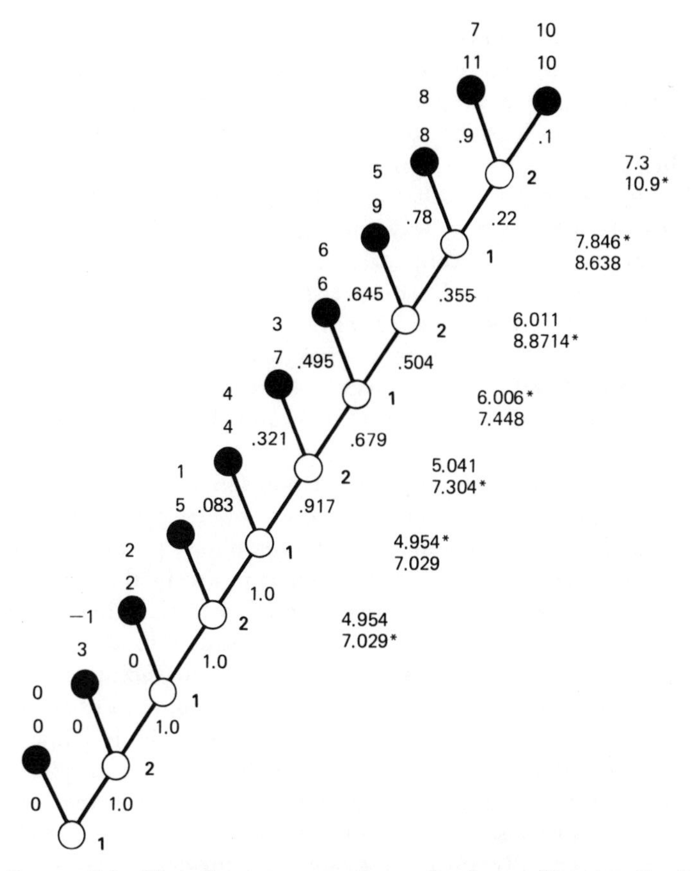

FIGURE 5.3 The effect on expected payoffs of some "irrationality."

equilibria. Section 6.2 takes up two equilibrium concepts for non-cooperative games that differ from the equilibrium point concept that dominates the field and dominates Chapters 2 through 5 of this book.

6.1 Some restrictions on equilibrium points

Three properties that have been suggested for noncooperative equilibria are that they be (a) *coalition-proof,* (b) *renegotiation-proof,* and (c) *subgame consistent.* The property of being coalition-proof will have immediate appeal for anyone familiar with the *core,* a well known cooperative game solution, covered in Chapters 7 and 8. Consider the equilibria that are enabled by the folk theorem and its relatives and think about the outcomes that such equilibria might be expected to support. Many would say that the particular outcomes that one really expects are those that give each player at least as great a payoff as she can command on her own and that cannot be Pareto dominated by some other supportable outcome. Thus, no individual loses and the group as a whole cannot do better.

Coalition-proofness adds a kindred condition. Consider a collection of players, $K \subset N$, that contains between 2 and $n-1$ players. The same common-sense view that expects individual rationality and Pareto optimality might expect a supported outcome to give to the members of K payoffs that are at least as favorable as anything the members could achieve by themselves. As an example, suppose each individual can be held to 0 (his minimax level) and that a total payoff of 100 can be attained and can be apportioned in any way among the, say, $n = 10$ players. Now suppose that the coalition consisting of players 1 and 2 can guarantee a total payoff to themselves of 8. However hard they are squeezed by players $\{3, \ldots, 10\}$, they can still be certain to obtain 8. Coalition-proofness requires that an acceptable outcome give them a total of 8, in addition to ensuring that each individual receives at least 0. This topic is explored in Bernheim, Peleg, and Whinston (1987) and Bernheim and Whinston (1987).

Renegotiation arises in the context of punishment phases of equilibria. Imagine a cooperative outcome supported by trigger strategies or by punishments that drive deviating players to minimax payoffs. Suppose that a player has, in fact, deviated and the players are in the beginning of a punishment phase. It could occur to some of them that they could all be better off by forgiving the deviator and returning to the actions they originally chose. On the one hand, they have an incentive to do this because they all gain. On the other hand, this possibility can be anticipated from the start and, if it is plausible, it undermines the discipline that the punishment provisions are meant to provide. If this discipline is undermined, then the original equilibrium is no longer viable. The first person formally to consider this problem was apparently Joseph Farrell. Solution of this difficulty apparently calls for punishments yielding payoffs that cannot, themselves, be improved upon. Recent working papers on this topic include Farrell and Maskin (1987) and Pearce (1987).

6.2 Iterated dominance and rationalizability

The Nash (1951) notion of equilibrium has unified noncooperative game theory for nearly 40 years. Despite the challenges it has encountered, Nash's notion remains the principal equilibrium concept. The challenge represented by both iterated dominance and rationalizability is that a player can face *strategic uncertainty*; that is, a player need not be able to predict exactly what the other players in a game will choose to do. This is obvious when there are multiple equilibria; however, even when the noncooperative equilibrium is unique, as in the centipede game, there may be reason to doubt whether other players will analyze the game *on the assumption that all players will adhere to the Nash standard.* The point of view taken under iterated dominance and rationalizability is that strategies can be ruled out if they can be shown to be patently foolish, but not otherwise. The belief that a player will certainly not use a given strategy requires a relatively strong justification.

Iterated dominance works like this. Suppose a game (N, S, P). For player i, let $S_i^0 \equiv S_i$ and $S^0 = \times_{i \in N} S_i^0$. A nested sequence of strategy spaces $\{S_i^k\}$ is developed by successively discarding dominated strategies. The strategies in S that are dominated are removed. Then any dominated strategies from the surviving set are removed. This process continues until no dominated strategies remain. Thus, $s_i \in S_i^k$ is (weakly) dominated if there is $s_i' \in S_i^k$ such that $P_i(s \backslash s_i') \geqslant P_i(s \backslash s_i)$ for all $s \in S^k$ (where $S^k = \times_{i \in N} S_i^k$). Let S_i^{k+1} be the undominated members of S_i^k, $S_i^* = \cap_{k=0}^{\infty} S_i^k$, and $S^* = \times_{i \in N} S_i^*$. The game (N, S, P) is *dominance solvable* if, for all s, $s' \in S^*$, $P(s) = P(s')$ [see Moulin (1979)]. The payoffs under iterated dominance in a dominance solvable game need not be the same as the payoffs at a noncooperative equilibrium.

Rationalizability is based on the iterated use of a best-reply criterion. As above, let $S_i^0 \equiv S_i$ and $S^0 = \times_{i \in N} S_i^0$. A nested sequence of strategy spaces $\{S_i^k\}$ is developed by successively discarding strategies that are never best replies. That is, S_i^{k+1} is a subset of S_i^k having the property that a member of S_i^{k+1} is a best reply to at least one element of (the convex hull of) S^k. Then $S_i^* = \cap_{k=0}^{\infty} S_i^k$, and $S^* = \times_{i \in N} S_i^*$. Any element of S^* is a rationalizable outcome of the game (N, S, P). Any such strategy combination, s, consists of plausible strategies in the sense that each s_i is a best reply to something in S^*. Of course, such an s need not be a noncooperative equilibrium, but a noncooperative equilibrium is rationalizable. The pioneering papers are Bernheim (1984) and Pearce (1984). Interestingly, a unique noncooperative equilibrium does not imply a unique rationalizable outcome (Bernheim, 1989, Figure 1).

7 Applications of time-dependent supergames

Two oligopoly applications are sketched in this section. The first is one in which the intertemporal link is a firm's capital stock. In the second, time dependence stems from advertising. Both are simple examples of time-dependent (nonstochastic) supergames.

7.1 A model of oligopoly with capital

Imagine a market in which the firms make differentiated products but are regarded as choosing output levels rather than prices. If one variable is easily changed within a time frame during which the second is not, then it is reasonable to take the second as the decision variable, to assume the time period is long enough that the second can be changed freely in each period, and to suppose the first variable is finely tuned often during the period in response to market conditions. For example, suppose the period is 1 day, and that the firms are individual fishermen who fish early in the morning and sell their catch during the rest of the day. Price is easily adjusted throughout the day, but quantity (the number of fish brought to market) is decided once per day. Output would then be the decision variable.

Although both variables are decided by the firm, it is really output that has the major strategic importance.

The inverse demand function of a firm is $p_{it} = f_i(q_t)$, where p_{it} is the price of the ith firm in period t and $q_t = (q_{1t}, \ldots, q_{nt})$ is the output vector of the n firms in the market. $f_i(q_t)$ is assumed continuous and differentiable with $f_t^i < f_t^j < 0$ $(j \neq i)$. An increase in any firm's output will lower the market clearing price for firm i, but firm i has a larger effect on p_{it} than any other single firm. The cost of production depends on output and capital, and is denoted by $C_i(q_{it}, K_{it})$, where K_{it} is the capital stock of firm i at time t. $C_i(q_{it}, K_{it})$ is a convex function that is increasing in q_{it}, marginal cost (i.e., $\partial C_i / \partial q_{it} = C_i^1$) declines as K_{it} rises. For any output level, there is an optimal (cost-minimizing) level of capital stock. This level rises as output rises. As K_{it} goes to 0, fixed cost goes to 0 and marginal cost, $C_i^1(q_{it}, K_{it})$, goes to infinity. Finally, the relationship between K_{it}, $K_{i,t+1}$, and the amount spent on new capital in period t (I_{it}) is $I_{it} = K_{it} g(K_{i,t+1}/K_{it})$. This is a convex function, increasing in $K_{i,t+1}$ and decreasing in K_{it}. That is, given the size of the present capital stock, the higher is tomorrow's level of capital, the more it will cost to attain. Conversely, to attain a specific level of $K_{i,t+1}$, the larger is K_{it}, the less it will cost. Convexity of $K_{it} g(K_{i,t+1}/K_{it})$ means that the cost of adding one more unit to the capital stock rises, or at best, remains constant, as the desired value of $K_{i,t+1}$ increases.

Thus, the single-period profit function is

$$\pi_{it} = q_{it} f_i(q_t) - C_i(q_{it}, K_{it}) - K_{it} g\left(\frac{K_{i,t+1}}{K_{it}}\right) \tag{5.66}$$

(For the details of this model, see Friedman (1983, Chapter 7)).

Formulating the strategy spaces for the players is cumbersome. In any period t, it is possible to specify an interval for output, $[0, q_i^+]$, where q_i^+ is chosen to be so high that production at this level would never be profitable. For example, q_i^+ might be high enough that $f_i(q \backslash q_i^+) = 0$ for all $q \geqslant 0$. The situation with $K_{i,t+1}$ is less straightforward. There is an upper bound, K_i^+, so high that it cannot be in a firm's interest to maintain such a large or larger capital stock; however, the minimum value of $K_{i,t+1}$ is not generally 0. It is the level satisfying the condition $0 = K_{it} g(K_{i,t+1}/K_{it})$, because the firm cannot, by assumption, sell off capital and, once bought, it is not in the firm's interest to throw away any capital.

Letting α_i be the ith firm's discount parameter, denoting strategies by $\sigma_i = (q_{i0}, K_{i1}, q_{i1}, K_{i2}, \ldots)$, and letting I' be a fixed start-up cost for the firm that buys an initial capital of $K_{i0} = K'$, the firm's supergame payoff function is

$$G_i(\sigma) = -I' + \sum_{t=0}^{\infty} \alpha_i^t \left[q_{it} f_i(q_t) - C_i(q_{it}, K_{it}) - K_{it} g\left(\frac{K_{i,t+1}}{K_{it}}\right)\right] \tag{5.67}$$

Note that in period t, firm i chooses q_{it} and $K_{i,t+1}$. Choosing $K_{i,t+1}$ is equivalent to choosing I_{it}. Turning to a specific example, let $n = 2$, $\alpha_i = .9$, i = 1, 2, and

$$p_{1t} = 140 - 1.5q_{1t} - q_{2t} \tag{5.68}$$

$$p_{2t} = 140 - 1.5q_{2t} - q_{1t} \tag{5.69}$$

$$C_i(q_{it}, K_{it}) = .05K_{it} + \frac{10 + 3q_{it} + q_{it}^2}{K_{it}}, \qquad i = 1,2 \tag{5.70}$$

$$I_{it} = \frac{K_{i,t+1}^2}{K_{it}} \tag{5.71}$$

Thus, the supergame payoff for player 1 is

$$G_1(\sigma) = -I' + \sum_{t=0}^{\infty} .9^t \Big[q_{it}(140 - 1.5q_{1t} - q_{2t}) - .05K_{1t} - \frac{10 + 3q_{1t} + q_{1t}^2}{K_{1t}} - \frac{K_{1,t+1}^2}{K_{1t}} \Big] \tag{5.72}$$

Looking for a steady-state equilibrium, and taking advantage of the symmetry of the model to find a symmetric steady state,

$$\frac{\partial G_1(\sigma)}{\partial q_{1t}} = .9^t \frac{\partial \pi_{1t}}{\partial q_{1t}} = .9^t \Big(140 - 3q_{1t} - q_{2t} - \frac{3 + 2q_{1t}}{K_{1t}} \Big) \Big] = 0 \tag{5.73}$$

$$\frac{\partial G_1(\sigma)}{\partial K_{1t}} = .9^t \Big[\frac{-2K_{1t}}{K_{1,t-1}} + .9\Big(-.05 + \frac{10 + 3q_{1t} + q_{1t}^2 + K_{1,t+1}^2}{K_{1t}^2} \Big) \Big] = 0 \tag{5.74}$$

Denoting steady-state values by q^* $(=q_{1t} = q_{2t})$ and K^* $(=K_{1t} = K_{2t})$, equations (5.73) can be expressed as

$$q^* = \frac{140K^* - 3}{4K^* + 2} \tag{5.75}$$

and equation (5.74) reduces to

$$-2 + .9\Big(-.05 + \frac{10 + 3q^* + q^{*2}}{K^{*2}} + 1 \Big) = 0 \tag{5.76}$$

Using equations (5.75) and (5.76), $q^* = 34.438$, $K^* = 31.96$, and steady-state, single-period profits are $\pi = 1783.77$.

In this model, the element of time dependence is very restricted in the sense that π_{it} depends only on the firm's own previous choice of capital and on no previous choices of the other firm. In addition, the only variables of other firms that directly enter a firm's payoff function are the q_{jt}. Other models of oligopoly with capital are Prescott (1973), from which the investment function $I_{it} = K_{it}g(K_{i,t+1}/K_{it})$ is borrowed, and Flaherty (1980). The Flaherty model has the unusual feature that it is symmetric in structure, has both symmetric and asymmetric steady states, and only asymmetric steady states are stable.

7.2 A model of oligopoly with advertising

The model presented here is similar in many ways to the capital model. Advertising treated like capital in the sense that an expenditure on advertising from time t (y_{it}) affects sales at t and all future periods. The effect is captured in a variable called *goodwill* (Y_{it}), whose value depends on the firm's past path of advertising. Of course, goodwill enters into a firm's demand function rather than its production cost function. An important difference between this model and the capital example is that here the demand facing firm i will depend on the goodwill levels of all firms in the market. The effect of an increase in the goodwill of firm j can be either to raise or to lower demand. Products and industries seem to differ with regard to whether the advertising of a firm aids or hurts competing firms.

The demand structure is given by an inverse demand function, $p_{it} = f_i(q_t, Y_t)$; cost of production depends on current output only, $C_i(q_{it})$; and the relationship between the current expenditure on advertising (y_{it}), the goodwill level of the preceding period ($Y_{i,t-1}$), and that of the current period (Y_{it}) is $y_{it} = \theta_i(Y_{it}, Y_{i,t-1})$. Analogously to $K_{it}g(K_{i,t+1}/K_{it})$, θ_i is convex, increasing in Y_{it} and decreasing in $Y_{i,t-1}$. In period t, the firm chooses q_{it} and Y_{it}. Single-period profit is

$$\pi_{it} = q_{it}f_i(q_t, Y_t) - C_i(q_{it}) - \theta_i(Y_{it}, Y_{i,t-1}) \tag{5.77}$$

and the supergame payoff, letting $\sigma_i = (q_{i0}, Y_{i0}, q_{i1}, Y_{i1}, q_{i2}, Y_{i2}, \ldots)$, is

$$G_i(\sigma) = q_{i0}f_i(q_0, Y_0) - C_i(q_{i0}) - \theta_i(Y_{i0}, 0) + \sum_{t=1}^{\infty} \alpha_i^t[q_{it}f_i(q_t, Y_t) - C_i(q_{it}) - \theta_i(Y_{it}, Y_{i,t-1})] \tag{5.78}$$

As a specific symmetric example of this model, let $n = 2$, production costs be nil, and

$$p_{1t} = 140 - 1.5q_{1t} - q_{2t} + Y_{1t} + wY_{2t} \tag{5.79}$$

$$p_{2t} = 140 - 1.5q_{2t} - q_{1t} + Y_{2t} + wY_{1t} \tag{5.80}$$

$$y_{it} = Y_{it}^2 + 1.5Y_{it}Y_{i,t-1} + Y_{i,t-1}^2, \qquad i = 1,2 \tag{5.81}$$

Single-period profits for firm 1 are

$$\pi_{1t} = 140q_{1t} - 1.5q_{1t}^2 - q_{1t}q_{2t} + q_{1t}Y_{1t} + wq_{1t}Y_{2t} - Y_{1t}^2 - 1.5Y_{1t}Y_{1,t-1} - Y_{1,t-1}^2 \tag{5.82}$$

The parameter w controls the degree to which the advertising of one firm helps or hurts the rival firm. Two values of w are used for sample computations: $w = +.8$ and $w = -.8$. The first figure causes advertising to be highly cooperative between firms, while the second causes it to be highly predatory. Letting $\alpha_i = .9$, supergame payoffs for firm 1 are

$$G_1(\sigma) = \sum_{t=0}^{\infty} .9^t[(140 - 1.5q_{1t} - q_{2t} + Y_{1t} + wY_{2t})q_{1t} - Y_{1t}^2 - 1.5Y_{1t}Y_{1,t-1} - Y_{1,t-1}^2] \tag{5.83}$$

where $Y_{1t} = 0$ for $t = -1$

The first-order conditions for equilibrium, which define in implicit form the equilibrium correspondence (see Definition 5.2), are

$$\frac{\partial G_1}{\partial q_{1t}} = .9^t[140 - 3q_{1t} - q_{2t} + Y_{1t} + wY_{2t}] = 0 \tag{5.84}$$

$$\frac{\partial G_2}{\partial q_{2t}} = .9^t[140 - 3q_{2t} - q_{1t} + Y_{2t} + wY_{1t}] = 0 \tag{5.85}$$

$$\frac{\partial G_1}{\partial Y_{1t}} = .9^t[q_{1t} - 2Y_{1t} - 1.5Y_{1,t-1} + .9(-1.5Y_{1,t+1} - 2Y_{1t})] = 0 \tag{5.86}$$

$$\frac{\partial G_2}{\partial Y_{2t}} = .9^t[q_{2t} - 2Y_{2t} - 1.5Y_{2,t-1} + .9(-1.5Y_{2,t+1} - 2Y_{2t})] = 0 \tag{5.87}$$

For $w = .8$, the equilibrium correspondence is

$$q_{1t} = 39.7015 - .121Y_{1,t-1} - .0805Y_{2,t-1} - .1089Y_{1,t+1} - 0.7243Y_{2,t+1} \tag{5.88}$$

$$q_{2t} = 39.7015 - .0805Y_{1,t-1} - .121Y_{2,t-1} - .07243Y_{1,t+1} - .1089Y_{2,t+1} \tag{5.89}$$

$$Y_{1t} = 10.4478 - .4266Y_{1,t-1} - .0212Y_{2,t-1} - .3839Y_{1,t+1} - .01906Y_{2,t+1} \tag{5.90}$$

$$Y_{2t} = 10.4478 - .0212Y_{1,t-1} - .4266Y_{2,t-1} - .01906Y_{1,t+1} - .3839Y_{2,t+1} \tag{5.91}$$

and the steady-state values are $q^* = 37.54$, $Y^* = 5.65$, $p^* = 56.31$, and $\pi^* = 2002.5$. For $w = -.8$, the equilibrium correspondence is

$$q_{1t} = 35.467 - .24276Y_{1,t-1} + .22276Y_{2,t-1} - .21848Y_{1,t+1} + .20048Y_{2,t+1} \tag{5.92}$$

$$q_{2t} = 35.467 + .22276Y_{1,t-1} - .24276Y_{2,t-1} + .20048Y_{1,t+1} - .21848Y_{2,t+1} \tag{5.93}$$

$$Y_{1t} = 9.333 - .45862Y_{1,t-1} + .05862Y_{2,t-1} - .41276Y_{1,t+1} + .052759Y_{2,t+1} \tag{5.94}$$

$$Y_{2t} = 9.333 + .05862Y_{1,t-1} - .45862Y_{2,t-1} + .052759Y_{1,t+1} - .41276Y_{2,t+1} \tag{5.95}$$

and the steady-state values are $q^* = 35.265$, $Y^* = 5.303$, $p^* = 52.9$, and $\pi^* = 1767.03$. As one would expect, where advertising is more cooperative, output, goodwill, price, and profit are all higher. To check whether the equilibrium correspondence is a contraction, add the absolute values of the coefficients of $Y_{1,t-1}$, $Y_{2,t-1}$, $Y_{1,t+1}$, and $Y_{2,t+1}$ for each equation to see if their sum is, in each case, less than 1. These sums are .38283 for equations (5.88) and (5.89), .85076 for equations (5.90) and (5.91), .88448 for equations (5.92), and (5.93), and .98726 for equations (5.94) and (5.95); therefore, the equilibrium correspondence (at least for the chosen parameter values) is a contraction, which means that the steady state is unique and stable. For more detail on this model, see Friedman (1983, Chapter 6).

8 Concluding comments

A connection between the time-dependent supergames of Section 1 and the stochastic games of Section 3 is worth noting. If the state space in the stochastic game model were generalized to be an uncountably infinite space, then the model of Section 1 would become a special case of a stochastic game. The state at time t would be defined as s_{t-1}, and the definition of a stochastic game would be satisfied. Such a stochastic game would be degenerate in the sense that there would be no stochastic elements; however, the two sections would then fit within a single framework. This approach is not taken here because Section 1 contains some results on uniqueness and stability of equilibrium that have not been developed, as far as I know, in the stochastic games literature. In addition, where one is dealing with a time-dependent supergame and there are no stochastic elements, it may be simpler to use an explicitly nonstochastic model.

The stochastic game model has a natural application to oligopoly. Developing such a model sufficiently fully takes too much space for present purposes, but the nature of the model in outline form can easily be given. (For a complete description, see Kirman and Sobel (1974) and Miller (1982).) The following description is closer to Miller. Suppose there are two firms, and, in each time period, each firm selects its price and output level. Demand for each firm is subject to random shocks from a distribution known to the firms. Thus, at the start of each period the state is described by two numbers, x_{1t} and x_{2t}. If x_{1t} is positive, it is the starting inventory of firm 1; if it is negative, it is the amount of backlogged orders. A similar interpretation is applied to x_{2t} for firm 2. Both backlogging and carrying inventories are costly; the latter represents storage costs, and the former represents penalties for being unable to satisfy some demand. The realized values of the random variables that affect demand in period t are not known to the firms until after they have chosen their prices and output levels for period t. Miller proves that the model has an equilibrium in which the firms eventually reach a region in which the firms choose the same prices in all periods and they choose output levels so that the sum of inventory and production would be the same in all periods. Inventory plus production is the stock available for sale in the period, and the equilibrium policies are constant stock in the sense that this sum is constant.

Bounded rationality has been around for at least several decades; Simon (1957) being an early contributor. The topic is difficult and appealing, and, I think it fair to say, results have not yet gotten very far. The topics relating to the chain store paradox and reputation are of recent origin and promise to develop quite a bit more in coming years. These investigations promise to resolve some situations in which the noncooperative equilibrium has appeared inappropriate, or in which a conventional application of it has been unsatisfactory.

Exercises

1. Let the single-period payoff functions of two players be

$$P_1(s_t) = -5s_{1,t-1}^2 - 5s_{1,t-1}s_{1t} - 10s_{1s}^2 + s_{2,t-1}s_{1t} + 5s_{1t}$$

$$P_2(s_t) = 10s_{2t} - 5s_{2,t-1}^2 - 4s_{2t}^2 - 6s_{1,t-1}s_{2t} - 2s_{1t}s_{2t}$$

   Let the single-period strategy spaces be [0, 1] for both players and suppose that the discount parameter is .8 for both players. What is the equilibrium correspondence? What is the steady-state value of s?

2. Suppose that two players will play the following prisoners' dilemma game a finite known number of times. There is no discounting (i.e., $\alpha_i = 1$).

		Player 2	
		C	NC
Player 1	C	0, 0	3, −1
	NC	−1, 3	2, 2

   a. Suppose this game will be played exactly twice and player 1 believes that player 2 will adhere to a strategy under which she plays NC with probability p in the first iteration; NC with probability p in the second iteration if (NC, NC) was observed in the first iteration; and C otherwise. What is the best reply for player 1 to this strategy as a function of p? Suppose that player 2 has precisely the same beliefs about player 1 that player 1 has about player 2. What will they actually play in each period, what will be their realized payoffs, and what will be their expected payoffs?
   b. Now suppose that this same game will be iterated T times and that the beliefs held by one player about the other are these. The player is believed to select NC with probability p in the first iteration; select NC with probability p in any later iteration when (NC, NC) has been observed in *all* previous iterations; and select C otherwise. Answer the questions posed in part (a).
   c. For $T = 100$, compare expected and realized payoffs for values of p close to 1, such as $p = .9$, .99, and .999.

Notes

1. Existence of a steady-state equilibrium, ensured by existence of a fixed point of the correspondence $\psi(s', s)$ when s' and s'' are restricted to equal one another, is implied by a weaker condition than Assumption 5.3. It is enough that ψ satisfy the conditions for the Kakutani fixed-point theorem. The only condition not implied already is that the sets $\psi(s, s)$ be convex for all $s \in S$. Note that (s', s'') need not be convex if $s' \neq s''$.
2. The condition that a contraction map a set into itself can be weakened. If a contraction $f(x): A \to B(A, B \subset R^n)$ satisfies a Lipschitz condition with ratio k, and if $\|x^0 - f(x^0)\| \leq \beta$, then any fixed point x^* of f would have to be within a

distance $\beta/(1-k)$ of x^0. Therefore, if B, the range of f, contains the set $\{x \in R^m \mid \|x - x''\| \leq \beta/(1-k)\}$, then f has a fixed point. Note that this result does not require compactness of A or of B. See Dieudonné (1960; page 261).

3. Blackwell (1965) treats dynamic programming, not games; however, if all players except player i are using policies, finding a best-reply for player i amounts to solving a dynamic programming problem of the sort Blackwell treats. The lemma is stated here in terms of the stochastic game context of this section.
4. Throughout this section, θ_i is referred to as either a policy or a strategy. θ_i is not literally, a strategy; however, the plan to use the policy θ_i in each and every period is a strategy, and this strategy is meant when θ_i is called a strategy. This slight abuse of terminology eliminates some cumbersome wording and should cause no confusion. No notational distinction is made in writing column and row vectors. The use of a prime or superscript T would add to an already heavy load of notation, and what is meant should be obvious from the context.
5. It is no restriction to require that $S_{it} \subset R^m$ for all i and k. Suppose $S_{ik} \subset R^{m_{ik}}$ that m_{ik} is finite for all i and k. Now let $m = \max_{i,k} \{m_{ik}\}$ and, clearly, $S_{ik} \subset R^m$.

6

Two-person cooperative games

This chapter, the first of three devoted to cooperative games, covers two-person cooperative games. The study of cooperative games can be simplified either by limiting attention to two persons, which eliminates any complications that arise when a myriad of coalitions is allowed, or by assuming *transferable utility*. Transferable utility is explained fully in Chapter 7, which is devoted entirely to n-person transferable utility games. In brief, the restriction allows a much simpler statement of the relevant aspects of a game. Chapter 8 takes up n-person nontransferable utility games. This chapter begins with a short overview of cooperative games.

1 Introduction to cooperative games

1.1 Comparison of cooperative and noncooperative games

The fundamental distinction between cooperative and noncooperative games is that cooperative games allow binding agreements while noncooperative games do not. Historically, however, the two branches of game theory have developed along different lines based upon a distinction in the technique of modeling. Noncooperative games have approached modeling from a strategic standpoint, as the preceding four chapters bear ample witness. The focus is on the decisions made by the players and on the way that their "rational" decision making affects the outcome of the game. Until less than a decade ago cooperative games avoided dealing with individual decisions as such and tended to focus on the set of possible outcomes and the conditions that might be "reasonable" or "appealing" or "fair" to impose on an acceptable outcome. This approach is sometimes referred to as *axiomatic bargaining theory*.

In principle there is no reason that one cannot analyze cooperative games using the tools and the approach of noncooperative game theory. Doing so requires that the possibilities for binding agreement be built into the formulation of the game; that is, that the payoff functions and strategy spaces reflect these possibilities. Such an approach is taken in the original

Nash (1950, 1953) articles; however, there Nash analyzes two-person cooperative games in a variety of ways, including via axiomatic bargaining. It was his axiomatic bargaining approach that caught on and really held sway until the early 1980s. Of course, Shapley (1953b), Harsanyi (1959), and others also use the axiomatic bargaining approach. With Rubinstein (1982) the noncooperative bargaining approach was revived and it has received considerable attention since then. In several provocative and enlightening chapters in Binmore and Dasgupta (1987), Binmore reconsiders the work of Nash.

In the remainder of this section, noncooperative games are contrasted with cooperative games analyzed from the axiomatic point of view. Points of comparison include (a) the fairness of outcomes, (b) the naturalness of outcomes, (c) the scope for players to make active choices, and (d) the levels on which players can interact.

The Nash noncooperative equilibrium is not generally heralded as a fair outcome; however, various cooperative game solutions are sometimes put forth as being fair outcomes on the assertion that the axioms characterizing them embody fairness. These assertions do not have universal appeal and often merely reflect the enthusiasm of their inventors and supporters. The lack of consensus about which solution is most desirable stimulates partisans of one or another solution to make claims for their favorites. Explicit concern with fair division schemes is outside the scope of this book; however, the interested reader might look at Crawford (1979, 1980), where additional references can be found.

By the "naturalness" of an outcome, I refer to whether it seems reasonable to expect a particular solution or equilibrium to be realized in practice. The noncooperative equilibrium seems natural in the sense that it is an outcome under which each player is, in an appropriate sense, doing the best he can. Put another way, it is intuitively plausible that players would behave in accord with such an equilibrium. Regarding cooperative games, it is easy to visualize players agreeing on a *core* outcome, because core outcomes satisfy a set of conditions guaranteeing that no individual or group could possibly fare better. The core, however, can be very large, or it can be empty. A core consisting of one point is unusual.

For other cooperative outcomes, it is not clear how they might naturally occur unless a pregame arrangement is made to select one of them. A pregame agreement that is made for only a single game would be difficult to imagine. Because different solutions generally treat players differently, each player would favor the solution that maximized his own payoff. The situation is different for a group of people who know that, from time to time, subsets of them will be in games, and no one knows in advance which games will arise or who the players will be. Agreeing in advance to use a particular cooperative solution is like establishing a clause in the constitution, or basic legal framework, of the group. If no one knows how and when he will enter a game, then each player can discuss the merits of various solutions from the standpoint of his overall sense of fairness. If an agreement is reached on a solution concept, the individuals reap two advantages over playing an unknown sequence of isolated single-period noncooperative games. First, they can assure in advance that each game will have a Pareto optimal outcome, which makes it likely that each player will, over time, have a larger total payoff than he would without the agreement. Second, a player's payoffs would be regarded as being ex ante fair. When a cooperative game solution is used by a prearrangement that lies outside the game proper, the solution is called an *arbitration scheme.*

The scope for players to make active choices is always great in noncooperative games. Were there not such scope, a noncooperative game would hold no interest. In cooperative games the situation varies. For example, if the Nash bargaining model is adopted, individual action is precluded. That is, if two players are in a fixed-threat game in which they decide to accept the Nash bargaining solution, then there is no further action for either of them to take. Of course, this modoel can apply only to special situations (see Section 2); however, the Shapley value, which can be used in a large class of games, has the same feature. In contrast, the Nash variable-threat solution and the bargaining set, among others, leave scope for individual actions aimed at improving a player's final outcome.

By the levels on which players interact, I refer to the scope for various subgroups to discuss possible joint actions. In Chapters 4 and 5, it was shown that all players could jointly make self-enforcing agreements. The possibility of two or more, but fewer than n, players making a self-enforcing agreement has not received much attention, probably because many interesting results are obtainable without bringing in groups of intermediate size; however, there is no reason in principle why such possibilities could not be explored. With cooperative games, it is often of central importance that groups of any size be allowed to form. Again, this is not a difference of principle; it is a difference in emphasis.

1.2 The various kinds of cooperative games

It was noted at the beginning of the chapter that Chapters 6, 7, and 8 are divided along lines that are pedagogically convenient. Relative to two-person cooperative games n-person cooperative games provide much more scope for coalition formation that is allowed when there are 2 players. A *coalition* is simply a subset of N that is allowed to make a binding agreement, and it is usually assumed that any subset of N can form for this purpose; therefore, there are 2^n possible coalitions if the *coalition of the whole* (i.e., the coalition consisting of N itself) and the *empty coalition* (consisting of no one) are allowed, and there are $2^n - 2$ coalitions consisting of between 1 and $n - 1$ players. The number of possible coalitions clearly grows much faster than n.

Intuitively, restricting n to 2, would seem to rid the analysis of some complications, and it is true that certain solution concepts for cooperative games have been developed specifically for $n = 2$. Some of these are examined in this chapter. These solution concepts apply to models with both transferable and nontransferable utility. *Transferable utility* means that each coalition can achieve a certain total amount of utility that it can freely divide among its members in any mutually agreeable fashion. The concept is discussed further in Chapters 7 and 8. To grasp what it involves assume that for all individuals utility is proportional to money, and one unit of money equals one unit of utility. Then, transferable utility exists in a game where all outcomes are measured in money and any amount of money that is achievable by a coalition can be divided arbitrarily among the coalition members. Transferable utility is a very restrictive assumption that forces a strong relationship onto the utility functions of the players in a game.

Certain solution concepts have been developed for n-person transferable utility games. Although some of them can be generalized to nontransferable utility games, there are losses, such as the uniqueness of a solution, in the process.

Another distinction in cooperative games is that between core-type and value-type solutions. This distinction is not always that clear cut; however, it is still useful. The core is a set of payoff vectors that passes a minimal criterion of acceptability: The payoff vector $x \in R^n$ is in the core if it is achievable by the members of N and if no coalition could, on its own, obtain more for all of its members than what it gets under x. The *von Neumann–Morgenstern solution* (also called the *stable set*) is similar to the core in singling out an acceptable set of outcomes, and in rarely achieving a unique outcome. In contrast, the Nash cooperative solution, and the relatives of it that are the subject of the present chapter, and the Shapley and Banzhaf values, and the generalizations of the Shapley values, are all value solutions. For many games, these solution concepts give unique outcomes. It is for this reason that they may be termed *value* solutions. Recall that von Neumann proved that any finite two-person zero-sum game has a (unique) value. Some of the motivation behind the value solutions appears to have been the desire to find a unique value for each game in a given class.

1.3 Outline of the chapter

Two-person cooperative games are covered from several points of view. The (fixed-threat) bargaining model of Nash (1950) is discussed in Section 2, following a brief discussion of Edgeworth's early (1881) contribution. This model deals with two-person situations in which the players each obtain fixed utility levels if they fail to make an agreement. There is a feasible set of outcomes that they can achieve if they make an agreement; however, in the absence of agreement, there is nothing a player can do to help or hurt either himself or the other player. Section 3 is concerned with precisely the same basic game situation, but the rules defining a solution concept, that is, an appropriate, fair, or desirable agreement, are different from those proposed by Nash. The models of Sections 2 and 3 do not really examine the bargaining process as such, nor is the element of time considered. These topics are examined in Section 4, where an explicit bargaining process is outlined that leads to Nash's outcome. A model that places an explicit value on time is also explored. A good source on the bargaining models in the present chapter is Roth (1979). The means of doing this is to suppose that each step of bargaining, offer or counteroffer, takes a fixed length of time and that the value of any particular outcome shrinks with the passing of time. Another model is examined in which players make demands of one another at the first stage of the game, and, in the second, they discover whether their demands are simultaneously feasible. Section 5 takes up variable-threat models. In these models the players have strategy sets from which to choose actions in the absence of agreement. Thus, the disagreement outcomes are not fixed; they depend on the chosen actions. It is natural to approach this cooperative game by styling it as a noncooperative game in which the disagreement actions determine disagreement payoffs and the disagreement payoffs determine the cooperative agreement. Then a player's choice of a disagreement action is guided by the effect of that action on a final cooperative outcome. Section 6 contains some applications and concluding comments appear in Section 7.

2 Fixed threat bargaining: the models of Edgeworth and Nash

Imagine two people isolated on a desert island. Suppose that each has a fixed supply of various commodities, and each has a utility function assigning a personal value to any conceivable commodity bundle. The two players' utility functions need not be the same, nor need their initial holdings be the same. To think of this situation as a game, define a player's payoff as her utility. Suppose the two players may engage in trade with one another, exchanging anything from their respective holdings that both agree on. The possibilities for them are depicted in two different ways in Figures 6.1 and 6.2. In Figure 6.1, the axes are labeled with the two players' utility levels, or payoffs. The point marked d is the outcome that obtains in the absence of trade and the set H is the set of feasible trades (in payoff space), assuming that the two players trade by reallocating their joint total holdings. That is, the possibility of discarding some holdings is not considered, but the possibility that one player gives resources to the other, receiving nothing in return, is allowed. The point d is sometimes called the *status quo* or the *no trade* or *threat point.*[1]

2.1 Edgeworth two-person bargaining and the core

This trading situation was first studied by Edgeworth (1881). The dimensionality of the *Edgeworth box,* shown in Figure 6.2, is equal to the number of commodities existing in the two-person society; Figure 6.1 is applicable for any number of commodities. Figure 6.2, then, shows two

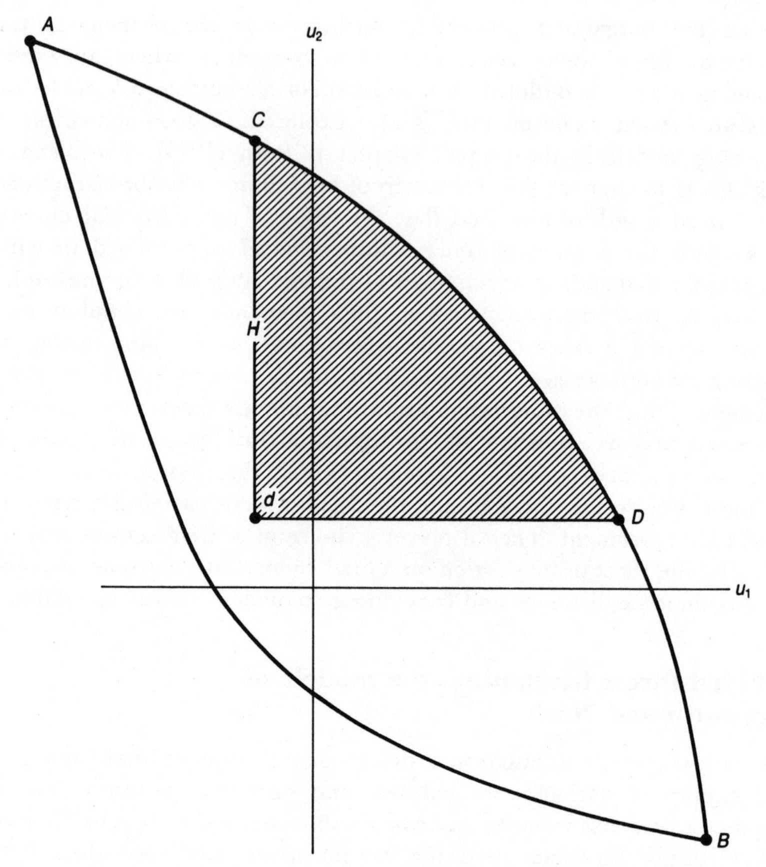

FIGURE 6.1 The attainable payoffs in a fixed-threat game.

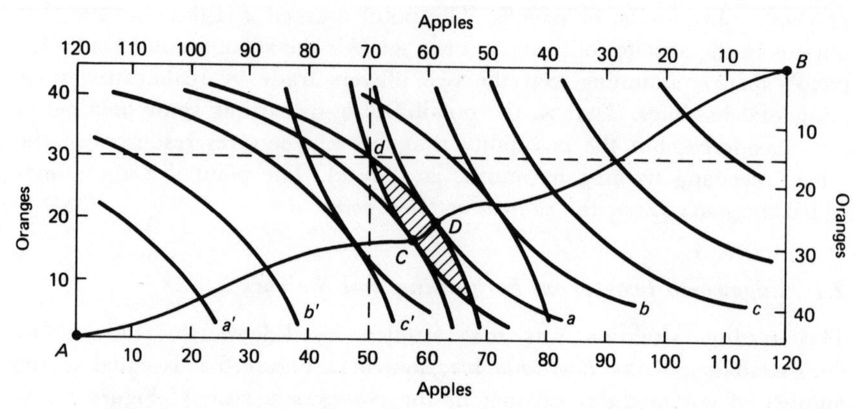

FIGURE 6.2 The Edgeworth box representation of a fixed-threat game.

goods, apples and oranges, with units of apples measured on the horizontal axis and oranges on the vertical. Curves such as a, b, and c are *indifference curves* for player 1, whose origin is at A, and for whom larger amounts are upward and to the right. An indifference curve for a player is a curve on which each point corresponds to exactly the same payoff for that player as any other point on the curve. The curves labeled a', b', and c' are indifference curves for player 2, whose origin is at B and for whom quantities and utility increase as one moves down and to the left in the diagram. The overall size of the box is determined by the total holdings of the two players. Thus, the point d, which shows their initial holdings, indicates that player 1 has 50 apples and 30 oranges and player 2 has 70 apples and 15 oranges.

Edgeworth's analysis of this trading situation is an important forerunner of cooperative game theory. He notes two limitations on any reasonable agreement: (1) Any agreement should be efficient in the sense that no alternative agreement is feasible that yields higher payoffs simultaneously to both players. Such agreements are the points on the curve going from A to B in both Figures 6.1 and 6.2. (2) No player can be expected to agree to a trade that leaves him with a lower payoff than he would get if no trade were made. Such agreements are the shaded regions of Figures 6.1 and 6.2. The curve from C to D in Figure 6.1 satisfies both criteria and is known in game theory as the *core*. The curve from C to D in Figure 6.2 is the set of allocations of goods (in commodity space) that yields payoffs in the core.

Of course, in a game of three or more players, a third limitation would have to be added in defining the core that stipulates that a core outcome would have to give each coalition as much as the coalition could obtain on its own. This issue is taken up in Chapter 7.

The curve from A to B represents the set of trades yielding outcomes on the payoff possibility frontier. It is often called the *Pareto optimal set*, for the obvious reason that each point on the curve is Pareto optimal for the players in the game. The requirement of Pareto optimality is also known in cooperative game theory as *group rationality*, because it seems intuitively unreasonable (irrational) that the set of all players would ever settle for a non-Pareto optimal outcome. Condition (2) is often called *individual rationality*, because it would be irrational for a single player to agree to a joint outcome under which he receives a lower payoff than he could assure himself by his own individual efforts.

2.2 The Nash fixed-threat bargaining model

Frequently, the core consists of many points, as in the illustration in Figures 6.1 and 6.2. The Nash solution, in contrast, always gives a unique outcome. The Nash solution satisfies several intuitively appealing conditions. In bargaining models, $H \subset R^2$ denotes the *set of* attainable *payoff pairs* and $d \in H$ denotes the *threat point*. If a payoff point $u = (u_1, u_2) \notin H$, then it is impossible for the players to achieve it, but if $u \in H$, then there are (joint) actions open to them that will result in u being the payoff. The salient feature of the threat point, d, is that the players will receive the payoffs $d = (d_1, d_2)$ if they fail to achieve an agreement. In Sections 2 and 3, the models are all characterized by such fixed-threat points. This means that there is no way that one player can take unilateral action that hurts the other. For example, if the game is an Edgeworth game of pure trade in which each player has an endowment of goods and the utility (payoff) of each player depends only on his own personal consumption, then a player cannot be forced to a payoff below the level he achieves by consuming the endowment of goods with which he begins.

2.2.1 The game and conditions defining the Nash bargaining solution

The class of games studied in Sections 2 and 3 is characterized by a pair (H, d) and by the rule that the players will attain any single payoff point in H upon which they jointly agree. In the absence of an agreement, they attain d. Definitions 6.1 and 6.2 formally describe these games, Definition 6.3 defines the meaning of *solution*, and Conditions 6.1 to 6.4 define the Nash solution to the class of games specified in Definitions 6.1 and 6.2.

DEFINITION 6.1 *The pair* $\Gamma = (H, d)$ *is a* **two-person fixed-threat bargaining game** *if* $H \subset R^2$ *is compact and convex,* $d \in H$*, and* H *contains at least one element,* u*, such that* $u \gg d$.

DEFINITION 6.2 *The* **set of two-person fixed-threat bargaining games** *is denoted* W.

Recall that the von Neumann–Morgenstern utility axioms are assumed throughout this book. Consequently, there could be a bargaining game in which there was a nonconvex set (including a finite set) of unrandomized outcomes. Then the set H would consist of the utility pairs associated with all possible lotteries defined on elements from the original set of outcomes. Such a set H would be convex. It would also be compact if the set of original outcomes were compact.

A solution to some game $\Gamma = (H, d)$ is a particular element of H, which is the payoff pertaining to the solution concept under discussion, for example, the Nash solution. Because (H, d) can be any game drawn from a large set of games, a particular solution concept can be conveniently described as a function of the game, $f(H, d) \in H$.

DEFINITION 6.3 *A* **solution to** $(H, d) \in W$ *is a function* $f(H, d)$ *that associates a unique element of* H *with the game* $(H, d) \in W$. $f(H, d) = (f_1(H, d), f_2(H, d))$.

The conditions defining the Nash solution are as follows:

CONDITION 6.1 $f(H, d) \gg d$ *for all* $(H, d) \in W$.

CONDITION 6.2 *Let* $a_1, a_2 \in R_{++}$, $b_1, b_2 \in R$, *and* $(H,d), (H',d') \in W$ *and define* $d'_i = a_i d_i + b_i$, $i = 1, 2$, *and* $H' = \{x \in R^2 \mid x_i = a_i y_i + b_i, i = 1, 2, y \in H\}$. *Then* $f_i(H', d') = a_i f_i(H, d) + b_i$, $i = 1,2$.

CONDITION 6.3 *If* $(H,d) \in W$ *satisfies* $d_1 = d_2$ *and* $(x_1, x_2) \in H$ *implies* $(x_2, x_1) \in H$, *then* $f_1(H,d) = f_2(H,d)$.

CONDITION 6.4 *If* $(H,d), (H',d') \in W$, $d = d'$, $H \subset H'$, *and* $f(H',d') \in H$, *then* $f(H,d) = f(H',d')$.

Condition 6.1 stipulates that the solution payoff to each player should be at least as large as the payoff the player would get if no agreement were reached. That is, the solution must be *individually rational.*

Condition 6.2 requires that the solution should be invariant to positive affine utility transformations. A *positive affine transformation* of utility is a transformation of the form $x_i = a_i y_i + b_i$, defined for all y_i, that moves the zero point of utility in an arbitrary way (b_i can be positive or negative), and changes the scale of units on which utility is measured (a_i must be strictly positive). The essential feature of such transformations is that if $y_i^0 - y_i^1 > y_i^2 - y_i^3$, then $(a_i y_i^0 + b_i) - (a_i y_i^1 + b_i) > (a_i y_i^2 + b_i) - (a_i y_i^3 + b_i)$. Two games related in the manner of (H,d) and (H',d') in Condition 6.2 are like two games that are identical in terms of physical possibilities but differ only in an arbitrary aspect of their utility representation. For such a pair of games, their solutions should be related by precisely the same utility transformation that relates the games themselves. If the two games embody the same set of physical outcomes and the two threat points correspond to the same physical outcome, then the same physical outcome is the solution to both. Condition 6.2 also states that the outcome does not depend on which particular von Neumann—Morgenstern utility function is used for a player (among those that, in fact, represent her preferences).

Condition 6.3 imposes symmetry: If the attainable set H is symmetric about a 45° line through the origin and $d_1 = d_2$, then $f_1(H,d) = f_2(H,d)$. This does not imply comparability of the two utility scales. Given symmetry, if $f_1(H,d) \neq f_2(H,d)$, it would appear that one player was being favored over the other.

Condition 6.4 is called *independence of irrelevant alternatives.* It compares two games (H,d) and (H',d') that are related in three ways: (1) they have the same threat point ($d = d'$), (2) the attainable set of one game is contained in the attainable set of the other ($H \subset H'$), and (3) the solution of the larger game is an attainable point in the smaller game ($f(H',d') \in H$). Thus the game (H,d) can be regarded as (H',d') with some of the attainable set of (H',d') removed, but with $f(H',d')$ remaining available. Condition 6.4 requires that $f(H',d')$ be the solution to (H,d). The heuristic justification is that, if $f(H',d')$ is to be preferred as the outcome for (H',d'), then it should be preferable to any alternative in $(H,d') = (H,d)$ because $H \subset H'$.

Nash also assumed that $f(H,d)$ is Pareto optimal. That is, if $u \in H$, then

$u \not> f(H, d)$; however, Pareto optimality is implied by Conditions 6.1 to 6.4, as the following lemma, due to Roth (1977), shows:

LEMMA 6.1 *Let* $(H, d) \in W$ *and let* $u^* = f(H, d)$ *satisfy Conditions* 6.1 *to* 6.4. *Then if* $u \in H$ *and* $u \neq u^*$ *either* $u_1^* > u_1$ *or* $u_2^* > u_2$.

Proof. Without loss of generality, we may let $d = 0$.[3] Assume the lemma false; then there is some $y^* \in H$ such that $y^* > u^*$. Define a game $(H', 0)$ by letting

$$H' = \left\{ x \in R^2 \;\middle|\; x_1 = \frac{u_1^*}{y_1^*} y_1, \quad x_2 = \frac{u_2^*}{y_2^*} y_2, \quad y \in H \right\} \tag{6.1}$$

Clearly, $H' \subset H$, $H' \neq H$, and $u^* \in H'$, because $y^* \in H$. By Condition 6.4, u^* is the solution to $(H', 0)$, but by Condition 6.2, $[(u_1^*/y_1^*)u_1^*, (u_2^*/y_2^*)u_2^*] \neq u^*$ is the solution to $(H', 0)$. This contradiction implies that the solution of $(H, 0)$ must be Pareto optimal. QED

2.2.2 *Characterization and existence of the Nash solution*

The Nash solution can be characterized in a very simple way: The element of H that maximizes the product of gains from agreement, $(u_1 - d_1)(u_2 - d_2)$, is the unique outcome satisfying Conditions 6.1 to 6.4. This is proved in Theorem 6.1 and the Nash solution is illustrated in Figure 6.3. The

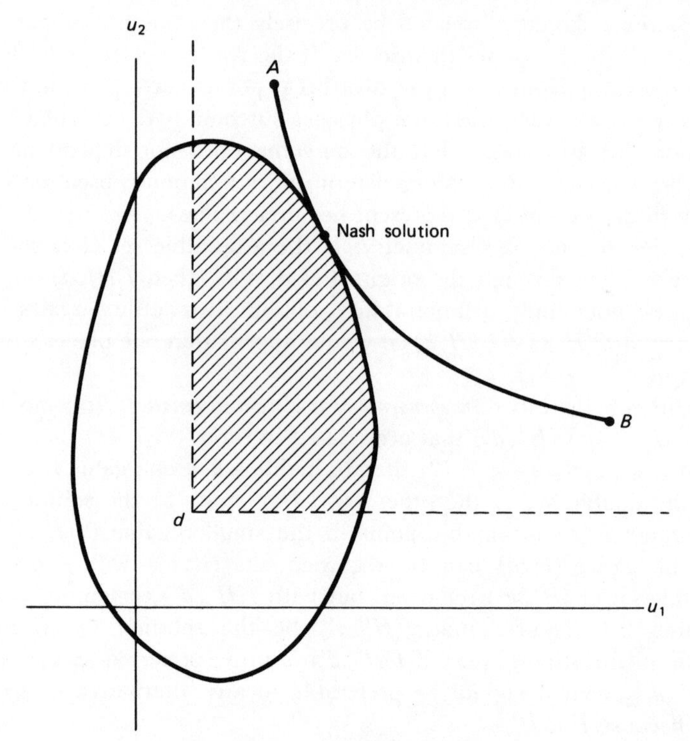

FIGURE 6.3 The Nash solution

shaded area of H is the region of outcomes that are individually rational. A rectangular hyperbola, as shown in Figure 6.3, that is asymptotic to the broken lines through d is a curve along which the product $(u_1 - d_1)(u_2 - d_2)$ is constant; therefore, the point of H at which the product of gains from agreement is maximized is the point of tangency between the upper boundary of H and the most highly placed rectangular hyperbola asymptotic to axes through d that touches H. The result is proved in two steps. First, Lemma 6.2 establishes that $(u_1 - d_1)(u_2 - d_2)$, called the *Nash product*, has a unique maximum, u^*, on the part of H lying above d; second, Theorem 6.1 shows that u^* is the unique point in H satisfying Conditions 6.1 to 6.4.

LEMMA 6.2 *$(u_1^* - d_1)(u_2^* - d_2) = \max_{u \in H, u \gg d} (u_1 - d_1)(u_2 - d_2)$ is unique.*

Proof. Uniqueness of the Nash product maximizer can be seen by supposing the utility product is maximized at both (u_1, u_2) and $(u_1 + \delta, u_2 - \gamma)$, where γ and δ have the same sign. then

$$u_1 u_2 = (u_1 + \delta)(u_2 - \gamma) \tag{6.2}$$

and $u_2 = \gamma + \gamma u_1/\delta$. Convexity of H implies that $(u_1 + \delta/2, u_2 - \gamma/2) \in H$ and calculation shows its utility product to

$$\left(u_1 - d_1 + \frac{\delta}{2}\right)\left(u_2 - d_2 + \frac{\gamma}{2}\right)$$

$$(u_1 - d_1)(u_2 - d_2) + \frac{\delta(u_2 - d_2)}{2} - \frac{\gamma(u_1 - d_1)}{2} - \frac{\delta\gamma}{4}$$

$$= (u_1 - d_1)(u_2 - d_2) + \frac{\delta\gamma}{4} > (u_1 - d_1)(u_2 - d_2) \tag{6.3}$$

which is a contradiction. thus the Nash product maximizer is unique. QED

THEOREM 6.1 *A game $(H, d) \in W$ has a unique Nash solution $u^* = f(H, d) \in H$ satisfying Conditions* 6.1 *to* 6.4. *The solution u^* satisfies Conditions* 6.1 *to* 6.4 *if and only if*

$$(u_1^* - d_1)(u_2^* - d_2) > (u_1 - d_1)(u_2 - d_2) \tag{6.4}$$

for all $u \in H$, $u \gg d$, and $u \neq u^*$.

Proof. It first proved that $(u_1^* - d_1)(u_2^* - d_2) = \max_{u \in H, u \gg d} (u_1 - d_1)(u_2 - d_2)$ satisfies Conditions 6.1 to 6.4. Then it is shown that if $u' \in H$ and $(u_1' - d_1)(u_2' - d_2) < (u_1^* - d_1)(u_2^* - d_2)$, then u' violates one of the four conditions. Condition 6.1 is satisfied by confining product maximization to the region in H in which $u \gg d$. For Condition 6.2, let (H', d') be defined by $d_i' = a_i d_i + b_i$, $i = 1, 2$ and $H' = \{y \in R^2 \mid y_i = a_i u_i + b_i,\ i = 1, 2,\ u \in H\}$. Let w be an arbitrary element of H', let u be the point in H that transforms into w, and let $w^* \in H'$ be the point to which u^*, the Nash product maximizer for (H, d), transforms. Comparing w^* and w,

$$(w_1^* - d_1')(w_2^* - d_2') - (w_1 - d_1')(w_2 - d_2') = a_1 a_2(u_1^* u_2^* - u_1 u_2) > 0 \quad (6.5)$$

To see that Condition 6.3 is satisfied, suppose H is symmetric about a 45° line through d and let u' be the point in H that is on the 45° line at the point at which it intersects the upper right boundary of H. Then $u_1' - d_1 = u_2' - d_2$ and, for any $u \in H$, $u_1 + u_2 \leq u_1' + u_2'$; that is, the set H is bounded above by a line of slope -1 through u'. For any $a \neq 0$, $(x-a)(x+a) < x^2$; therefore, for any $u \in H$, $u \neq u'$, we have

$$(u_1 - d_1)(u_2 - d_2) < (u_1' - d_1)(u_2' - d_2) \quad (6.6)$$

and $u^* = u'$.

That Condition 6.4 is satisfied is trivial, for if u^* is the product maximizer in H, and H' is derived from H by paring away some regions of H, then $(u_1^* - d_1)(u_2^* - d_2)$ must exceed the corresponding product for any point in H'.

A constructive argument is used to prove that if u' is not the product maximizer, it violates at least one of the four conditions. As above, let u^* be the product maximizer for the game (H, d). Transform (H, d) into $(H^1, 0)$ by letting $H^1 = \{v \in R^2 \mid v_1 = u_1 - d_1,\ v_2 = u_2 - d_2, \text{ and } u \in H\}$. This transforms u^* into $v^* = (u_1^* - d_1, u_2^* - d_2)$. Now transform $(H^1, 0)$ into $(H^2, 0)$ by letting $H^2 = \{w \in R^2 \mid w_1 = v_2^* v_1 / v_1^*,\ w_2 = v_2, \text{ and } v \in H^1\}$. This transforms v^* into $w^* = (v_2^*, v_2^*) = (u_2^* - d_2, u_2^* - d_2)$. Next transform $(H^2, 0)$ into $(H^3, 0)$ by adding points to H^2. Let β the largest value attained in H^2 by $|w_1| + |w_2|$.

$$H^3 = \{y \in R^2 \mid y_1 + y_2 \leq 2w_2^* \text{ and } y_1 + y_2 \leq \beta\} \quad (6.7)$$

The upper right boundary of H^3 is a line of slope -1 through $y^* = v^* = (u_2^* - d_2, u_2^* - d_2)$ and H^3 contains H^2; therefore, by Condition 6.3 (symmetry) v^* is the solution of $(H^3, 0)$; by Condition 6.4 (independence of irrelevant alternatives) v^* is the solution of $(H^2, 0)$; and by Condition 6.2 (invariance to affine transformations) v^* is the solution to $(H^1, 0)$ and u^* is the solution to (H, d).

The preceding construction rules out the possibility that some $u' \neq u^*$ is a solution to (H, d) in the following way. In the transformations above, u' is transformed into v', w', and y' in the three games, respectively. Condition 6.2 dictates that if u' is the solution to (H, d), then v' is the solution of $(H^1, 0)$ and w' is the solution of $(H^2, 0)$. However, we know by Condition 6.3 that $y^* \neq y'$ is the solution of $(H^3, 0)$; therefore, y' as the solution to $(H^3, 0)$ violates Condition 6.3. QED

2.2.3 *Zeuthen's anticipation of the Nash solution*

An interesting precursor to the Nash bargaining solution that also attempts to deal with the bargaining process itself was proposed by Zeuthen (1930) as a method of resolving labor–management disputes. Harsanyi (1956)

discovered that it yields an outcome formally identical to Nash's solution. Zeuthen envisaged a bargaining process in which each player, at any point in time, had a proposal on the table. Each proposal could be stated as a point on the payoff possibility frontier. Depending on how the two proposals were related, it would be incumbent upon one of the two players to replace his proposal with a new one that made a concession. Letting the proposal of player 1 be $u^1 = (u_1^1, u_2^1)$, that of player 2 be u^2, and $d = 0$, the values $(u_1^1 - u_1^2)/u_1^1$ and $(u_2^2 - u_2^1)/u_2^2$ are compared. If the first is smaller than the second, then Zeuthen directed that player 1 should make a concession large enough to reverse the inequality. Zeuthen reasoned that $(u_i^i - u_i^j)/u_i^i$ measures the cost to player i of making a concession to player j, and the player facing the smaller cost should make a concession. It is easily seen that $(u_1^1 - u_1^2)/u_1^1 < (u_2^2 - u_2^1)/u_2^2$ is equivalent to $u_1^1 u_2^1 < u_1^2 u_2^2$; therefore, the player whose proposal has the smaller Nash product makes a concession of sufficient size that his Nash product becomes larger than that of the other player. Clearly this process ends at the Nash solution. Although the proposed bargaining process seems ad hoc, it is interesting to see this totally independent route to Nash's outcome.

3 Other approaches to fixed-threat bargaining

Several alternatives to the Nash approach have been put forth, some of which are reviewed in this section. The first of these takes up a suggestion of Raiffa's described briefly in Luce and Raiffa (1957, page 136). Raiffa's model has been axiomatized by Kalai and Smorodinsky (1975). The basic intuitive notion is that each player naturally aspires to receive the largest payoff available in the game that is consistent with individual rationality. These two individual maximum payoffs are, in general, not attainable simultaneously and the proposed solution is to settle at the largest attainable payoff point that is proportional to them. The Raiffa–Kalai–Smorodinsky model is examined in Section 3.1. The other solutions reviewed in Section 3.2 are members of a family to which the Nash solution also belongs. In this family, Condition 6.4, independence of irrelevant alternatives, is replaced with a parallel condition in which the place of the threat point d is taken by another point that lies below the payoff possibility frontier.

3.1 The Raiffa–Kalai–Smorodinsky solution

For the Nash model the threat point, d, is a *reference point* with respect to which the solution is found. All the models in Section 2 and 3 use reference points, but the Raiffa–Kalai–Smorodinsky solution is unique among them because it is the only solution for which there are two reference points. One is d and the other is above and to the right of the Pareto optimal curve.

3.1.1 A description of the Raiffa–Kalai–Smorodinsky solution

As in the discussion of the Nash solution, a game is characterized by (H, d) as specified in Definitions 6.1 and 6.2. The solution concept is easily described with the aid of Figure 6.4. Locate point A, where player 1 has a payoff M_1, the highest payoff among all points in H that are individually rational. This payoff of M_1 is regarded as an aspiration level for player 1, at least in the sense that the larger is M_1, the more the player thinks he should obtain at a final settlement. Point B and M_2 are analogously defined for player 2. The point $M = (M_1, M_2)$, which generally lies beyond the attainable set H, is called the *ideal point* and is a reference point. A straight line is then drawn from d, the threat point, to (M_1, M_2) and the solution occurs at μ, where the straight line intersects the payoff possibility frontier. This simple characterization yields the only outcome that satisfies a set of conditions including Conditions 6.2 and 6.3, Pareto optimality, and monotonicity. These latter two conditions are stated formally, following

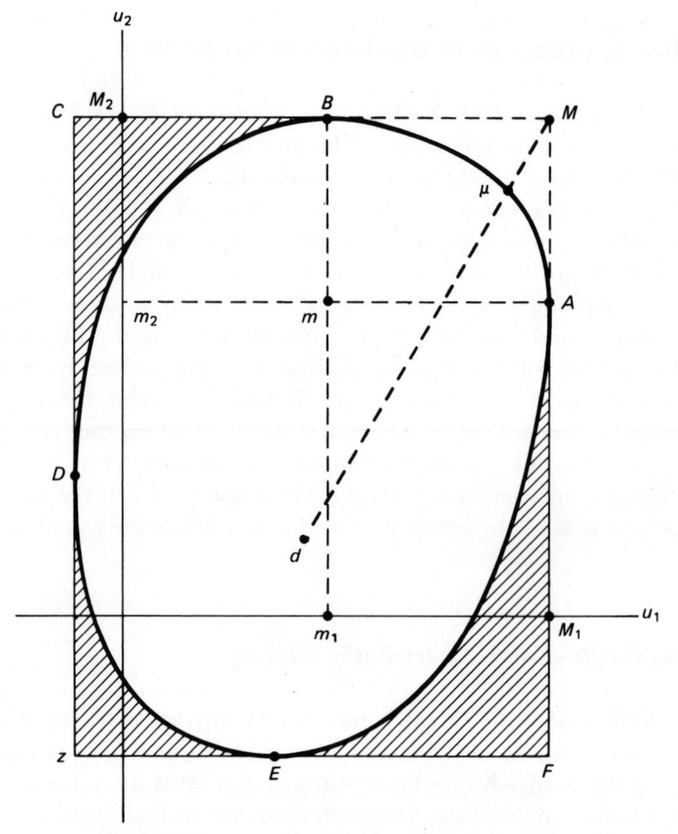

FIGURE 6.4 The Raiffa–Kalai–Smorodinsky solution.

some definitions.

$$M_1(H, d) = \max\{x_1 \in R \mid (x_1, x_2) \in H \text{ and } x_2 \geqslant d_2\} \tag{6.8}$$

$$m_2(H, d) = \max\{x_2 \in R \mid (M_1(H, d), x_2) \in H\} \tag{6.9}$$

$$M_2(H, d) = \max\{x_2 \in R \mid (x_1, x_2) \in H \text{ and } x_1 \geqslant d_1\} \tag{6.10}$$

$$m_1(H, d) = \max\{x_1 \in R \mid (x_1, M_2(H, d)) \in H\} \tag{6.11}$$

$M_1(H, d)$ is the largest payoff in H for player 1 among all outcomes at which player 2 obtains at least d_2. $m_2(H, d)$ is the payoff player 2 receives when player 1 gets $M_1(H, d)$. If the payoff for player 2 corresponding to $M_1(H, d)$ is not unique, then $m_2(H, d)$ is the largest of the corresponding payoffs. $M_2(H, d)$ and $m_1(H, d)$ are defined analogously.

DEFINITION 6.4 *The point* $M(H, d) = (M_1(H, d),\ M_2(H, d))$ *is called the* **ideal point.**

DEFINITION 6.5 *The point* $m(H, d) = (m_1(H, d),\ m_2(H, d))$ *is called the* **point of minimal expectations.**

Note that, in general, the ideal point is not an element of H. Indeed, if it were, there would be nothing to bargain over (i.e., no conflict in the interests of the two players).

It will prove convenient to work with a set somewhat larger than H that is obtained by adding a *comprehensiveness* assumption to the model. To that end, let $z_i = \min\{y_i \in R \mid (y_1, y_2) \in H\}$ for $i = 1,2$. Thus, z_i is the smallest payoff in H that player i could conceivably receive, and $z = (z_1, z_2)$. The set $\bar{H}$ is defined as the set of payoff points that both weakly dominate z and that are weakly dominated by an element of H. That is,

$$\bar{H} = \{y \in R^2 \mid y \geqslant z \text{ and, for some } x \in H,\ x \geqslant y\} \tag{6.12}$$

In Figure 6.4, the set H is the unshaded set enclosed by the curve $ABDEA$, and the set $\bar{H}$ consists of H plus the three shaded areas (i.e., the set enclosed by the curve $ABCzFA$). The Raiffa–Kalai–Smorodinsky (RKS) solution satisfies Conditions 6.5 and 6.6 below.

CONDITION 6.5 *If* $x > f(H, d)$ *then* $x \notin H$.

CONDITION 6.6 *Let* (H, d) *and* (H', d') *satisfy* (a) $d = d'$, (b) $M_1(H, d) = M_1(H', d')$, and (c) $\bar{H} \subset \bar{H}'$. *Then* $f_2(H, d) \leqslant f_2(H', d')$.

Condition 6.5, *strong Pareto optimality,* states that there is no alternative in H that gives more to one player than $f(H, d)$ without giving less to the other. *Weak Pareto optimality* states that $u \in H$ is Pareto optimal if there is no $u' \in H$ for which $u' \gg u$. That is, a point is not weakly Pareto optimal if there is another point that gives strictly more to each player. Both concepts are used in Chapters 6 to 8, and the choice between them is usually based on mathematical tractability. Condition 6.6, *monotonicity,* states that if two games yield the same maximum payoff to player 1 ($M_1(H, d) = M_1(H', d')$), the second game affords at least as large as a maximum payoff as the first to player 2, they have the same threat point, and the Pareto set of the second game lies on or above the Pareto set of the first game ($\bar{H} \subset \bar{H}'$), then the solution payoff to player 2 in the second game must at least as large as in the first game. The conditions on M_2 and the payoff frontier can be stated as stipulating that, for fixed x_1, the largest value of x_2 is at least as large in $\bar{H}'$ as in $\bar{H}$. The point here is that the game (H', d') is at least as favorable to player 2 as the game (H, d), while the two games retain the same value of M_1. In this situation, monotonicity requires that player 2 get at least as large a payoff from the second game as from the first.

Figure 6.5 shows an example in which the Nash solution violates Condition 6.6. For both (H, d) and (H', d') the threat point is at d. The set $\bar{H}$ is the four-sided figure whose vertices are at d, A, B, and D. The set $\bar{H}'$ is the four-sided figure whose vertices are at d, A, C, and D. Note that $M_1(H, d) = M_1(H', d')$ (marked at M_1 in Figure 6.5), and $\bar{H} \subset \bar{H}'$. The curves aa' and bb' are each curves along which $(u_1 - d_1)(u_2 - d_2)$ is constant; therefore, the Nash solution to (H, d) is at B, the Nash solution to (H', d') is at C, and player 2 receives a smaller Nash solution payoff from (H', d') than from (H, d). This is contrary to Condition 6.6.

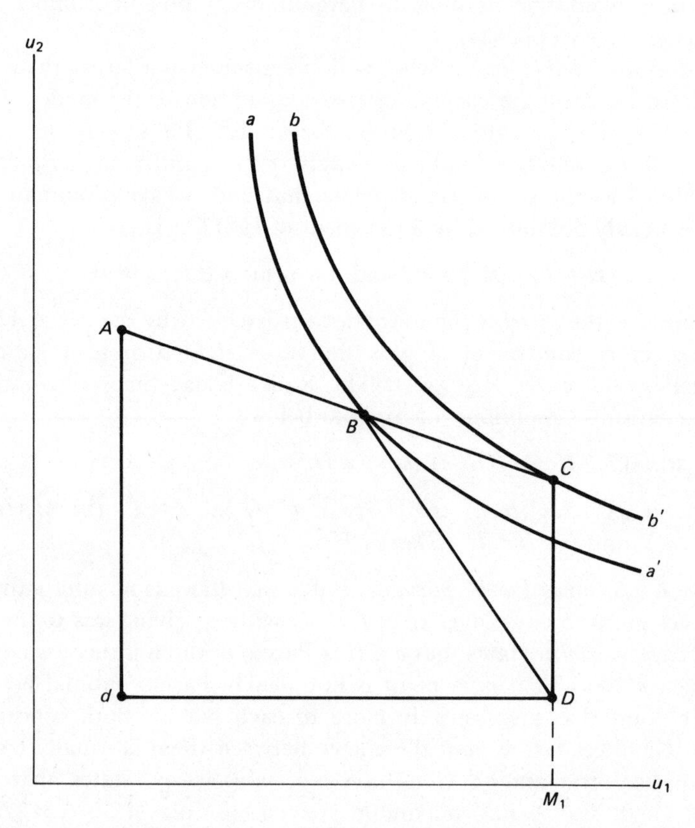

FIGURE 6.5 An example showing that the Nash solution can violate Condition 6.6.

Now let the solution $f(H, d)$ be defined by the following two conditions:
(a)

$$\frac{f_2(H, d) - d_2}{f_1(H, d) - d_1} = \frac{M_2(H, d) - d_2}{M_1(H, d) - d_1} \tag{6.13}$$

(b) if

$$\frac{x_2 - d_2}{x_1 - d_1} = \frac{M_2(H, d) - d_2}{M_1(H, d) - d_1} \tag{6.14}$$

and $x > f(H, d)$, then $x \notin H$. Condition (a) states that the amount each player receives over and above his threat point payoff is proportional to his ideal point payoff, with the proportionality constant being parallel for both players. Then condition (b) requires that the proportionality constant must be the largest value that can be achieved within the feasible set H.

3.1.2 *Existence of the Raiffa–Kalai–Smorodinsky solution*

The main result of Kalai and Smorodinsky (1975) is that the function $f(H, d)$, defined by (a) and (b) in Section 3.1.1, is well defined and is the only function that satisfies Conditions 6.1 to 6.3, 6.5, and 6.6. This is proved in Theorem 6.2 with the aid of the following two lemmas.

LEMMA 6.3 *A nonnegatively sloped ray (i.e., straight line) through d passes through a strongly Pareto optimal point of H if and only if if the slope of the ray lies in the interval*

$$\left[\frac{m_2(H, d) - d_2}{M_1(H, d) - d_1}, \frac{M_2(H, d) - d_2}{m_1(H, d) - d_1}\right] \tag{6.15}$$

Proof. A nonnegatively sloped ray whose slope lies in the interval in equation (6.15) must pass through a point on the (weak) upper right frontier of H. (A point $u \in H$ is on this frontier if there is no $u' \in H$ such that $u' \gg u$.) To fail to do so would contradict the convexity of H, and for the point to fail to be Pareto optimal would, likewise, contradict the convexity of H. A ray of slope steeper than the upper bound in equation (6.15) could intersect the upper right boundary of H only at a point where $x_1 < m_1(H, d)$ and $x_2 \leq M_2(H, d)$; therefore, x could not be strongly Pareto optimal. A parallel argument can be made for nonnegatively sloped rays that are too flat to lie in the interval in equation (6.15). QED

LEMMA 6.4 *The function $f(H, d)$ defined by equations* (6.13) *and* (6.14) *has a unique value for every* (H, d) *and corresponds to a strongly Pareto optimal element of* H.

Proof. $f(H, d)$ is that point in H that lies on a ray through d of slope $[M_2(H, d) - d_2]/[M_1(H, d) - d_1]$. This slope lies in the interval in equation (6.15); therefore, it passes through a strongly Pareto optimal element of H. The ray could not pass through two or more Pareto optimal elements, because its slope is positive and hence, one of the supposed Pareto optimal points would strongly dominate the other. QED

THEOREM 6.2 *The function* $f(H, d)$ *is well defined, satisfies Conditions* 6.1 *to* 6.3, 6.5, *and* 6.6, *and is the only function to satisfy these conditions.*

Proof. First it is shown that $f(H, d)$ satisfies the several conditions and is well defined; then it is shown that it is the only function to satisfy the conditions. It is proved in Lemma 6.4 that $f(H, d)$ is well defined and satisfies Condition 6.5. That Condition 6.1 is satisfied is obvious, that Condition 6.2 is satisfied is left as an exercise for the reader, and that Condition 6.3 is satisfied follows from the definition of $f(H, d)$ and the definition of a symmetric game.

To see that Condition 6.6 is satisfied, suppose that (H, d) and (H', d') are two games satisfying $d = d'$, $M_1(H, d) = M_1(H', d')$, and $\bar{H} \subset \bar{H}'$. The latter implies that $M_2(H, d) \leqslant M_2(H', d')$. The ray from a threat point d through a point $f(H, d)$ is called the *defining ray for* $f(H, d)$. Note the following facts: (a) Because $M_1(H, d) = M_1(H', d')$ and $M_2(H, d) \leqslant M_2(H', d')$, the defining ray for $f(H', d')$ has a slope at least as great as that for $f(H, d)$. (b) The set of Pareto optimal points of H is identical to that of $\bar{H}$, and the same holds for H' and $\bar{H}'$. (c) Let y denote the point on the upper right boundary of $\bar{H}$ that the defining ray of $f(H, d)$ passes through, and let $\bar{y}$ denote the point on the upper right boundary of $\bar{H}$ through which the defining ray of $f(H', d')$ passes. Then $\bar{y}_2 \geqslant y_2$. (d) Let y' denote the point on the upper boundary of $\bar{H}'$ through which the defining ray of $f(H', d')$ passes. Then $y'_2 \geqslant \bar{y}_2$. Now note that y is on the upper right boundary of H; hence $y = f(H, d)$. Similarly, y' is on the upper right boundary of H', implying that $y' = f(H', d)$. Therefore, $f_2(H', d') \geqslant f_2(H, d)$, which establishes that $f(H, d)$ satisfies Condition 6.6.

To see that only $f(H, d)$ satisfies the various conditions, it is first shown to hold for games in which $d = 0$ and $M_1(H, 0) = M_2(H, 0) = 1$. Extension to the full class of games follows from the requirement that the solution satisfy invariance to affine transformations (Condition 6.2). Denote the true solution by $f^*(H, d)$, and let H' be the convex hull of the set of points $\{(0, 0), (0, 1), (1, 0), f(\bar{H}, 0)\}$. Monotonicity requires that $f^*(\bar{H}, 0) \geqslant f^*(H, 0)$ and that $f^*(\bar{H}, 0) \geqslant f^*(H', 0)$, because $H \subset \bar{H}$ and $H' \subset \bar{H}$. At the same time, the definition of $\bar{H}$ and the Pareto optimality of $f(H, 0)$ imply

$$f^*(H, 0) = f^*(\bar{H}, 0) \tag{6.16}$$

Symmetry ensures that

$$f^*(H', 0) = f(\bar{H}, 0) \tag{6.17}$$

Meanwhile, $f(\bar{H}, 0)$ is in the Pareto optimal set of both H and $\bar{H}$; hence,

$$f(\bar{H}, 0) = f(H, 0) \tag{6.18}$$

and

$$f^*(\bar{H}, 0) = f^*(H', 0) \tag{6.19}$$

Equations (6.16) to (6.19) imply that $f^*(H, 0) = f(H, 0)$. QED

3.1.3 A comparison of the Raiffa–Kalai–Smorodinsky and Nash solutions

Before moving on to other bargaining solutions, it is worthwhile to compare Condition 6.4 with Condition 6.6. Nash's condition says that two games sharing the same threat point must have the same solution if one feasible set is contained in the other ($H \subset H'$) and the solution to the larger game is available in the smaller game ($f(H', d') \in H$). This is analogous to saying that, if the best tennis player in Europe is from Sweden, then the best tennis player in Europe must be the best tennis player in Sweden. Condition 6.6 says, approximately, that if the payoff frontier is moved outward to increase $M_2(H, d)$ without increasing $M_1(H, d)$, then the solution payoff to player 2 should be increased. The former condition states when a change in the game should have no effect on the solution, while the latter states when a change should benefit a particular player. In Section 3.2, a class of solutions is examined to which the Nash solution belongs, but to which the RKS solution does not. For this class of solutions a condition similar to Condition 6.4 is used.

3.2 Bargaining solutions based on invariance with respect to a reference point

In the game (H, d), the threat point, d, plays the role of reference point for the Nash solution. Although Condition 6.4 has traditionally been called independence of irrelevant alternatives, it might be more aptly named dependence on the threat point. In Section 3.2.1, a condition of *dependence on the reference point* is formulated and a large class of reference points is considered. Section 3.2.2 establishes the existence and uniqueness of solutions based on the class of reference points from Section 3.2.1. In Section 3.2.3 an example is presented for which several solutions are computed and compared, each for a different reference point. Finally, comparisons of these solutions with noncooperative equilibria are made in Section 3.2.4.

3.2.1 Conditions defining a class of reference points

The concept of reference point can be formulated abstractly, following Thomson (1981). Let $g(H, d)$ be an element of H that is not in the Pareto set of H (i.e, there is some $y \in H$ such that $y \gg g(H, d)$). Two examples of reference points are the point of minimal expectations, $m(H, d)$, and a convex combination of d and $m(H, d)$, $g(H, d) = kd + (1 - k)m(H, d)$ for $k \in [0, 1]$. The reference point $g(H, d)$ is required to obey two conditions that are obviously satisfied by d. They are

CONDITION 6.7 *Let* (H, d), $(H', d') \in W$, a_1, $a_2 \in R_{++}$, b_1, $b_2 \in R$, $d'_i = a_i d_i + b_i$, $i = 1, 2$, *and* $H' = \{x \in R^2 \mid x_i = a_i y_i + b_i,\ i = 1, 2,\ y \in H\}$. *Then* $g_i(H', d') = a_i g_i(H, d) + b_i$, $i = 1,2$.

This condition states that when two games are related by an affine transformation, their reference points are related by the same affine transformation.

CONDITION 6.8 *Let* (i) $\delta \in R$ *be defined by the conditions that* $(\delta, \delta) \in H$ *and* $(\delta + \varepsilon, \delta + \varepsilon) \notin H$ *for all* $\varepsilon > 0$, (ii) $(H, d) \in W$, (iii) $g_1(H, d) = g_2(H, d)$, *and* (iv) $x \notin H$ *if* $x_1 + x_2 > 2\delta$. *Then there exists* $(H', d') \in W$ *satisfying* (a) $H \subset H'$, (b) *if* $(x_1, x_2) \in H'$ *then* $(x_2, x_1) \in H'$, (c) *if* $x \in H'$ *then* $x_1 + x_2 \leq 2\delta$, *and* (d) $g(H, d) = g(H', d')$.

For a game (H, d), suppose a line of slope -1 is drawn that is tangent to the upper right boundary of H. If this tangency point, x, is symmetric (i.e., $x_1 = x_2$), and the reference point is symmetric (i.e., $g_1(H, d) = g_2(H, d)$), then the game is *almost symmetric*. Conditions 6.8 states that the game (H, d) can be enlarged into the symmetric game (H', d') with the two games having the same reference point $(g(H, d) = g(H', d'))$ and with $[u_1 - g_1(H, d)][u_2 - g_2(H, d)]$ maximized on H at the same point u^* where $[u_1 - g_1(H', d')][u_2 - g_2(H', d')]$ is maximized on H', if the original game (H, d) is almost symmetric.

Condition 6.9 is parallel to Condition 6.4; it is condition of independence of irrelevant alternatives in which the reference point d is replaced with the reference point $g(H, d)$. Let

$$W_g = \{(H, d) \in W \mid g(H, d) \in H \text{ and } y \gg g(H, d) \text{ for some } y \in H\} \quad (6.20)$$

W_g is the subset of W in which the reference point of a game $(H, d) \in W_g$ is not Pareto optimal.

CONDITION 6.9 *If* (H, d), $(H', d') \in W_g$, $g(H, d) = g(H', d')$, $H \subset H'$, *and* $f(H', d') \in H$, *then* $f(H, d) = f(H', d')$.

Results parallel to Lemma 6.1 and 6.2 and Theorem 6.1 hold, but with W_g replacing W and $g(H, d)$ replacing d. The counterparts to Lemma 6.1 and Theorem 6.1 are stated here as Lemma 6.5 and Theorem 6.3; the proofs are omitted because they trivially repeat the earlier proofs. In parallel to Condition 6.1.

CONDITION 6.10 $f(H, d) \gg g(H, d)$.

3.2.2 *Existence and uniqueness of solutions to games based on reference points*

With the conditions defining a class of reference points stated, attention can now be turned to existence and uniqueness of solutions based on them. This is done in Theorem 6.3, but first an intermediate result is needed.

LEMMA 6.5 *Let* $(H, d) \in W_g$, *let* $u^* = f(H, d)$ *satisfy Conditions* 6.2, 6.3, 6.9 *and* 6.10, *and let* $g(H, d)$ *satisfy Conditions* 6.7 *and* 6.8. *Then if* $u \in H$ *and* $u \neq u^*$ *either* $u_1^* > u_1$ *or* $u_2^* > u_2$.

THEOREM 6.3 *Let* $(H, d) \in W_g$ *be a game with reference function* $g(H, d)$ *that satisfies Conditions* 6.7 *and* 6.8. *Then* (H, d) *has a unique solution* $u^* = f(H, d)$ *satisfying Conditions* 6.2, 6.3, 6.9 *and* 6.10. *The solution satisfies*

$$(u_1^* - g_1(H, d))(u_2^* - g_2(H, d)) > (u_1 - g_1(H, d))(u_2 - g_2(H, d)) \quad (6.21)$$

for all $u \in H$, $u \geq g(H, d)$, *and* $u \neq u^*$.

Thomson (1981) cites several functions that satisfy Conditions 6.7 and 6.8, including (a) the point of minimal expectations, $m(H, d)$, (b) the *point of minimal compromise*, given by

$$\left(\frac{M_1(H, d) + m_1(H, d)}{2}, \frac{M_2(H, d) + m_2(H, d)}{2} \right) \quad (6.22)$$

(c) the center of the smallest rectangle containing H, and (d) the center of gravity of H. These reference functions share a common characteristic: Each depends on the shape of H, but has no dependence on the threat point d. It is proved in Thomson (1981; page 438). that a reference function that is a convex combination of reference functions satisfying Conditions 6.7 and 6.8 will also satisfy these conditions. Thus, many reference functions can be constructed by combining d and one or more of the four foregoing reference functions.

3.2.3 Comparison of different solutions by example

Suppose two players can attain any utility pair that lies on or below the curve $u_2 = 7.5 + u_1 - .1u_1^2$ and is also on or above the u_1 axis. The threat point is fixed at the origin. Figure 6.6 depicts this set. In the figure, five

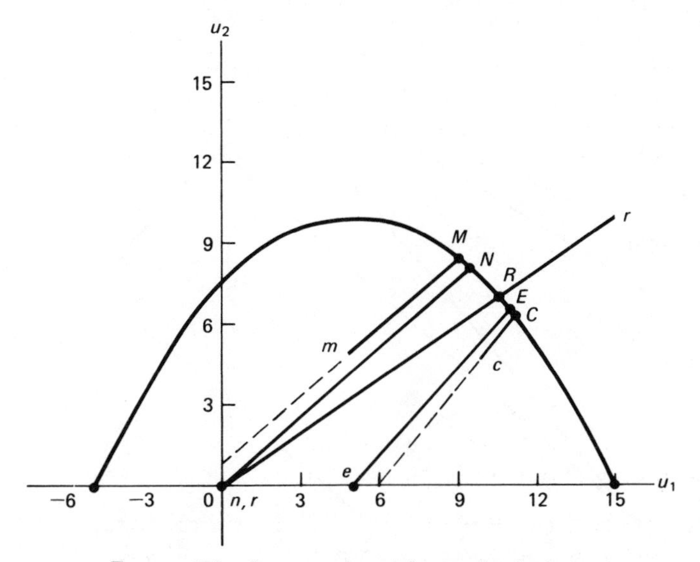

FIGURE 6.6 A comparison of several solutions.

solutions are shown and each is denoted by a letter. The letter appears in uppercase on the payoff frontier at the solution point itself and in lowercase at the reference point for the solutions. The solutions and their accompanying letters are (1) The Nash solution (N), (2) the Raiffa–Kalai–Smorodinsky solution (R), (3) the minimal expectations reference point solution (E), (4) the minimal compromise reference point solution (C), and (5) the middle of the smallest rectangle containing H reference point solution (M). Note that two reference points are specified for the Raiffa–Kalai–Smorodinsky solution. This is because the threat point and the point (15, 10), giving the ideal point for the two players, are both reference points. The solution is on the frontier where the straight line from the threat point to (15, 10) intersects it.

Contrast the game just described with the game shown in Figure 6.7. For this game, the set H is bounded below and on the left by the two axes. The upper right boundary of H is given by $u_2 = 7.5 + u_1 - u_1^2$, as before, from $u_1 = 8$ to $u_1 = 12$. Going leftward from $u_1 = 8$ and $u_2 = 9.1$, the boundary is a straight line with slope $-.6$ until it reaches the u_2 axis at $u_2 = 13.9$. From $u_1 = 12$ and $u_2 = 5.1$, the boundary goes downward and to the left at a slope of $+1$ until the u_1 axis is reached at $u_1 = 6$. The solutions and reference points for both games are shown in Table 6.1. Note the change in the

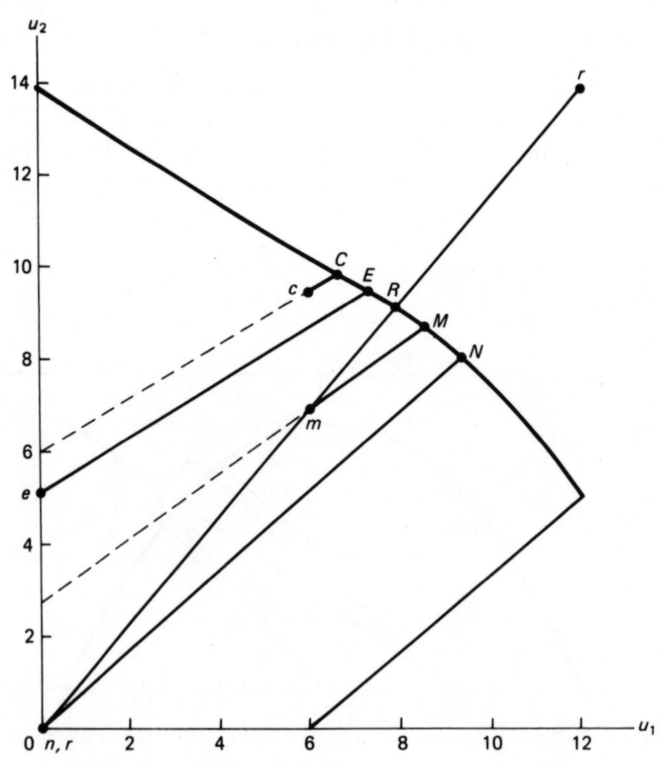

FIGURE 6.7 The order of solutions can be reversed.

TABLE 6.1 Location of reference and solution points for models in Figures 6.6 and 6.7

	Figure 6.6				Figure 6.7			
	Reference point		Solution point		Reference point		Solution point	
	Player		Player		Player		Player	
Solution name	1	2	1	2	1	2	1	2
Nash	0	0	9.34	8.11	0	0	9.34	8.11
Raiffa–Kalai–Smirodinsky	0	0	10.49	6.99	0	0	7.91	9.16
	15	10			12	13.9		
Minimal expectation	5	0	10.77	6.67	0	5.1	7.33	9.5
Minimal compromise	10	5	11.08	6.31	6	9.5	6.67	9.9
Middle of rectangle	5	5	9.08	8.33	6	6.95	8.54	8.75

relative positions of the solutions between the two games. Except for the middle of the smallest rectangle solution favoring player 2 relative to the Nash solution in both games, the solutions appear in reversed order. The best three for player 1 in Figure 6.6 are C, E, and R, in that order, but in the game in Figure 6.7, C, E, and R are the best for player 2 in that order.

3.2.4 Comparison of cooperative solutions with noncooperative games

It is instructive now to compare the bargaining models described above with noncooperative games. Within a bargaining game to which a specific solution is applied, there is no action for a player to take. The structure of the game (specification of H and d) along with the definition of the solution concept decide the outcome. Why, then, should these solution concepts be of any interest? There are three answers explored below: fairness, efficiency, and descriptive accuracy.

Fairness was discussed in Section 1.1. Clearly, if the members of a group regard a particular solution as fair, they have an incentive to adopt it as an arbitration scheme in order to achieve efficiency (Pareto optimal outcomes) in a generally acceptable way. As noted earlier, this makes sense as an aspect of a group's operating rules when it is believed that a sequence of games will be encountered in the future and there is no idea in advance of which games they will be. In general noncooperative equilibria are not efficient.

Concerning descriptive accuracy, it is possible that the empirical investigation of cooperative game situations will reveal that certain solutions are persistent outcomes. Such a finding would be very important and would be a compelling reason to give a special place of importance to these solutions; as far as I know, however, results of this type have not yet surfaced.

This discussion is not likely to have compelled the reader to accept a particular solution. Game theory presently offers many cooperative game solution concepts; in addition to the bargaining solutions covered above, additional solutions are discussed in the remainder of this chapter and in the next chapter. The solutions covered in this book do not exhaust the literature. None is widely accepted as preeminent, and perhaps the brief discussion of fairness, efficiency and descriptive accuracy makes clear why no single solution has conquered the field.

4 Bargaining over time

In this section a model of the bargaining process due to Rubinstein (1982) is discussed. The problem Rubinstein takes up is very simple: Two players can split between them a unit of a good that both value. If player 1 gets $s_1 \in [0, 1]$, then player 2 gets $s_2 = 1 - s_1$, but neither player receives anything unless the two players come to an agreement. The game is modeled as a noncooperative game, played over time, in strategic form. In the first period, player 1 proposes a division of the good, which player 2 either accepts or rejects. If player 2 accepts, the game ends, with player 1 getting s_1 and player 2 getting $1 - s_1$. If player 2 rejects the proposal, then the game continues with player 2 making the next proposal, which player 1 can accept or reject. Each period consists of a proposal being made by one player followed by the acceptance or rejection of it by the other player. The game ends as soon as one player accepts the other's offer. Prior to termination, player 1 makes an offer in each even-numbered period $t = 0, 2, 4, \ldots$, and accepts or rejects offers of player 2 in each odd-numbered period $t = 1, 3, \ldots$. Player 2 makes offers in each odd-numbered period and passes on offers of player 1 in each even-numbered period. The value of an outcome to a player depends on what he receives and on the time period in which agreement is reached. Naturally, the value rises with the amount received and falls with time. There are no rules concerning how one offer is related to another; in particular, the offer of player i in period t need not concede as much as the offer he made in period $t - 2$. The game is cooperative in the sense that if one player states an offer and the other accepts, the offer is a binding agreement for both.

4.1 The strategy spaces and payoff functions

In presenting the assumptions of Rubinstein's model, I introduce one alteration. Rubinstein does not give an explicit payoff function for each player; rather, he defines an ordering on outcomes for each player. In the presentation that follows, each player has a payoff function.

ASSUMPTION 6.1 *$u_i(s_i, t)$ is the payoff function of player i. It is continuous and strictly increasing in s_i, strictly decreasing in t, defined for $s_i \in [0, 1]$ and $t \in \{0, 1, 2, \ldots\}$, and satisfies*

(a) $u_i(0, \infty) < u_i(s_i, t)$ *if* $s_i > 0$ *and* $t < \infty$

(b) $u_i(s_i, t) = u_i(s_i', t + 1)$ *if and only if* $u_i(s_i, t') = u_i(s_i', t' + 1)$.

(c) *if* $u_i(s_i + \varepsilon, 1) = u_i(s_i, 0)$, $u_i(s_i' + \varepsilon', 1) = u_i(s_i', 0)$, *and* $s_i < s_i'$, *then* $\varepsilon < \varepsilon'$.

Assumption 6.1 asserts that the good to be divided is strictly valuable, the passage of time is strictly costly, and the payoff of a player is continuous in the amount of the good received. In addition, (a) to (c) stipulate that (a) any positive amount of the good received at any finite time is preferred to obtaining nothing, (b) the payoff functions are stationary with respect to time, and (c) the amount of compensation required for a one-period delay in receiving the good increases with the amount of the good.

The game is played noncooperatively in the sense that each player has a strategy set from which she chooses a strategy. Rubinstein is interested in subgame perfect noncooperative equilibria for the game. For player 1, a strategy has the form $\sigma_1 = (f_{10}, f_{11}, f_{12}, f_{13}, \ldots)$, where

$f_{10} = s_{10}$, the offer of player 1 at time 0
f_{11} is a function whose argument is $(s_{10}, s_{21}) \in [0, 1]^2$ and whose value is *accept* or *reject*
f_{1t} is a function whose argument is $(s_{10}, s_{21}, s_{12}, s_{23}, \ldots, s_{2,t-1}) \in [0, 1]^t$ *and whose value is* $s_{1t} \in [0, 1]$ when t is even, and when t is odd, the argument is $(s_{10}, s_{21}, \ldots, s_{2t}) \in [0, 1]^{t+1}$ and the value is *accept* or *reject*

Strategies for player 2, $\sigma_2 = (f_{20}, f_{21}, \ldots)$, are analogous to strategies for player 1. For even values of t, f_{2t} has $(s_{10}, s_{21}, \ldots, s_{1t}) \in [0, 1]^{t+1}$ as its argument and takes the value *accept* or *reject*. For odd values of t, f_{2t} has $(s_{10}, s_{21}, \ldots, s_{1,t-1}) \in [0, 1]^t$ as an argument and takes a value $s_{2t} \in [0, 1]$.

4.2 Equilibrium and perfect equilibrium outcomes

In this game, any division of the good can be associated with a noncooperative equilibrium. This is unfortunate in the sense that one usually hopes for equilibrium conditions greatly to narrow the field of possible outcomes. Ideally, that field is narrowed to one. If any outcome can be associated with equilibrium, then the model lacks any predictive power. In the present case, σ^* is an equilibrium point where $\sigma_i^* = (f_{i0}^*, f_{i1}^*, \ldots)$

For even t

$$f_{1t}^* = s_1^*$$
$$f_{2t}^* = \textit{accept} \text{ if } s_{1t} \leq s_1^*$$
$$f_{2t}^* = \textit{reject} \text{ if } s_{1t} > s_1^*$$

For odd t

$$f_{1t}^* = \textit{accept} \text{ if } s_{2t} \leq s_2^* = 1 - s_1^*$$
$$f_{1t}^* = \textit{reject} \text{ if } s_{2t} > s_2^* = 1 - s_1^*$$
$$f_{2t}^* = s_2^* = 1 - s_1^*$$

It is obvious that these equilibria are not perfect equilibria and are supported by a knife-edge quality built into the terms of acceptance.

Player 1 proposes s_1^* and, if this is rejected, will accept in the future only agreements giving himself s_1^* or more. Player 2 will, likewise, accept $1 - s_1^*$ or more and will propose this amount in any odd-numbered period. Clearly, neither player can unilaterally increase his payoff by the use of an alternate strategy.

To be a subgame perfect equilibrium, the following conditions should be met at any time t, following a sequence of offers and rejections. (a) The player whose turn it is to make an offer has no superior alternative to offering what his strategy prescribes. (b) The other player, if she planned to accept the offer at time t, cannot do better by rejecting it. (c) If she planned to reject, she could do no better than reject. As an example of an equilibrium that is not subgame perfect, consider the equilibrium discussed by Rubinstein (1982; page 103). Let $u_1(s_i, t) = s_i - c_i t$ with $c_1 = .1$ and $c_2 = .2$, and let $s_1^* = .5$. If player 1 offered $s_1^* = .6$ player 2 would reject it, expecting settlement in the next period at $s_1^* = .5$. But the payoff that player 2 rejects is $(1 - .6) = .2t = .4 - .2t$ and what she would receive instead is $(1 - .5) - .2(t + 1) = .3 - .2t$. Player 2 would lose .1 by this behavior, which violates (c) above.

Rubinstein proves that the final outcome of a perfect equilibrium strategy combination must be a pair $(s_1^*, 1 - s_1^*)$ where s_1^* is the first member of a pair (s_1^*, s_2^*) that simultaneously satisfies two equations, $s_2 = d_1(s_1)$ and $s_1 = d_2(s_2)$ that are defined as follows:

$$\begin{aligned} d_1(s_1) &= 1 \text{ if, for all } s_2 \in [0, 1],\ u_1(1 - s_2, 0) > u_1(s_1, 1) \\ &= s_2 \text{ if } u_1(1 - s_2, 0) = u_1(s, 1) \end{aligned} \tag{6.23}$$

$$\begin{aligned} d_2(s_2) &= 1 \text{ if, for all } s_1 \in [0, 1],\ u_2(1 - s_1, 0) > u_2(s_2, 1) \\ &= s_1 \text{ if } u_2(1 - s_1, 0) = u_2(s_2, 1) \end{aligned} \tag{6.24}$$

Any such $(s_1^*, 1 - s_1^*)$ can be a perfect equilibrium realization, and any game satisfying Assumption 6.1 has at least one such realization. An example of a perfect equilibrium is

$$\left.\begin{aligned} f_{1t}(s_{10}, \ldots, s_{2,t-1}) &= s_1^* \\ f_{2t}(s_{10}, \ldots, s_{1t}) &= \textit{accept} \text{ if } s_{1t} \leq s_1^* \\ &= \textit{reject} \text{ if } s_{1t} > s_1^* \end{aligned}\right\} \quad t = 0, 2, 4, \ldots$$

$$\left.\begin{aligned} f_{1t}(s_{10}, \ldots, s_{2t}) &= \textit{accept} \text{ if } s_{2t} \leq s_2^* \\ &= \textit{reject} \text{ if } s_{2t} > s_2^* \\ f_{2t}(s_{10}, \ldots, s_{1,t-1} &= s_2^* \end{aligned}\right\} \quad t = 1, 3, 5, \ldots$$

where $s_1^* = d_2(s_2^*)$ and $s_2^* = d_1(s_1^*)$. Note that the equilibrium allocation is $(s_1^*, 1 - s_1^*)$; however, this is because player 1 is the first to propose an offer. If player were to be the first to make an offer, then the equilibrium allocation would be $(s_2^*, 1 - s_2^*)$, which, in general, would be better for player 2 than $1 - s_1^*$ would be.

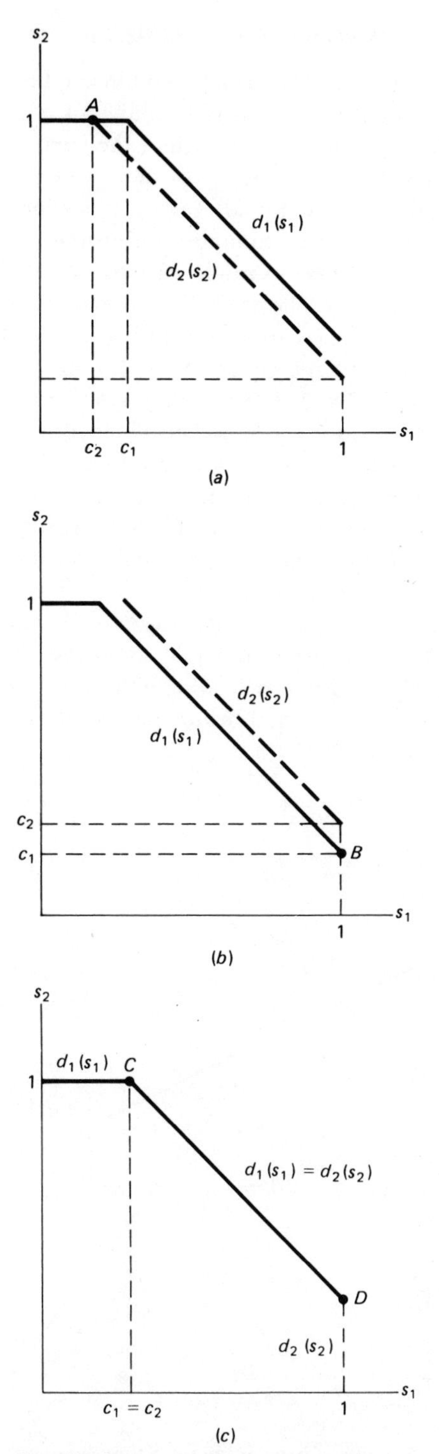

FIGURE 6.8 Equilibrium in a game of bargaining over time.

4.3 Examples of the model of bargaining over time

It appears that bargaining will usually end in the first round ($t=0$), that a player will do better if he is to first to make an offer, and that a player's equilibrium amount of the good is higher, the smaller is the drop in payoff he suffers by postponing receipt of the good by one period. These conclusions are not proved for all games satisfying Assumption 6.1, but they are borne out by two examples Rubinstein discusses and that are examined here. In the first example, the payoff functions are $s_1 - c_1 t$ for player 1 and $s_2 - c_2 t$ for player 2. The functions d_1 and d_2, defined in equations (6.23) and (6.24), are shown in Figure 6.8. In Figure 6.8*a*, $c_1 > c_2$ and the only point satisfying equations (6.23) and (6.24) simultaneously is A, where $s_1^* = c_2$ and $s_2^* = 1$. Thus, with player 1 making the first offer, the outcome would be $(c_2, 1 - c_2)$; however, if player 2 were able to make the first offer, the outcome would be $(0, 1)$. In Figure 6.8*b*, $c_1 < c_2$, and the only point satisfying the equilibrium requirements is B. With player 1 going first, the outcome is $(1, 0)$ and with player 2 going first, it is $(1 - c_1, c_1)$. Finally, if $c_1 = c_2$, as in Figure 6.8*c*, any point on the line from C to D can be associated with an equilibrium point.

For the second example, let δ_1, $\delta_2 \in (0, 1)$ and let the two payoff functions be $s_1 \delta_1^t$ for player 1 and $s_2 \delta_2^t$ for player 2. Equation (6.23) and (6.24) are plotted in Figure 6.9, where the only point giving rise to an equilibrium is at A. The equilibrium payoffs are $(s_1^*, 1 - s_1^*)$ if player 1

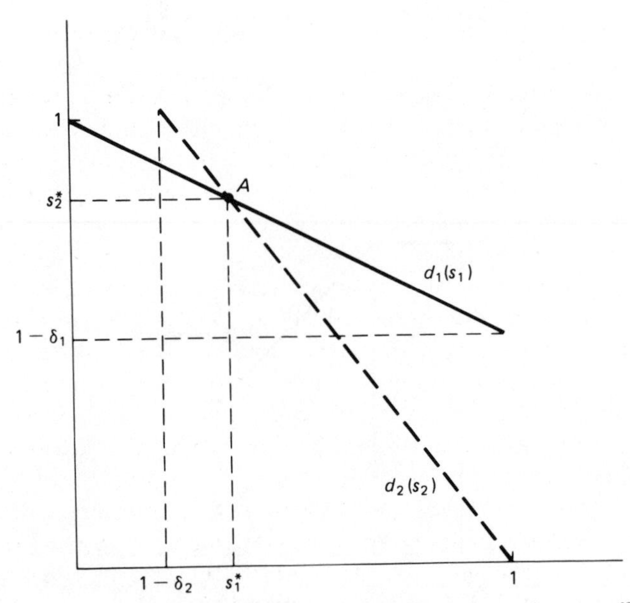

FIGURE 6.9 A game of bargaining over time having a unique equilibrium.

offers first and $(1 - s_2^*, s_2^*)$ if player 2 offers first. Note that with $\delta_1 < \delta_2$, player 1 places less weight on the future relative to the present than player 2. This favors player 2 in the sense that $s_1^* < s_2^*$. That is, if player 1 were to offer first, his equilibrium payoff would be less than if player 2 were to offer first.

It is difficult to imagine the process of bargaining as anything but a noncooperative procedure, with the essential distinction from most noncooperative games being that the players are allowed to make binding agreements. As in the Rubinstein model, each player here forms a strategy that dictates his move at each stage of the process, and the strategy is chosen to maximize his payoff at the final cooperative outcome to which the bargaining process leads. Were it found that actual bargaining processes tended toward a particular cooperative solution, it would become very important to discover noncooperative game formulations of bargaining that would lead to the same solution.

Rubinstein's model does give an explicit value to time that may differ for the two players and may differ for a particular player according to the size of the player's payoff. A value for time may enter for either of two reasons. First, the actual quantity of good to be divided may shrink over time. This can occur because the good is subject to deterioration or, as in labor–management negotiations taking place during a strike, because each day that passes without an agreement is a day in which the productive power of both sides is wasted. Second, time is likely to be discounted: a given reward received today is valued higher than the same reward would be valued today if it were to be received tomorrow.

An important phenomenon on which the Rubinstein model does not touch is the failure to make agreements or the making of agreements at a later than optimal time. In labor-management negotiations, strikes are observed, and in most instances it would undoubtedly be better for both parties if the contract that was finally agreed on would have been accepted in time to avert a strike. If the bargainers have incomplete or inaccurate information concerning one another's payoff functions, coming to an agreement might be made more difficult and strikes might be understandable. Indeed, even the failure to agree at any time might be understandable. The labor–management example has a feature that goes beyond the scope of cooperative games studied in this book: The bargaining over a contract may be seen by both players as a single episode in an ongoing sequence of contract negotiations. Therefore, the outcome of one particular contract would naturally be evaluated for its effect in the future on the remaining contracts. Seen in this light, one side or both may be quite willing to have an occasional strike to avoid the appearance of weakness.

Between Nash (1953) and Rubinstein (1982) attempts were made to deal with the bargaining process, most notably by Ståhl (1972). The subject did not take hold, however, until Rubinstein's work came out. The reader interested in delving further can look at Binmore and Dasgupta (1987) and the surveys by Rubinstein (1987) and Sutton (1986).

5 Variable threat games

It is easy to imagine two-person games in which binding agreements are possible, but in which each player has considerable scope for action in the absence of an agreement, and in which the decision of each player affects both of them. Nash (1953) dealt with this class of games from several angles, all of which provided justification for a revised version of his bargaining solution. The approach taken here combines arbitration with a noncooperative game. Suppose each of the two players has a strategy set, S_i for player i, that is compact and convex, and assume that $P_i(s_1, s_2)$ is the payoff function for player i in the absence of an agreement. Thus the game $\{(1, 2\}, S, P)$ is a default noncooperative game that the two players must play if they cannot agree. There is no threat point as such. Suppose further that there is a compact convex set $H \subset R^2$ consisting of all the payoff points that the two players can reach by means of binding agreements. H would naturally contain as a subset all those points attainable in the default game.

DEFINITION 6.6 $\Gamma = (N, S, P, H)$ *is a* **variable-threat two-person cooperative game** *in which* (N, S, P) *is a two-person noncooperative game that is played if no agreement is reached and the compact, convex set* H *is the cooperative attainable payoff set.* H *contains* $\{P(s) \in R^2 \mid s \in S\}$.

The Nash arbitration game proceeds by having the two players simultaneously choose strategies for the default game, $s_i^T \in S_i$ that are used to determine a threat point, $P(s^{\mathrm{T}})$. The arbitrated outcome to the game is the Nash bargaining solution to $(H, P(s^T))$. Thus, the players are not interested in the payoffs $P_i(s)$ for their own sakes; they care about the effect on the final outcome that is due to their choices of s_i^T. Potentially, any point on the Pareto optimal frontier of H could be an arbitrated outcome. Figure 6.10 shows a useful and interesting relationship between points on the Pareto optimal frontier of H and the points in H that could serve as threat points. The slope of a line from the threat point to the Nash solution is equal in absolute value and opposite in sign to the slope of the payoff possibility frontier at the Nash solution point. This property is stated formally and proved in the following lemma.

LEMMA 6.6 *Let* (H, d) *be a two-person bargaining game whose Nash bargaining solution is* u^*. *Then for all* $u \in H$ *such that* $u \geqslant d$ *and* $u \neq u^*$. $(u_1 - d_1)(u_2 - d_2) < (u_1^* - d_1)(u_2^* - d_2)$ *implies that no points of* H *lie above a line of slope* $-(u_2^* - d_2)/(u_1^* - d_1)$ *through* u^*. *The slope of the line from* d *to* u^* *is* $(u_2^* - d_2)/(u_1^* - d_1)$.

Proof. Consider the rectangular hyperbola $(x_1 - d_1)(x_2 - d_2) = (u_1^* - d_1)(u_2^* - d_2)$. The slope of this curve at u^* is $-(u_2^* - d_2)/(u_1^* - d_1)$; therefore any point on the curve other than u^* lies above the straight line with this slope that passes through u^* (i.e., the tangent to the rectangular hyperbola at u^*). If some point y of H other than u^* lay above this tangent line, then, by the convexity of H, there would be a point on the line connecting u^* and y that lay on or above the rectangular hyperbola. This contradicts the

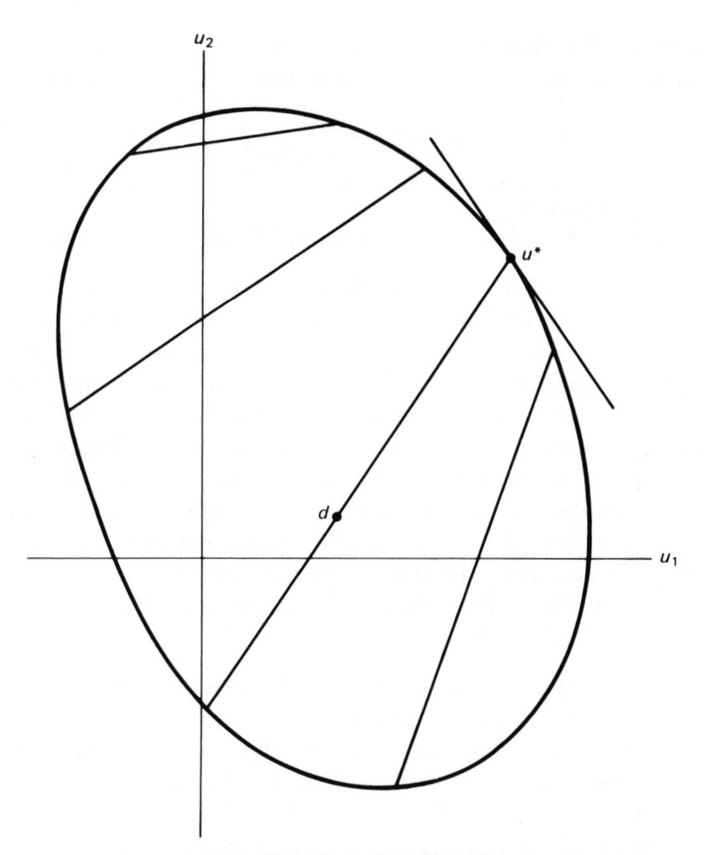

FIGURE 6.10 The Nash variable-threat game.

condition that u^* is the unique Nash product maximizer. That the slope of the line from d to u^* is $(u_2^* - d_2)/(u_1^* - d_1)$ follows from the definition of the line. QED

With the help of Figure 6.10, it is easy to see the nature of a Nash variable-threat cooperative equilibrium and to see the nature of the threat strategies that would be associated with it. Suppose the strategy combination s' results in d in Figure 6.10 as the threat point. That is, suppose the players choose s' and, by the arbitration procedure, u^* is the resulting cooperative outcome. Whether s_1' is an optimal choice of threat strategy for player 1 depends on whether there is some $s_1'' \in S_1$ such that $P(s_1'', s_2')$ would lie below the line connecting d and u^*; for any point below this line, as a threat point, would lead to a cooperative outcome that was more favorable to player 1. Therefore, if s' is an optimal threat point, all points $P(s_1', s_2'')$ lie on or below the line through d and u^* for all s_1''. Furthermore, if the game has a solution, which is the same as there being an optimal threat, then the solution payoff is unique. Uniqueness of the solution payoff follows from Theorem 3.2, because the noncooperative game whose outcome is the cooperative solution is strictly competitive. These facts are summarized in the following theorem.

THEOREM 6.4 *Let (N, S, P, H) be a two-person variable threat game. If u^* is the Nash solution to the game and s^T is the optimal threat strategy combination, then*
(a)

$$\frac{u_2^* - P_2(s_1^T, s_2)}{u_1^* - P_1(s_1^T, s_2)} \geqslant \frac{u_2^* - P_2(s^T)}{u_1^* - P_1(s^T)} \geqslant \frac{u_2^* - P_2(s_1, s_2^T)}{u_1^* - P_1(s_1, s_2^T)} \tag{6.25}$$

for all $s_1 \in S_1$ and $s_2 \in S_2$, and
(b) *all cooperative Nash variable-threat solutions yield the same payoff pair.*

Proof. The first part of the proof follows from the fact that the arbitration scheme requires the final outcome to be the Nash bargaining solution to $(H, P(s^T))$. To see part (b), note first that, given the arbitration scheme, the cooperative payoffs can be regarded as functions of s^T, thus the payoff functions $P_i^*(s)$ may be understood to be the Nash cooperative payoffs associated with the threat strategy combination s. Clearly, $P^*(s)$ is continuous in s, although this is not needed to prove uniqueness of the equilibrium payoffs. Uniqueness follows using precisely the same argument that establishes uniqueness of the equilibrium payoffs in two-person strictly competitive games. The crucial point in both instances is that the payoff structure is perfectly competitive. In the present case, $P_1^*(s) < P_1^*(s')$ if and only if $P_2^*(s) > P_2^*(s')$. QED

Note that there is no assurance of existence of a Nash cooperative solution; Theorem 6.4 applies to games having such equilibria. Existence is assured if the strictly competitive game whose payoff function is $P^*(s)$ satisfies the usual concavity condition that $P_i^*(s)$ is concave in s_i.

6 Applications of two-person cooperative games

Contract negotiations between a union and a firm have been cited several times as an example of a two-person cooperative game. This is strictly correct if the union is the only source of labor for the firm and the firm is the only source of employment for the union members. In Section 6.1, this situation is examined using a simple model. then, in Section 6.2 a two-country arms race is modeled.

6.1 Labor–management: An example of bilateral monopoly

Imagine a firm that is a monopolist in the market for its output and a monopsonist in the labor market. Suppose a labor union is a monopolist in the labor market. Thus, the situation in the labor market is one of bilateral monopoly. Letting L denote the level of employment and w, the wage rate, suppose the union has a utility function $u(L, w) = (Lw)^{.5}$. In other words, the union values both numbers employed and the rate of pay per unit of work. The firm values profits, and measures utility by profit. The inverse demand function facing the firm is $p = 100 - q$, where p and q are price and output, respectively. The firm's production function is $q = L$; that is, output is proportional to employment. Thus, the firm's profit function is $\pi = L(100 - L) - wL$.

In this situation, the most natural threat on the part of the firm is to cease production, which means the union members will all be without jobs. Similarly, the union can refuse to supply any labor to the firm, and, again, there will be neither production nor jobs. Looking for a Nash solution, then (0, 0) is the threat point, and the game is a Nash bargaining game. The Pareto optimal curve can be found by maximizing

$$F(L, w) = \rho\pi + (1 - \rho)u = \rho(100 - L - w)L + (1 - \rho)(Lw)^{.5} \quad (6.26)$$

with respect to L and w for values of ρ between 0 and 1. Carrying out the maximization, the first-order conditions are

$$\frac{\partial F}{\partial L} = \rho(100 - 2L - w) + \tfrac{1}{2}(1 - \rho)\frac{(Lw)^{1/2}}{L} = -\rho L + \frac{\rho\pi}{L} + \frac{(1 - \rho)u}{2L} = 0 \quad (6.27)$$

$$\frac{\partial F}{\partial w} = -\rho L + \tfrac{1}{2}(1 - \rho)L(Lw)^{-1/2} = -\rho L + \frac{(1 - \rho)L}{2u} = 0 \quad (6.28)$$

From $\pi = 100L - L^2 - wL$ and $u = (Lw)^{1/2}$ we have $\pi = 100L - L^2 - u^2$ or $\pi + u^2 = 100L - L^2$. From equation (6.28), $\rho = 1/(2u + 1)$, which can be used in equation (6.27) to obtain $\pi + u^2 = L^2$. Therefore, $100L - L^2 = L^2$, $L = 50$, and

$$\pi = 2500 - u^2 \quad (6.29)$$

which is the equation of the Pareto optimal curve. Now the Nash solution is easily found by maximizing

$$u\pi = 2500u - u^3 \quad (6.30)$$

with respect to u. The outcome is $u = 16.67(3)^{.5}$ and $\pi = 1666.67$. These payoffs are associated with employment of $L = 50$ and a wage rate of $w = 16.67$.

6.2 An arms race

A usual presumption in an arms race is that each party values the extent to which it is stronger than its rival. This sounds like a zero-zum situation, and it would be if arms were costless. Let x denote the arms level of player 1 and y the arms level of player 2. For each player, utility consists of two terms. The first is the contribution to utility that comes from the relative arms strength of the players; the second is a negative term that results from the cost of the arms that the player purchases. The two players' utility levels are

$$u_1 = 20(x - y) - x^{1.5} \quad (6.31)$$

$$u_2 = 10(y - x) - y^{1.5} \quad (6.32)$$

The Pareto optimal curve is found in the same way as in the bilateral monopoly example; Maximize $\rho u_1 + (1-\rho)u_2$ with respect to x and y. The first-order conditions are

$$(30 - 1.5x^3)\rho - 10 = 0 \tag{6.33}$$

$$-30\rho - 1.5(1-\rho)y^{.5} + 10 = 0 \tag{6.34}$$

Solving for u_2 as a function of u_1 is not easy in this case, so instead, equations (6.33) and (6.34) can be used to obtain x and y as functions of ρ. These give

$$x = \left(\frac{60\rho - 20}{3\rho}\right)^2 \tag{6.35}$$

$$y = \left(\frac{20 - 60\rho}{3(1-\rho)}\right)^2 \tag{6.36}$$

Both x and y must be nonnegative, so equation (6.35) is correct for $\rho \geqslant 1/3$ and $x = 0$ otherwise. Likewise, equation (6.36) holds for $\rho \leqslant 1/3$ and $y = 0$ otherwise. This makes sense, because it can never be Pareto optimal for *both* players to have positive arms expenditure. It will always be possible to reduce both players' arms expenditure by the amount of the smaller expenditure of the two. This keeps the value of $x - y$ constant and reduces costs for both of them.

Figure 6.11 shows the Pareto optimal frontier for this game. Unlike the bilateral monopoly example, here players can spend on arms whether or not they make an agreement; therefore, the threat situation may be more complicated. There are two ways to look at their threat behavior. In the absence of agreement, they may simply play noncooperatively. If so, they

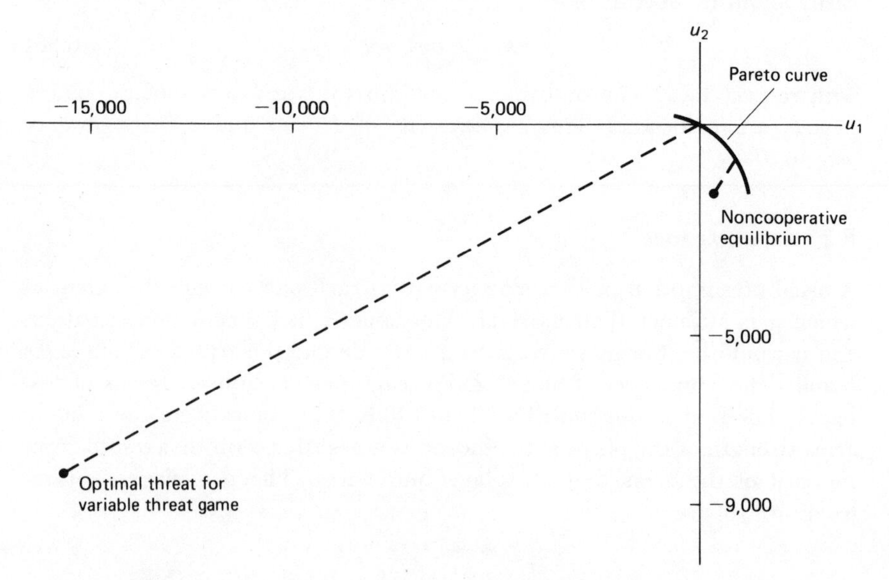

FIGURE 6.11 Optimal threats in a variable-threat arms race.

may take the Nash noncooperative equilibrium as the threat point for a Nash bargaining game. The noncooperative equilibrium payoffs are $u_1 = 296.3$ and $u_2 = -1629.6$, and the Nash bargaining solution, based on the noncooperative equilibrium as a threat point, is $u_1 = 870.7$ and $u_2 = -779.5$. Whether there is a Nash variable-threat outcome depends on the ability of the players to choose threat outcomes at will. Choosing arbitrary threat outcomes is possible if the players are able to make commitments, or if they have agreed to a binding arbitration under which they must take the Nash cooperative solution that is based on whatever threat point results from threat strategies. The latter is not plausible in the context of an arms race, because some higher authority must be present to enforce the binding arbitration. The former can be imagined. Each country might set in motion internal policies that cannot be changed and that force selection of arbitrary threat strategies should there fail to be an agreement. Supposing this is the case, it is easily checked that the optimal threats are $x^T = 711.1$ and $y^T = 177.8$, with threat payoffs being $u_1^T = -15{,}592.6$ and $u_2^T = -8{,}296.3$. The associated Nash solution is $x = 0$ and $y = 0$, where the payoffs are $u_1 = 0$ and $u_2 = 0$.

7 Concluding comments

The foregoing models are often called bargaining models, as if the outcomes they prescribe are the likely outcomes of a bargaining process. In fact, as the reader can easily see, most of these models describe a solution, or equilibrium, without showing any kind of actual sequential bargaining process. The two exceptions are Harsanyi's interpretation of Zeuthen and Rubinstein's model. The Zeuthen process is not convincing as a process that bargainers, acting in their own interests, are likely to follow. Rubinstein's model is more convincing on this score. Of course, the absence of a convincing process to lead to a particular outcome does not mean that outcome cannot be the end point of such a process; it only means that nothing has yet been found.

The various solutions are of interest apart from bargaining. Any one of them could be used by an arbitrator to settle a dispute and any could be built into the laws of a society or into the rules of an organization to deal with situations fitting the model.

Perhaps the greatest limitation of these models is their restriction to two players. This objection can be met, on the one hand, by noting that some of these models can be generalized to more than two players, and, on the other, by noting that many gamelike situations are comprised of two-persons. Thus two-person theory has considerable scope for practical application. Indeed, the word *bargaining* brings to mind, above all, a pair of people (or organizations) confronting one another: the union versus the company, the owner of a house versus a potential purchaser, and so forth.

Exercises

1. Suppose Barbara and Don can trade apples for oranges, but they must trade integer quantities. Barbara has two apples and no oranges, and Don has no apples and two oranges. The utility each attaches to apple–orange commodity bundles is listed as follows.

		Utility for	
Apples	Oranges	Barbara	Don
0	0	0	5
1	0	4	10
0	1	6	8
2	0	10	11
0	2	12	15
1	1	14	25
1	2	20	36
2	1	16	35
2	2	21	37

   a. Describe the situation that Barbara and Don face as a Nash bargaining game.
   b. What is the Nash solution of the game?
2. a. For the game in problem 1, find the point of minimal expectations and the ideal point.
   b. What is the Raiffa–Kalai–Smorodinsky solution of the game?
3. a. For the game in problem 1, find the solution and reference point associated with the center of the smallest rectangle.
   b. Find the solution and reference point associated with the point of minimal compromise.
4. Suppose the set H consists of the points lying on and within a circle of radius 2, having a center at $(2, 2)$. If the threat point, d, is at $(2, 2)$, what are the Nash, Raiffa–Kalai–Smorodinsky, minimal expectations, minimum compromise, and smallest rectangle solutions? What are the reference points?
5. Suppose H is as in problem 4, except that $d = (0, 2)$. What are the Nash, Raiffa–Kalai–Smorodinsky, minimal expectations, minimum compromise, and smallest rectangle solutions? What are the reference points?

Notes

1. It is possible to distinguish between the status quo, denoted by a pair of utility levels that were achieved prior to the start of bargaining, and the threat point, where the players will be if they bargain and fail to agree. Given that they will bargain, it is not clear that the status quo concept should be relevant to the outcome. Brito, Buoncristiani, and Intriligator (1977) analyze a model in which the status quo and threat point are different.
2. It is, in fact, possible that reasonable persons will be able to make binding agreements and fail to achieve a Pareto optimal outcome. This could occur when a bargaining process allows players to make mutually incompatible demands and, when such demands are made, leaves the players at the threat point. Where

solutions are explored in the absence of a bargaining process, the Pareto optimality requirement is easy to accept.

3. If $d = 0$ were not assumed, equation (6.1) would use

$$x_i = \frac{u_i^* - d_i}{y_i^* - d_i}(y_i - d_i)$$

to define H', and suitable changes would be required in the remainder of the proof. The details are left to the reader.

7

n-person cooperative games with transferable utility

In this chapter and Chapter 8, n-person cooperative games are examined and several solution concepts are reviewed. This chapter assumes *transferable utility*, whereas Chapter 8 does not. This chapter is more general than Chapter 6 in that it deals with n- rather than two-person games, but it is less general in that the models of Chapter 6 did not require transferable utility.

To assume transferable utility is to require that the payoffs attainable by any particular coalition (subset of N) consist of all individual payoffs that sum to no more than a particular number. For example, suppose a five-player game and a coalition consisting of players 1 and 4. Each player has his own utility scale; however, if the transferability condition is met, then there will be a number, $v(\{1,4\})$, that describes the payoff possibilities of the coalition $\{1,4\}$. They can achieve any payoff (x_1, x_4) that satisfies the condition $x_1 + x_4 \leq v(\{1,4\})$. This condition is dropped in Chapter 8, but, in the meantime, it greatly simplifies the analysis in this chapter. If the players in the game are firms in a market, it may seem reasonable to assume transferable utility on the ground that income in money measures utility for each firm, and does so in the same way for all firms. Additionally, many models from this chapter are useful in developing results for the nontransferable utility games covered in Chapter 8.

1 An overview of the chapter

The *core,* due to Gillies (1953), is the first solution studied. It is an appealing solution; although it suffers from two faults: Some games have no outcomes in the core and others have too many outcomes in the core. The groups of players that may form and act in concert are called *coalitions.* In general, a coalition is a subset of players that has the right to make binding agreements with one another, and it is usually assumed that any subset of players can do this. Another aspect of the core is that it does not appear to take into consideration which particular coalitions will actually form. In this respect, a closely related solution concept, the *stable set* (also called the von Neumann–Morgenstern (1944) solution) at least partly recognizes actual coalition formation. Relative to the core, the stable set has the advantage that many games with empty cores have (nonempty) stable sets.

Similar advantages are associated with the *bargaining set,* a solution that exists for a large class of games and that takes explicit account of the way players divide themselves into coalitions. A payoff vector is considered in conjunction with a specific division of the players into coalitions. Such a division of players is called a *coalition structure.* A particular payoff vector–coalition structure pair is in the bargaining set (i.e., is a solution) if no subset of players who are already members of the same coalition can find a way of improving their payoffs. There are specific rules for improving payoffs that involve surmounting two hurdles. The first is that these players, who are said to form an *objection,* are able to do better for themselves with the aid of a few other players, called their *partners,* who also benefit from the change. The second is that the other players cannot mount a successful *counterobjection.* A counterobjection is successful if these other players can do as well as they did originally and can ensure more to the partners of the objectors. Closely related to the bargaining set are the *kernel* and the *nucleolus.*

A solution concept that yields a unique outcome in a large class of games is the *Shapley* (1953*b*) *value* and the somewhat similar solution proposed by Banzhaf (1965) and axiomatized by Owen (1978). The Shapley value differs markedly from the solutions covered earlier in the chapter in that it is unique. It is defined by several conditions, called the Shapley axioms, and is characterized by a specific function that gives the payoff to each player as a function of the characteristic function of the game. All of these solutions are discussed in this chapter.

Section 2 contains basic concepts used in the study of characteristic function games. Section 3 introduces the core and gives conditions for a nonempty core. Section 4 takes up the stable set, and Section 5 is devoted to the bargaining set and its cousins, the kernel and the nucleolus. The Shapley and Banzhaf values are discussed in Section 6, an example of political power is examined in Section 7, and concluding comments are made in Section 8.

2 Basic concepts for cooperative games

2.1 Coalitions

As in previous chapters, $N = \{1, 2, \ldots, n\}$ is the set of players. Subsets of players, called *coalitions,* are denoted K, L, M, and so forth. The corresponding lowercase letters denote the number of players; thus, K has k players, L has l players, and M has m players. The actual members of, say, K are $\{i_1, \ldots, i_k\}$, where each i_j is distinct from all the others and is an integer between 1 and n. For example, suppose $n = 50$ and $K =$

$\{1, 3, 5, 8, 12, 24, 46\}$. Then $k = 7$, $i_1 = 1$, $i_2 = 3, \ldots, i_7 = 46$. The coalition consisting of all players other than those in K, called the *complementary coalition of* K, is denoted by $N \backslash K$ and also by $\tilde{K}$. The notation $K \backslash L$ denotes the players who are in K but are not in L. Thus, $K \backslash L = K \cap \tilde{L}$.

DEFINITION 7.1 *A* **coalition** *is a subset of the set of players, N, that is able to make a binding agreement.*

ASSUMPTION 7.1 *Any subset of N, including N itself, can form a coalition.*

2.2 Characteristic functions

There is an important sense in which our point of view naturally changes when examining cooperative games: The actual strategies available to the players recede into the background; instead, attention focuses on what payoffs the players and coalitions are able to achieve for themselves. These possibilities are summarized in the *characteristic function,* which is defined as follows.

DEFINITION 7.2 *The* **transferable utility characteristic function** *of a game having the set of players N is a scalar valued function, $v(K)$, that associates $v(K) \in R$ with each $K \subset N$. The characteristic function value for the empty coalition is 0. That is, $v(\phi) = 0$.*

In the rest of this chapter, the term *characteristic function* always means *transferable utility characteristic function.* In Chapter 8, the characteristic function for nontransferable utility games is defined. The quantity $v(K)$ is interpreted in this chapter as the maximum payoff to members of the coalition K that the coalition can guarantee to itself.

Describing the way that characteristic functions can be related to games in strategic form requires that some useful concepts be developed. First, a notion of what a player can guarantee herself alone is needed. This is commonly done in one of two ways. Both are described in Section 2.2.1. Section 2.2.2 defines the characteristic function form of the game.

2.2.1 *The α and β characteristic functions*

A characteristic function is usually derived from a game in strategic form in one of two ways. The first, giving rise to the *α-characteristic* function, is based on what a player (or coalition) can guarantee herself when the remaining players act to minimize her payoff. The second, called the *β-characteristic* function, is based on the payoff to which the remaining players can hold a player (or coalition).

The α-characteristic function is the more frequently encountered concept, and is defined first. Let $\Gamma = (N, S, P)$ be a game in strategic form, and let the joint strategy space of the players in a coalition K be $S^K = \times_{i \in K} S_i$. Elements of S^K are denoted s^K. It will be convenient to use the notation $(s \backslash t^K)$ to denote the strategy combination in which player i is using s_i if $i \notin K$ and t_i if $i \in K$. Also, $(s\backslash t_i)$ is defined to be $(s\backslash t^{\{i\}})$. A player can be utterly certain of no more than her *maximin* value in this game. That is, imagine that player i is in a situation in which the remaining players, $N\backslash\{i\}$, form a coalition with the express purpose of minimizing the payoff of player i. Then the largest payoff that player i can assure herself is

$$y_{\alpha i} = \max_{t_i \in S_i} \min_{s^K \in S^K} P_i(s\backslash t_i) \qquad \text{for } K = N\backslash\{i\} \tag{7.1}$$

Let u^K denote a payoff vector in R^k giving payoffs for members of the coalition K, and let y^K be the vector of maximin payoffs of the members of K.

ASSUMPTION 7.2 *If $u^K \in R^K$ can be achieved by the coalition K, then K can achieve any u'^K satisfying*

$$\sum_{i \in K} u_i' \leq \sum_{i \in K} u_i \tag{7.2}$$

Assumption 7.2 embodies two conditions. The first is the transferable utility condition that, if K can achieve u^K, then it can achieve any other outcome (u'^K) in which the players' payoffs sum to the same total $(\sum_{i \in K} u_i' = \sum_{i \in K} u_i)$. The second condition is free disposal. If u^K can be achieved and $u'^K < u^K$, then u'^K can also be achieved. Assumption 7.2 allows a very compact way to represent the payoffs obtainable by a coalition. For a given game, each coalition has associated with it the largest total payoff that it is capable of attaining. To define the α-characteristic function, these are maximin payoffs for the coalition. The α-characteristic function is

$$v_\alpha(K) = \max_{t^K \in S^K} \min_{s^{\tilde{K}} \in S^{\tilde{K}}} \sum_{i \in K} P_i(s\backslash t^K) \tag{7.3}$$

By definition, $v_\alpha(\{i\}) = \underline{u}_{\alpha i}$; however, equation (7.3) is defined for all coalitions K.

Equations parallel to (7.1) and (7.3) that define the β-characteristic function are

$$\underline{u}_{\beta i} = \min_{s^K \in S^K} \max_{t_i \in S_i} P_i(s\backslash t_i) \qquad \text{for } K = N\backslash\{i\} \tag{7.4}$$

$$v_\beta(K) = \min_{s^{\tilde{K}} \in S^{\tilde{K}}} \max_{t^K \in S^K} \sum_{i \in K} P_i(s\backslash t^K) \tag{7.5}$$

In contrasting the α and β forms, it is helpful to perform a thought experiment in which K for α (or $N\backslash K$ for β) moves first. In defining v_α, suppose that K must choose s^K and announce it to $N\backslash K$, and then, knowing s^K, $N\backslash K$ can select $s^{N\backslash K}$. With this procedure, K knows that, for any s^K it chooses it can be held to

$$\min_{s^{\tilde{K}} \in S^{\tilde{K}}} \sum_{i \in K} P_i(s\backslash s^K) \tag{7.6}$$

K can guarantee itself the maximum over s^K of equation (7.6), which is equation (7.3). By contrast, to define v_β, suppose that $N\backslash K$ chooses $s^{N\backslash K}$, announces it to K, and then K selects s^K. Then $N\backslash K$ can hold K down to

$$\max_{s^K \in S^K} \sum_{i \in K} P_i(s \backslash s^K) \tag{7.7}$$

In general, $v_\alpha(K) < v_\beta(K)$ for all K. This is entirely analogous to the security levels of the two players in a two-person zero-sum game. In defining $v_\alpha(K)$ and $v_\beta(K)$, the technique is to treat K and $N\backslash K$ as the two players in a zero-sum game in which the payoff to player K is $\sum_{i \in K} P_i(s)$ and to player $N\backslash K$ is $-\sum_{i \in K} P_i(s)$.

2.2.2 *The characteristic function form*

The following results and assumptions are stated for characteristic function games in general, and hold whether the α or β form is used. To simplify notation, $\underline{u}$ is used in place of $\underline{u}_\alpha$ and $\underline{u}_\beta$, and v is used in place of v_α and v_β. It is generally assumed that coalitions can achieve at least as much as the sum of what their members can achieve. This condition is called *superadditivity,* and is defined in terms of any partition of a subset. That is, if K and L are subsets of N with $K \cap L = \varnothing$, then $v(K \cup L) \geqslant v(K) + v(L)$.

ASSUMPTION 7.3 *The characteristic function, $v(K)$, for a game (N, S, P) is superadditive. That is, for any disjoint coalitions, K and L contained in n, $v(K \cup L) \geqslant v(K) + v(L)$.*

It is convenient to refer to the characteristic function and set of players, (N, v), as a game rather than mention (N, S, P), because $\Gamma = (N, v)$ contains all the needed information. This description of the game is called the *characteristic function form,* or the *coalitional form.*

DEFINITION 7.3 *For transferable utility games, the* **characteristic function form** *of a game, also called the* **coalitional form,** *is given by $\Gamma = (N, v)$. It is characterized by the set of players, N, and the characteristic function, v.*

2.3 Imputations and essential games

Certain payoff vectors are naturally singled out for study as the set containing all reasonable outcomes for a cooperative game; this is called the *set of imputations.* An *imputation* is a payoff vector that gives each player at least as much as he can guarantee himself and gives all players together $v(N)$.

DEFINITION 7.4 *A payoff vector, $x \in R^n$, is an* **imputation** *in the game $\Gamma = (N, v)$ if $x_i \geqslant \underline{u}_i$ for all $i \in N$ (i.e., x is individually rational) and $\sum_{i \in N} x_i = v(N)$ (i.e., x is group rational). The set of imputations is denoted $I(N, v)$.*

In this chapter and Chapter 8, a payoff vector is sometimes called an *allocation.* This terminology is natural in the sense that an imputation, in particular, is a payoff vector in which $v(N)$ is allocated among the n players.

Characteristic function games are naturally divided into two categories, depending on whether there is anything of interest in them to analyze. A game satisfying Assumptions 7.1 to 7.3 in which $\sum_{i\in N} v(\{i\}) = v(N)$ is a game in which no coalition can possibly achieve more than the individual members can do on their own. Such a game is called *inessential.* Games in which $v(N)$ is strictly greater than $\sum_{i\in N} v(\{i\})$ are called *essential,* and it is these games that are the center of attention in the next section.

2.4 Domination

A key concept in the study of n-person cooperative games is *domination,* which refers to the power a coalition can exert through its ability to go it alone. Recalling that $I(N, v)$ denotes the set of imputations, suppose that x, $y \in I(N, v)$. Then *x is dominated by y via the coalition K* if y gives more to the members of K than does x and y^K is achievable by K.

DEFINITION 7.5 *For $x, y \in I(N, v)$, y* **dominates** *x* **via** *K if $y^K > x^K$ and $\sum_{i\in K} y_i \leq v(K)$.*

The definition of domination is

DEFINITION 7.6 *y* **dominates** *x if, for some $K \subset N$, y dominates x via K.*

If y dominates x via K, it is sometimes said that K can *improve on x.*[1] The concept of domination is used to define the core.

3 The core and the ε-core

Many games have empty cores, a drawback of the core as a solution concept. Facing a game with an empty core, one is tempted to ask the question of how "close" the game is to having a nonempty core. This query naturally leads into the study of the ε-core, introduced in Section 3.3. Section 3.1 includes an interesting historical example that is an early instance of the core; the core itself is taken up in Section 3.2. Finally, Section 3.4 is concerned with a special class of games called *simple games.*

3.1 Edgeworth and the core

The core is a cooperative game solution concept that predates formal game theory, having appeared a century ago in the economics literature in Edgeworth (1881). Edgeworth discusses equilibrium for a pure trade economy in which traders are not forced to transact with one another through conventional markets utilizing competitive prices over which they have no individual control. Instead, Edgeworth places no restrictions on the institutions of the economy, and any group of traders can engage in any trade that falls within the resources at their command. That is, a particular collection of traders (Mary, Harry, Barry, Larry, and Cary, for example) can agree to redistribute among themselves the resources that they own in any fashion they wish. Prices need not enter such an arrangement, and no trader need be able to transact in a conventional competitive market. With this setup, Edgeworth argued that certain trades could be ruled out and others could not as final trades for the economy. This set of possible equilibria consisted of trades that gave to each trader a level of utility no lower than he would achieve by consuming his own commodity bundle, and, in addition, gave to each subset of consumers utility levels that the group could not improve upon on its own.

This last point can be illuminated by an example. Suppose that trader i (i = Mary, Harry, and so forth), has an endowment of m goods, $w^i = (w_1^i, \ldots, w_m^i)$, that the utility of that endowment bundle is $u_i(w^i) = 0$, and that there is a trade available to Mary, Harry, and Larry as a group that gives them, respectively, utilities of 5, 2, and 8. Edgeworth argued that no acceptable final outcome could (a) give any trader a utility less than 0 and (b) simultaneously give Mary a utility of less than 5, Harry less than 2, and Larry less than 8. The reasoning behind (a) is that no one is forced to trade, and hence cannot be forced to end up worse off than $u_i(w^i)$. The reasoning behind (b) is that such a proposal would never get the agreement of the three of them, nor could they be forced to accept it; for, on their own, they could provide themselves with $(5, 2, 8)$.

Similar reasoning can be applied to any outcomes achievable by any subset of traders, including the subset consisting of all players. The trades that are left as possible outcomes constitute trades associated with payoff vectors in the core. Edgeworth believed that competitive equilibrium points were always in the core and that, as the number of traders became very large, the core converged to the set of competitive equilibria. Subsequent research has proved Edgeworth's conjecture to be correct (see Debreu and Scarf (1972)). Edgeworth's game is one of nontransferable utility. A model in the spirit of his is discussed in the applications at the end of Chapter 8.

3.2 The core

3.2.1 Two examples illustrating empty and nonempty cores

Given a game $\Gamma = (N, v)$, there is a well-defined sense in which the core, $C(N, v)$, is not empty if $v(N)$ is sufficiently large relative to the values of the $v(K)$ for other coalitions K. This is obvious after a little thought. Consider a family of games that are identical in every way except for the value of $v(N)$. That is, any two games (N, v) and (N', v') have the same set of players $(N = N')$, and for all K having fewer than n players, $v(K) = v'(K)$. If $v(N)$ is small relative to the $v(K)$, then there will be no core. As an example, look at the game in which three players are told they can divide \$100 any way they wish. An agreement signed by two or more players that gives a division of the money into three parts is binding on all three of them. Suppose that utility is measured by money in this game. Then $v(\{i\}) = 0$ for all i and $v(K) = 100$ for any coalition having two or three members. No matter what imputation is tentatively proposed, there is a way for two players to form a coalition that dominates it. For instance, if (50, 50, 0) is proposed, players 1 and 3 can dominate it with (75, 0, 25). Consider any imputation, (x, y, z). If, for example, x, $z > 0$, then players 2 and 3 dominate (x, y, z) with $(0, y + x/2, z + x/2)$; however, this can be dominated by players 1 and 2 proposing $(z/2, x + y + z/2, 0)$, and so it goes.

Now imagine a variant of this game, denoted (N, v'), in which contracts signed by two players remain binding and determine the allocation of payoffs to all three players, but suppose that a contract signed by all three players can distribute \$150. That is, the rules of this new game state that a pair of players can receive \$100, divided between the two of them however they agree, but that three players can receive \$150, divided among them however they agree. Thus, $v'(\{1,2\}) = v'(\{1,3\}) = v'(\{2,3\}) = 100$ and $v\{i\} = 0$ $(i \in N)$ as before, and $v'(\{1,2,3\}) = 150$. For this game, the imputation (50, 50, 50) is in the core, because each two-person coalition receives \$100, and each individual receives at least 0. Similarly, suppose the coalition of three can achieve $\$y \geq 0$, while, as previously, two-person coalitions can achieve \$100. Then an allocation (a, b, c) is in the core if a, b, $c \geq 0$, $a + b$, $a + c$, $b + c \geq 100$, and $a + b + c = y$.

3.2.2 *Existence of the core*

The core is formally defined in Definition (7.7).

DEFINITION 7.7 *The* **core** *of the game* $\Gamma = (N, v)$, *denoted* $C(N, v)$, *is a subset of the set of imputations consisting of the imputations that are not dominated.*

Another way of stating Definition 7.7 is that an imputation y is in the core if $\sum_{i \in K} y_i \geq v(K)$ for all coalitions K. It is possible to determine if a game $\Gamma = (N, v)$ has a nonempty core by solving a linear programming problem in which the objective function is the minimal value of $v(N)$ that allows a nonempty core. The form of the problem is

$$\text{minimize} \sum_{i \in N} x_i \tag{7.8}$$

$$\text{subject to} \sum_{i \in K} x_i \geq v(K) \text{ for all } K \subset N,\ K \neq N \tag{7.9}$$

For the family of three-person games discussed above, this linear programming problem is

$$\text{minimize } x_1 + x_2 + x_3 \tag{7.10}$$

Subject to

$$\left.\begin{array}{llll} x_1 & & & \geq 0 \\ & x_2 & & \geq 0 \\ & & x_3 & \geq 0 \\ x_1 & + x_2 & & \geq 100 \\ x_1 & & + x_3 & \geq 100 \\ & x_2 & + x_3 & \geq 100 \end{array}\right\} \tag{7.11}$$

Clearly, if Assumptions 7.1 to 7.3 hold for a game $\Gamma = (N, v)$, the problem posed in equations (7.8) and (7.9) can be solved. Any solution is bounded below by $\sum_{i \in N} \underline{u}_i$, and there are obviously large finite values for the x_i that will satisfy all constraints (see, for example, Hillier and Lieberman (1974)). Let Z denote the solution to the linear programming problem posed in equations (7.8) and (7.9). Then

LEMMA 7.1 *Let (N, v) be a game satisfying Assumptions 7.1 to 7.3. Then the core, $C(N, v)$, of the game is nonempty if and only if $Z \leq v(N)$.*

Proof. The proof is obvious from the structure of the programming problem. QED

3.3 The ε-core

The *epsilon core,* introduced by Shapley and Shubik (1966), consists of imputations that are within ε of being in the core. Thus, any game satisfying Assumptions 7.1 to 7.3 has an ε-core for a large enough ε. To be in the core, an imputation must give at least $v(K)$ to each coalition K, but to be in the ε-core, an imputation need only give $v(K) - \varepsilon$.

DEFINITION 7.8 *The* **ε-core** *of a game (N, v), denoted $C_\varepsilon(N, v)$, is a subset of the set of imputations. $y \in I(N, v)$ is an element of $C_\varepsilon(N, v)$ if $\sum_{i \in K} y_i \geq v(K) - \varepsilon$ for all coalitions K.*

As an example, recall the game in which three players divide \$100. If $\varepsilon = 33\frac{1}{3}$, then the ε-core is not empty and consists of the single point $(33\frac{1}{3}, 33\frac{1}{3}, 33\frac{1}{3})$.

There is nothing inherent in the definition of the ε-core that requires ε to be nonnegative. If negative values are also admitted, there is a unique value ε^* such that $C_\varepsilon(N, v) \neq \emptyset$ if and only if $\varepsilon \geq \varepsilon^*$.

LEMMA 7.2 *Let (N, v) be a game satisfying Assumptions 7.1 to 7.3. For this game there is a unique ε^* such that the ε-core of the game is not empty if and only if $\varepsilon \geq Z - v(N)$.*

Proof. By the definition of the ε-core, there is a finite value of ε sufficiently large that the ε-core is not empty. Similarly, there is a finite value, possibly negative, small enough that the ε-core is empty. It also follows that if the ε-core is not empty for ε', then it is not empty for any larger value of ε. Therefore, it is clear that there exists a value ε^* such that, for $\varepsilon > \varepsilon^*$, the ε-core is not empty and for $\varepsilon < \varepsilon^*$, the ε-core is empty. It remains to show that the ε-core is not empty for $\varepsilon = \varepsilon^*$. This follows from the inequalities defining the ε-core being weak inequalities. QED

The ε-core associated with the value ε^* from Lemma 7.2 is called the *least core,* because it is the smallest nonempty core associated with the game (N, v).

3.4 The zero–one normalized form and simple games

In some games a coalition is naturally described as either *winning* or *losing*. Such games are called *simple games,* and are the focus of most of this section. Prior to taking them up, a normalized form of the characteristic function, called the *zero–one normalized form,* is defined in Section 3.4.1. If two games have the same normalized form, then they are equivalent from the point of view of many solution concepts. For the study of simple games, the normalized form is particularly natural and convenient. Section 3.4.2 defines simple games, provides two examples, and contains results on the core of such games.

3.4.1 The zero–one normalized form

Two games may have different characteristic functions and be the same in their essentials. Whether this is true for a given pair of games is determined by transforming each game to its *zero–one normalized form.* If the two games reduce to the same zero–one normalized form (or *zero–one normalization*), then they are regarded as belonging to the same equivalence class and as being the same in their essentials. For instance, the cores, the bargaining sets, and the Shapley values of the two games will be related by the transformations that define their common zero–one normalization.

DEFINITION 7.9 *Let (N, v) be a game. Then the* **zero–one normalization of** *(N, v), denoted (N, v^*), is given by*

$$v^*(K) = \frac{v(K) - \sum_{i \in K} v(\{i\})}{v(N) - \sum_{i \in N} v(\{i\})}, \qquad K \subset N \tag{7.12}$$

Games having identical zero–one normalizations are called *S-equivalent.* S-equivalence is proved in Lemma 7.3 to be an equivalence relation. That is, a relation that is reflexive (Γ is S-equivalent to Γ), symmetric (if Γ is S-equivalent to Γ', then Γ' is S-equivalent to Γ), and transitive (if Γ is S-equivalent to Γ' and Γ' is S-equivalent to Γ'', then Γ is S-equivalent to Γ'').

DEFINITION 7.10 *Two games, $\Gamma = (N, v)$ and $\Gamma' = (N, v')$, that have the same zero–one normalized form are* **S-equivalent.**

It is proved in Lemma 7.4 that the core of (N, v) and the core of its zero–one normalization, (N, v^*), are related by the transformation in equation (7.12). That is, $u \in C(N, v)$ if and only if $u^* \in C(N, v^*)$, where

$$u_i^* = \frac{u_i - v(\{i\})}{v(N) - \sum_{i \in N} v(\{i\})} \tag{7.13}$$

LEMMA 7.3 *S-equivalence is an equivalence relation.*

Proof. That S-equivalence is reflexive and symmetric is obvious. For transitivity, note that, if Γ and Γ' have the same zero–one normalized form, and Γ' and Γ'' do also, then Γ and Γ'' must share the same zero–one normalized form. QED

LEMMA 7.4 *Let (N, v) be any game satisfying Assumptions 7.1 to 7.3 whose zero–one normalization is (N, v^*), and let $u^* \in I(N, v^*)$ be related to $u \in I(N, v)$ by*

$$u_i^* = \frac{u_i - v(\{i\})}{v(N) - \sum_{i \in N} v(\{i\})} \tag{7.14}$$

Then $u^* \in C(N, v^*)$ if and only if $u \in C(N, v)$.

Proof. u is in the core of (N, v) if and only if $\sum_{i \in K} u_i \geqslant v(K)$ for all K. Let $\sum_{i \in K} u_i = v(K) + a_K$. Then u is in the core if and only if none of the a_K is negative. From equation (7.14),

$$\sum_{i \in K} u_i^* = \frac{\sum_{i \in K} [u_i - v(\{i\})]}{v(N) - \sum_{i \in N} v(\{i\})} \tag{7.15}$$

Using $\sum_{i \in K} u_i = v(K) + a_K$ in equation (7.15) gives

$$\sum_{i \in K} u_i^* = \frac{v(K) - \sum_{i \in K} v(\{i\})}{v(N) - \sum_{i \in N} v(\{i\})} + \frac{a_K}{v(N) - \sum_{i \in N} v(\{i\})}$$

$$= v^*(K) + \frac{a_K}{v(N) - \sum_{i \in N} v(\{i\})} \tag{7.16}$$

Thus, $u^* \in C(N, v^*)$ if and only if all the a_K are nonnegative, which means that $u^* \in C(N, v^*)$ if and only if $u \in C(N, v)$. QED

3.4.2 Simple games

There is a category of games, called *simple games,* in which a coalition is either *winning* or *losing*. These games are characterized by a fixed gain from coalition formation that a winning coalition obtains.

DEFINITION 7.11 $\Gamma = (N, v)$ *is a* **simple game** *if the zero–one normalization of (N, v), denoted (N, v^*), satisfies (a) $v^*(K) = 1$ or $v^*(K) = 0$ for all $K \subset N$. (b) If $v^*(K) = 1$, then K is called a* **winning coalition**, *and if $v^*(K) = 0$, then K is called a* **losing coalition.** *(c) (N, v) is superadditive. (d) $v^*(N) = 1$.*

In the definition of a simple game, the notion that a coalition K is winning is captured by $v(K)=1$; the notion that is losing is captured by $v(K)=0$. If the game were not zero–one normalized, the payoff to a winning coalition would be $v(K)=a+\sum_{i\in K} v(\{i\})$ and, the payoff to a losing coalition would be $v(K)=\sum_{i\in K} v(\{i\})$. Winning would add a fixed amount, a, to the sum of what the individual members could achieve on their own. The definition implies (from superadditivity) that the complement of a winning coalition is a losing coalition; however, note that the converse need not hold, as an example will illustrate. The superadditivity condition means that if K is a winning coalition, then adding members to K will result in a coalition that is still winning. Therefore, if K is winning, then $N\backslash K$ must be losing. Otherwise, $v(K)=v(N\backslash K)=1$ and superadditivity would be violated. The definition implies that the game is essential; that is, that there exists at least one winning coalition, N, and at least one losing coalition. This is because, for $n\geqslant 2$, both $\{1\}$ and $\{2\}$ cannot be winning. If they were, then $v(\{1\})=v(\{2\})=v(N)=1$, which violates superadditivity.

As an example of a simple game, imagine a committee or a legislature in which a simple majority of the members has full power. Then $v(K)=1$ if $k>n/2$ and $v(K)=0$ if $k\leqslant n/2$. (For a study of politics under the assumption that some political processes are characterized by simple games, see Riker (1962).) Other examples of voting situations as simple games include majority games in which substantially more than half the votes are required to win, and weighted majority voting games in which player i has a weight $w_i>0$. If a vote total of w_0 $(>\sum_{i\in N} w_i/2)$ is required to win in a weighted majority game, $v(K)=1$ if $\sum_{i\in K} w_i\geqslant w_0$ and $v(K)=0$ otherwise. The winning majority could be anything larger than a half, including 100 percent of the weighted votes.

As an example of a game in which the complement of a losing coalition is a winning coalition, consider a simple majority voting game in which each player has one vote and the number of players is odd. Then $v(K)=1$ if and only if $v(N\backslash K)<n/2$; for either $k>n/2$ or $(n-k)>n/2$. Now consider a majority voting game in which each player has one vote, but a winning coalition requires 75 percent of the votes. If a coalition K has 60 percent of the votes and $N\backslash K$ has 40 percent, then both K and $N\backslash K$ are losing.

A key fact about simple games is that they always have empty cores if they are at all interesting. Roughly speaking, interesting simple games are those having some (nonempty) losing coalitions and at least two winning coalitions. A player i can be thought of as a *dictator* if she is essential to all winning coalitions. In other words, for i to be a dictator means that if i does not belong to the coalition K, then K cannot be winning. If N is the only winning coalition, then all players are dictators. It is proved below that a simple game that has no dictators has an empty core. The proof is carried out in two steps. The first, Lemma 7.5, establishes the result for normalized games. Then, Lemma 7.6 extends Lemma 7.5 to all simple games having no dictators.

LEMMA 7.5 *Let $\Gamma = (N, v)$ be a zero–one normalized simple game. If $v(N \backslash \{i\}) = 1$ for all $i \in N$, then the core of (N, v) is empty.*

Proof. Let $u \in I(N, v)$. Then at least one player has a positive payoff under u. Suppose, without loss of generality, that $u_1 > 0$. Then u is dominated by u', where $u_i' + u_i + u_1/(n-1)$ for $i > 1$ and $u_1' = 0$. QED

In a simple game, (N, v), the reader can easily verify that K is winning if $v(K) = a + \sum_{i \in K} v(\{i\})$ and losing if $v(K) = \sum_{i \in K} v(\{i\})$, where a is a positive constant.

LEMMA 7.6 *If (N, v) is a simple game in which $N \backslash \{i\}$ is a winning coalition for all i, then the core of the game is empty.*

Proof of this lemma is left to the reader.[2]

4 The stable set

The stable set, also known as the von Neumann–Morgenstern solution, was proposed by von Neumann and Morgenstern (1944), and is closely related to the core. Recall that the core is the set of undominated imputations. Thus, an allocation in the core is not dominated by another allocation in the core, nor is it dominated by an allocation outside of the core. Let I^* be a subset of $I(N, v)$. Then I^* is a *stable set* if no allocation in I^* is dominated by any other allocation in I^*, and each allocation outside of I^* is dominated by an allocation in I^*. Two important differences between the stable set and the core are, first, that the stable set is not a unique set, and, second, that many games with stable sets also have empty cores.

DEFINITION 7.12 *A* **stable set**, *$C^*(N, v)$, is a subset of the set of imputations that satisfies two conditions: (a) If $u \in C^*(N, v)$ then u is not dominated by any other member of $C^*(N, v)$. (b) If $u \notin C^*(N, v)$, then there exists $u' \in C^*(N, v)$ such that u' dominates u.*

As an example, recall the three-person game in which any two players can determine the division of \$100 among all three players. One stable set for this game is

$$\{u \in R_+^3 \mid u_1 + u_2 = 100 \text{ and } u_3 = 0\} \tag{7.17}$$

Other stable sets are

$$\{u \in R_+^3 \mid u_1 + u_2 = 100 - a \text{ and } u_3 = a\} \text{ for } 0 < a < 50 \tag{7.18}$$

Before Lucas (1969) found a game that had no stable set it was not known whether all games had stable sets. For more on this topic (see also Owen (1982).

In thinking about cooperative games, it is natural to think about coalition formation and to wonder which coalitions will, in fact, be formed. One might expect a *coalition structure*—that is, a partition of the players into coalitions to emerge. The underlying notion here is that the players divide themselves into groups, with each group carrying out some binding agreement. With respect to the core as a solution concept, the natural coalition structure often seems to be N all by itself. In other words, for a game having a core, and in which the core is the solution, the players would, in many instances, seem to form the coalition of the whole, N, and to choose some core point. Even if none of the smaller coalitions is formed, their potential plays a role in determining the outcome. The influence of the smaller coalitions is seen in the definition of the core, where an allocation is ruled out of being in the core if it does not provide to each coalition a total payoff at least as large as that coalition could guarantee itself.

The stable set looks as if it might be closer to saying something about actual coalitions, other than the coalition of the whole, that might form. In the game in which two or three players can divide $100, the stable set consisting of all allocations giving player 3 nothing and splitting the $100 between players 1 and 2 is a set that looks like the options that are open to the coalition $\{1, 2\}$, with player 3 taking it on the chin because he is left out by the others. Still, coalitions appear in the stable set in an implicit way. The definition of a stable set does not rest on stipulating a particular coalition structure and then determining an outcome, or set of outcomes, based on some "reasonable" criteria. The latter approach is taken in the bargaining set and the kernel, which are discussed in Section 5.

5 The bargaining set, the kernel, and the nucleolus

The stable set hints at taking into account coalitions that might actually form; however, such an interpretation is implicit, not explicit, in the solution concept. In contrast, the *bargaining set* is clearly concerned with the actual coalitions that might form, because an equilibrium outcome is defined in relation to a particular partition of the players into coalitions. There are several different bargaining set concepts, only one of which is described here.

Closely related to the bargaining set are the *kernel* and the *nucleolus*. These solutions can be used to show the bargaining set to be nonempty; the kernel can be proved to be contained in the bargaining set, which means a nonempty kernel implies a nonemtpy bargaining set. The nucleolus is always associated with a payoff configuration in the kernel (hence, also in the bargaining set). Thus, if the nucleolus exists, then the bargaining set is not empty. Connections among the bargaining set, the kernel, the nucleolus, and some other constructs are in Maschler, Peleg, and Shapley (1979).

The kernel and the nucleolus may have independent appeal as solution concepts in addition to their role in proving the bargaining set nonempty. For example, the nucleolus is generally unique, and its formulation may make it attractive as a fair division concept. The bargaining set is covered in Section 5.1, the kernel in Section 5.2, and the nucleolus in Section 5.3. Examples are given in Section 5.4.

5.1 The bargaining set

A feature of the bargaining set is that it is not enough for some players to be able to improve on a particular allocation in order to rule it out. To rule out an allocation, a coalition must be able to improve on it, and it must be impossible for members of that coalition to be enticed away by another coalition that can improve on what the first coalition proposed as an alternative to the originally proposed allocation. Furthermore, only certain coalitions are allowed to try to improve on an allocation.

For example, suppose an allocation x is under review, and imagine that a coalition K proposes the allocation y as an alternative to x, where y gives all members of K as much or more than they get under x and y^K is enforceable by K. It is next determined whether there exists an allocation z and coalition L having the properties that some, but not all, members of K can be in L and all members of L get as much under z as they did under x and y. The members of K who are in L must get at least what y afforded them and the remaining members of L must get at least what x afforded them. The proposal y is called an *objection* and the proposal z is called a *counterobjection*. An allocation x is in the bargaining set if any objection to x can be met with a counterobjection. There are actually several different bargaining set concepts that vary according to the exact manner in which the missing details are specified.

5.1.1 Coalition structures and payoff configurations

The bargaining set is intimately related to what various coalitions can achieve given that the set of players is partitioned into several coalitions. In other words, a payoff vector is examined in relation to a particular partition of the players into subsets. Such a partition is a *coalition structure* and is defined as follows.

DEFINITION 7.13 *A* **coalition structure** $\mathcal{T} = (T_1, \ldots, T_m)$ *is a partition of the set of players into coalitions. Each set* T_k *is nonempty,* $T_j \cap T_k = \varnothing$ *for all* $j, k \in \{1, \ldots, m\}$ *with* $j \neq k$, *and* $\bigcup_{k \in \{1,\ldots,m\}} T_k = N$.

To make it easier to think about payoff vectors in the presence of a given coalition structure, the *payoff configuration* is defined in Definition 7.14. A payoff configuration is merely a particular payoff vector paired with a particular coalition structure. The payoff vector must also give to each coalition in the associated coalition structure precisely as much as the coalition can assure itself.

DEFINITION 7.14 *A* **payoff configuration** *is a pair* $(x, \mathcal{T})$, *where* $x \in R^n$, $\mathcal{T}$ *is a coalition structure, and* $\sum_{i \in T_k} x_i = v(T_k)$ *for* $k = 1, \ldots, m$.

A payoff configuration is a payoff vector that is group rational in a local sense that is defined by its associated coalition structure. For a payoff configuration $(x, \mathcal{T})$, each coalition in $\mathcal{T}$ receives a total payoff that is equal to what that coalition can achieve. The next definition refines this concept by adding individual rationality.

DEFINITION 7.15 *An* **individually rational payoff configuration** *is a payoff configuration,* $(x, \mathcal{T})$ *for which* $x_i \geq v(\{i\})$ *for all* $i \in N$. *The set of all individually rational payoff configurations for the game* (N, v) *relative to a coalition structure* $\mathcal{T}$ *is denoted* $I_{\mathcal{T}}(N, v)$.

5.1.2 Objections and counterobjections

The bargaining set consists of individually rational payoff configurations that pass a certain test. Roughly speaking, the test is that no *allowable coalition* has an effective complaint against the proposed payoff configuration. An allowable coalition is one that is a subset of a coalition in the coalition structure. An effective complaint is an *objection* to which there is no *counterobjection.* On the way to defining these last two concepts, the notion of *partners* of a coalition with respect to a coalition structure is needed. Partners of K in $\mathcal{T}$ are all those players who are in coalitions that also contain members of K.

DEFINITION 7.16 *For a coalition structure* $\mathcal{T}$ *and coalition* K, *the* **partners of** K *in* $\mathcal{T}$ *are the members of the set* $P(K, \mathcal{T}) = \{i \mid i \in T_j, T_j \cap K \neq \emptyset\}$.

For example, let $\mathcal{T} = (\{1, 2, 5\}, \{3, 6, 9\}, \{4, 7, 8\})$ and let $K = \{1, 2, 6\}$. Then $P(K, \mathcal{T}) = \{1, 2, 3, 5, 6, 9\}$.

An *objection* of K against L is an individually rational payoff configuration $(y, \mathcal{U})$ under which the members of K all get more than they get at $(x, \mathcal{T})$ and all partners of K get at least as much as they get under $(x, \mathcal{T})$. K and L are both subsets of the same member, T_j, of $\mathcal{T}$. An intuitive notion underlying the objection is that the members of K believe they are getting less of the payoff going to their coalition (T_j) than is fair and that the members of L are getting more than K thinks they deserve.

DEFINITION 7.17 *Suppose that* $(x, \mathcal{T}) \in I_{\mathcal{T}}(N, v)$, $K, L \subset T_j \in \mathcal{T}$, *and* $K \cap L = \emptyset$. *Then an* **objection of** K **against** L *is* $(y, \mathcal{U}) \in I_{\mathcal{U}}(N, v)$ *such that* (*a*) $P(K, \mathcal{U}) \cap L = \emptyset$, (*b*) $y_i > x_i$ *for all* $i \in K$, *and* (*c*) $y_i \geq x_i$ *for all* $i \in P(K, \mathcal{U})$.

Note that the members of K are not restricted to choosing a $\mathcal{U}$ of which K is a member, nor must K be a subset of member of $\mathcal{U}$, but they must be sure that all other players in coalitions with them under $\mathcal{U}$ are sufficiently well treated.

Roughly speaking, a *counterobjection* is a proposal of L that is similar to an objection and that leaves all interested players at least as well off as they would otherwise have been. This means that the partners of L must get at least as much as they get under $(x, \mathcal{T})$, and those partners of K under $(y, \mathcal{U})$ that are partners of L under $(z, \mathcal{V})$ must get at least as much as they would get under $(y, \mathcal{U})$. In effect, the counterobjection is good enough to dissuade some of the partners of K from joining K in the objection and good enough to keep the allegiance of L's partners.

DEFINITION 7.18 *Suppose that* $(x, \mathcal{T}) \in I_{\mathcal{T}}(N, v)$, K, $L \subset T_k \in \mathcal{T}$, $K \cap L = \varnothing$, *and* $(y, \mathcal{U}) \in I_{\mathcal{U}}(N, v)$ *is an objection of* K *against* L. *Then a* **counterobjection of** L **against** K *is* $(z, \mathcal{V}) \in I_{\mathcal{V}}(N, v)$ *satisfying (a)* $K \not\subset P(L, \mathcal{V})$, *(b)* $z_i \geqslant x_i$ *for* $i \in P(L, \mathcal{V})$, *and* $z_i \geqslant y_i$ *for* $i \in P(K, \mathcal{U}) \cap P(L, \mathcal{V})$.

Provision (a) in Definition 7.18 indicates that some members of K can be partners of L under $(z, \mathcal{V})$, but at least one member of K is excluded.

5.1.3 *Definition of the bargaining set*

The bargaining set consists of all individually rational payoff configurations for which every objection can be met with a counterobjection. Loosely speaking, a plan is in the bargaining set if, when a group of dissidents tries to bribe a few others to go along with a second plan that benefits all of them, it is possible for another group profitably to offer a third plan containing a larger bribe to those whom the dissidents need.

DEFINITION 7.19 *The* **bargaining set** $\mathcal{M}_1^{(i)}$ *is the set of all individually rational payoff configurations* $(x, \mathcal{T}) \in I_{\mathcal{T}}(N, v)$ *such that whenever some coalition* K *has an objection against a coalition* L, *at least one member of* L *has a counterobjection.*

To recapitulate, imagine a particular coalition structure, $\mathcal{T}$, and a payoff vector, x. These form an individually rational payoff configuration $(x, \mathcal{T})$. thus, x gives to each coalition in $\mathcal{T}$ a payoff just equal to what the coalition can achieve for itself. If some coalition K is a subset of $T_j \in \mathcal{T}$, it may have an objection against a coalition L that is a subset of the same member of $\mathcal{T}$. K and L must be disjoint. That K has an objection against L can be interpreted as saying that the members of K believe they are not getting as much as they think they should and that the members of L are getting too much. The objection itself is a new coalition structure $\mathcal{U}$ and an individually rational payoff configuration $(y, \mathcal{U})$ under which all members of K get more than before ($y_i > x_i$ for i in K), and all partners of K get at least as much as before ($y_i \geqslant x_i$ for partners). Any player i is a partner of K if the coalition U_j to which i belongs also contains one or more members of K. In other words, the members of K need not be in the same single coalition in the structure $\mathcal{U}$ even though they had to be in the same single coalition under the structure $\mathcal{T}$. The payoff requirements for partners of K reflects the notion that K needs the active cooperation of all players who appear in coalitions with members of K under the new coalition structure $\mathcal{U}$ that K proposes.

Now turning to the counterobjection of L to K, L must propose $(z, \mathcal{V})$, *where* z gives all members of L and all partners of L at least what they received under x ($z_i \geqslant x_i$ for members of L and partners of L). In addition, any members of K who are also partners of L must receive at least as much as they were to get from y ($z_i \geqslant y_i$ for partners of L who are members of K). If L has a counterobjection for a particular objection of K, then the objection is not considered to be viable. The reason is that K cannot get its members and its partners to go along with the objection if L can offer a counterobjection under which the counterobjectors do sufficiently well, and, in so doing, deprive K of some of the supporting players it needs for its objection. If $(x, \mathcal{T})$ is in the bargaining set, then for any objection of any coalition K against any coalition L, the coalition L has a counterobjection.

As an example, look at the game in which three players divide $100, with two or three players able to dictate the payoff vector. First, $[(0,0,0), (\{1\}, \{2\}, \{3\})]$ is in the bargaining set. The reason is that no player is in a coalition with another player against whom she can make an objection. The payoff configuration $[(50,50,0), (\{1,2\}, \{3\})]$ is also in the bargaining set. Obviously, player 2 can make an objection against player 1 and propose $[(0, 50+a, 50-a), \{1\}, \{2,3\})]$, where $0<a\leqslant 50$. To this proposal player 1 can make the counterobjection $[(50,0,50), (\{1,3\}, \{2\})]$. Thus, any payoff configuration consisting of a coalition structure and an allocation that gives 50 to each of two players who form a coalition and zero to the remaining player, who is in a coalition alone, is in the bargaining set. In the same vein, it can be shown that $[(a,a,b), (\{1,2\},\{3\})]$ for $0\leqslant b<100$ and $2a+b=100$ is in the bargaining set. Interestingly, $[(50+a, 50-a, 0), (\{1,2\}, \{3\})]$ for $a>0$ is not in the bargaining set. To see this, notice that $[(0, 50-a/2, 50+a/2), (\{1\}, \{2,3\})]$ can be put forth by player 2 as an objection against player 1, but this cannot be countered by player 1. A successful counterobjection by player 1 must give at least $50+a$ to player 1 and at least $50+a/2$ to player 3. Clearly, this cannot be done within the constraint that there is $100 to divide. Similarly, any individually rational payoff configuration $[(a,b,c), (\{1,2,3\})]$ is in the bargaining set, because any objection can be met with a counterobjection.

The bargaining set is an appealing concept, because of the interpretations given to its definition. The notion that players are, in fact, sorted out into particular coalitions is attractive. Given a coalition structure and a payoff vector, objections can be mounted only by players who are inside the same coalition, and only against one or more other members of the same coalition. This makes sense from the standpoint that these are the players who are contracting with one another at the moment. But if an objection is to be made, the objectors can go outside the immediate coalition and propose a whole new global structure (i.e., a new coalition structure and payoff vector); however, the new proposal is successful only if it cannot be countered by the group against which the objection was originally made. Thus, it is not good enough to propose a change that benefits those taking part in it; the change must be impervious to a counterproposal. All these provisions have the flavor of bargaining processes, although they are not actual processes.

A simple alternative formulation of the bargaining set is based on requiring that both K, the objecting coalition, and L, the coalition against which K objects, each consist of one player. There remains the fundamental question of whether a given game has a bargaining set, and, more importantly, whether all games within a large class can be shown to possess a bargaining set. This question is handled indirectly by moving on to the kernel and the nucleolus.

5.2 The kernel

5.2.1 Basic definitions for the kernel

Two useful concepts are defined that are related to the gains from forming various coalitions and to the comparative importance of pairs of players in coalitions. The first of these, defined in Definition (7.20), is the *excess*. It measures the payoff potential of a coalition relative to a particular payoff vector.

DEFINITION 7.20 *In a game* (N, v) *the* **excess of the coalition** K **with respect to the payoff vector** u is

$$e(K, u) = v(K) - \sum_{i \in K} u_i \tag{7.19}$$

The *surplus of i against j* can be thought of as a measure of the bargaining power of player i relative to player j. In comparing i and j in this way, one can also look at the surplus of j against i.

DEFINITION 7.21 *In a game* (N, v) *the* **surplus of** i **against** j *is* $s_{ij}(u) = \max\{e(K, u) \mid K \subset N, i \in K, j \notin K\}$.

The excess $e(K, u)$ measures the amount by which the members of a coalition can improve their joint payoff over what they receive under a proposed payoff vector u. The surplus of i against j is the largest excess $e(K, u)$ over all coalitions that include i but exclude j. Relative to the payoff vector u, K is the most profitable coalition that has i as a member and that lacks j as a member. In thinking about players i and j, keeping other things equal, and evaluating the worth of player i by herself, without the help of j, relative to the payoff vector u, $s_{ij}(u)$ measures the largest contribution to a coalition that can be associated with i and has no connection with j. Two players i and j are in a kind of balance, relative to u, if $s_{ij}(u) = s_{ji}(u)$, and this balance is the basis of the *kernel*.

DEFINITION 7.22 *The* kernel, $\mathcal{K}$, *consists of all* $(u, \mathcal{T})$ *such that, if* $i, j \in T_k \in \mathcal{T}$, *then (a)* $s_{ij}(u) = s_{ji}(u)$ *or (b)* $s_{ij}(u) < s_{ji}(u)$ *and* $u_i = v(\{i\})$ *or (c)* $s_{ji}(u) < s_{ij}(u)$ *and* $u_j = v(\{j\})$.

The conditions in Definition (7.22) are stated for each pair of players i and j who are in the same coalition T_k. Either the surplus of i against j ($s_{ij}(u)$) equals the surplus of j against i ($s_{jk}(u)$), or, if this is not the case, the player whose surplus is smaller is already receiving the minimum possible payoff (i.e., $u_i = v(\{i\})$ if player i has the smaller surplus). The reasoning behind this specification is that if $s_{ij}(u) < s_{ji}(u)$ and $u_i > v(\{i\})$, then player j is in a position to make an objection to which i has no counterobjection. In doing this, j will increase her payoff at the expense of i. If their surpluses are equal, then a successful objection by either one is impossible. Also, if $s_{ij}(u) < s_{ji}(u)$ with $u_i = v(\{i\})$, it is impossible for j to make a successful objection because i is already at the lowest payoff to which he can be driven.

5.2.2 *The relationship between the kernel and the bargaining set*

THEOREM 7.1 *For a game (N, v) the kernel is contained in the bargaining set.*

Proof. This theorem is proved in several steps. The first shows that if a coalition K has an objection against another coalition L, then there must be a single member subset of K, $\{i\}$, that has an objection against a single member subset of L, $\{j\}$. After this is established, it is shown that, for a point in the kernel $(x, \mathcal{T})$ and for any objection of $\{i\}$ against $\{j\}$, (where i and j are in the same set T_k) $\{j\}$ has a counterobjection.

Suppose that $(x, \mathcal{T})$ is in the kernel, that K and L are disjoint subsets of the same $T_k \in \mathcal{T}$, and that K has an objection against L. Then by the definition of objection, precisely the same objection could be made by any $\{i\} \subset K$ against any $\{j\} \subset L$.

Now, let $(y, \mathcal{U})$ be an objection of $\{i\}$ against $\{j\}$. Clearly the partners of $\{i\}$ will coincide with the members of the set U_k to which i belongs. Consequently, $\sum_{l \in U_k} (y_l - x_l)$ can be no larger than $s_{ij}(x)$. But $\{j\}$ has a counterobjection if there is $(z, \mathcal{V})$ under which each partner, l, of j receives at least x_l and, among them, any that were partners of i under the objection receive at least y_l. It is possible to mount such a counterobjection if $s_{ji}(x) \geqslant s_{ij}(x)$, because this latter amount is the amount by which the payoff of a carefully chosen coalition containing j, but not containing i, can receive above what they received under the original proposal $(x, \mathcal{T})$. Furthermore, this is at least as large as the amount by which i can reward those who join her in her objection, and, thus j can win back any of those she needs and also sufficiently reward any other required players to sustain the counterobjection.

It remains to cover the case in which $s_{ij}(x) > s_{ji}(x)$. In this case, $x_j = v(\{j\})$ as well; therefore, in the face of an objection by $\{i\}$, $\{j\}$ can counterobject with a $(y, \mathcal{U})$ under which $y_j = v(\{j\})$ and $\{j\} \in \mathcal{U}$. QED

An appealing feature of the kernel is that it is stated in terms of pairs of players. An individually rational payoff configuration is in the kernel when no pair of players in the same coalition is in a position where one can make a successful objection against the other. Of course, it is important that the kernel is always in the bargaining set. This fact, in conjunction with the nucleolus being in the kernel, is useful in showing that the bargaining set is not empty.

5.3 The nucleolus

Like the bargaining set, the kernel can have many elements; however, for a large class of games the nucleolus consists of just one point. The nucleolus is defined in relation to a given set of payoff vectors that can be an arbitrary set. This differs from both the kernel and the bargaining set, which are defined in relation to sets $I_{\mathcal{T}}(N, v)$. The idea behind the nucleolus is very simple: A payoff vector is in the nucleolus if, in a sense to be defined presently, the excesses for all coalitions for that payoff vector are made as small as possible.

5.3.1 Definition of the nucleolus

The definition of the nucleolus rests on comparisons among the excesses associated with various payoff vectors. To make a comparison between various payoff vectors, a function is constructed, $\theta(u) = (e_1(u), e_2(u), \ldots, e_{2^n}(u)) \in R^{2^n}$ that has one coordinate for each of the 2^n coalitions contained in N (including the coalitions N and ϕ). Order the 2^n coalitions $K_1, K_2, K_4, \ldots, K_{2^n}$ so that $\theta_j(u) = e(K_j, u)$ and $\theta_j(u) \geqslant \theta_{j+1}(u)$ for $j = 1, \ldots, 2^n - 1$. This can be done for any u in the set, X, of payoff vectors being considered. Note that $\theta_j(u)$ and $\theta_j(u')$ may be associated with different coalitions, in general. In both instances, however, the jth coordinate is the jth largest excess relative to u and u', respectively.

$\theta(u)$ and $\theta(u')$ can be compared using the *lexicographic ordering*. Under this ordering, $\theta(u)$ is smaller than $\theta(u')$ if $\theta_1(u) < \theta_1(u')$ or, for $j > 1$, $\theta_j(u) < \theta_j(u')$ and $\theta_i(u) = \theta_i(u')$ for $i = 1, \ldots, j-1$. This relation is denoted $\theta(u) <_L \theta(u')$. If $\theta_j(u) = \theta_j(u')$ for all j, then $\theta(u) =_L \theta(u')$, and $\theta(u) \leqslant_L \theta(u')$ means $\theta(u) <_L \theta(u')$ or $\theta(u) =_L \theta(u')$. For example, suppose that $n = 3$ and let $\theta(u^0) = (100, 62, 60, 50, 40, 36, 32, 30)$, $\theta(u^1) = (100, 62, 60, 52, 0, 0, 0, 0)$, $\theta(u^2) = (99, 98, 97, 96, 95, 94, 93, 92)$, and $\theta(u^3) = (120, 50, 45, 40, 35, 30, 25, 20)$. The lexicographic ordering of these four elements is $\theta(u^2) <_L \theta(u^0) <_L \theta(u^1) <_L \theta(u^3)$.

DEFINITION 7.23 *For a set of payoff vectors X, the* **nucleolus over** *X is* $nuc(X) = \{u \in X \mid u' \in X \text{ implies } \theta(u) \leqslant_L \theta(u')\}$.

Thus, the nucleolus over X consists of those payoff vectors, $u \in X$, having the lexicographically smallest associated excesses.

5.3.2 *Existence and uniqueness of the nucleolus*

There are two results in this section. The first is that, if X is compact, then nuc(X) is nonempty. The second is that nuc(X) is a single point if X is convex as well as compact.

THEOREM 7.2 *Let X be a nonempty compact subset of R^n. Then nuc(X) is not empty.*

Proof. Let $A_1 = \{\theta(u) \in R^{2^n} \mid u \in X\}$ and let $z_1 = \inf_{y \in A_1} y_1$. Because X is compact, there must be elements of A_1 having z_1 as the first coordinate. Therefore $X_2 = \{u \in X \mid \theta_1(u) = z_1\}$ is compact and nonempty. Now define $A_2 = \{\theta(u) \in A_1 \mid u \in X_2\}$.

From here, the proof proceeds by induction. Suppose that $X_i \subset X_{i-1}$ is nonempty, compact, and consists of all payoff vectors in X_{i-1} for which $\theta_j(u) = \theta_j(u')$ for all u, $u' \in X_{i-1}$ and all $j < i$, and that, if $u \in X_i$, then $\theta_{i-1}(u) = z_{i-1}$ where z_{i-1} is the minimum value of $\theta_{i-1}(u')$ for all $u' \in X_{i-1}$. Let $A_i = \{\theta(u) \in R^{2^n} \mid u \in X_i\}$. Note that, for $\theta(u)$, $\theta(u') \in A_i$, $\theta_j(u) = \theta_j(u')$, $j = 1, \ldots, i-1$ and that $\theta_j(u)$, $\theta_j(u') \leqslant \theta_{i-1}(u)$ for $j \geqslant i$. Now let $z_i = \inf_{y \in A_i} y_i$. Because X_i is compact, there must be elements of A_i having z_i as the ith coordinate. Let $X_{i+1} = \{u \in X_i \mid \theta_i(u) = z_i\}$. As with X_2, X_{i+1} is compact and nonempty. A_{i+1} is defined recursively as $A_{i+1} = \{\theta(u) \in R^{2^n} \mid u \in X_{i+1}\}$, and, thus for $y \in A_{i+1}$, $y_j = z_j$, $j = 1, \ldots, i$.

Letting $X_1 = X$, it is clear nuc(X) $\subset X_i$ for $i = 1, \ldots, 2^n + 1$, that nuc(X) $= X_{2^n+1}$, and that $\theta(u) = (z_1, z_2, \ldots, z_{2^n})$ for $u \in$ nuc(X) $= X_{2^n+1}$. QED

COROLLARY *If X is also convex, then nuc(X) contains exactly one point.*

Proof. Suppose that u and u' are elements of nuc(X); hence, $\theta(u) =_L \theta(u')$. In the vector $\theta(u)$, there are s distinct values of the coordinates, with k_1 occurrences of the largest number, k_2 occurrences of the second largest, and so forth. Of course, $\sum_{i=1}^{s} k_i = 2^n$. With respect to $\theta(u)$, let the order of the coalitions be $K_1, K_2, \ldots, K_{2^n}$ (i.e., $\theta_j(u) = e(K_j, u)$ for all j), and, similarly, let the order of the coalitions with respect to $\theta(u')$ be $K'_1, K'_2, \ldots, K'_{2^n}$. Letting $\lambda \in (0, 1)$ and $u^\lambda = \lambda u + (1 - \lambda)u'$, note that $\theta(u^\lambda)$ can have, at most, k_1 occurrences of $\theta_1(u)$, and if $\theta(u^\lambda)$ has exactly k_i occurrences of $\theta_{k_1+\cdots+k_{i-1}+1}(u)$, $i = 1, \ldots, j-1$, then it can have, at most, k_j occurrences of $\theta_{k_1+\cdots+k_{j-1}+1}(u)$.

The reasons for this are as follows: Take the occurrences of $\theta_1(u)$. There are k_1 of these in $\theta(u^\lambda)$ if and only if the two sets of coalitions $\{K_1, \ldots, K_{k_1}\}$ and $\{K'_1, \ldots, K'_{k_1}\}$ have precisely the same members. If the two collections of coalitions are not identical in their membership, there is some coalition, K_i ($i \leqslant k_1$), from the first list such that $e(K_i, u) > e(K_i, u')$. If this inequality holds, then there are fewer than k_1 coordinates of $\theta(u^\lambda)$ having the value $\theta_1(u)$. There can be no coordinates of $\theta(u^\lambda)$ with a value higher than $\theta_1(u)$; therefore, $\theta(u^\lambda) <_L \theta(u)$.

If the collection of coalitions $K_1, \ldots, K_{k_1}$ and $K'_1, \ldots, K'_{k_1}$ are the same, then the argument can be repeated for the coordinates achieving $\theta_{k_1+1}(u)$. Either the collections of coalitions, $K_{k_1+1}, \ldots, K_{k_1+k_2}$ and $K'_{k_1+1}, \ldots, K'_{k_1+k_2}$ are the same or $\theta(u^\lambda) <_L \theta(u)$. This argument can be repeated inductively with the conclusion that $\theta(u^\lambda) =_L \theta(u)$ if and only if $u = u'$, and otherwise $\theta(u^\lambda) <_L \theta(u)$. The latter inequality implies the contradiction that neither u nor u' are elements of nuc(X); hence, nuc(X) consists of just one point. QED

5.3.3 *The nucleolus is contained in the kernel*

To see better what is going on, consider the individually rational payoff configurations, relative to the coalition structure $\mathcal{T}$, for the game $(N, v): I_{\mathcal{T}}(N, v)$. The payoff vectors associated with $I_{\mathcal{T}}(N, v)$ are the elements of $A_{\mathcal{T}}(N, v) = \{u \in R^n \mid (u, \mathcal{T}) \in I_{\mathcal{T}}(N, v)\}$, which is compact, so nuc$[A_{\mathcal{T}}(N, v)]$ is not empty and is in both the bargaining set and the kernel. In fact, this result not only establishes that the bargaining set is nonempty, but that there is at least one element in the bargaining set relative to each coalition structure.

THEOREM 7.3 *Let $\mathcal{K}(N, v)$ be the kernel of the game (N, v), let $\mathcal{T}$ be a coalition structure, and let $u \in$ nuc$[A_{\mathcal{T}}(N, v)]$. Then $(u, \mathcal{T}) \in \mathcal{K}(N, v)$.*

Proof. Suppose that $(u, \mathcal{T}) \in I_{\mathcal{T}}(N, v)$ *and* $(u, \mathcal{T}) \notin \mathcal{K}(N, v)$. Then there is a pair of players, $i, j \in N$ such that $s_{ij}(u) > s_{ji}(u)$ and $u_j > v(\{j\})$. Let δ equal the smaller of $(s_{ij}(u) - s_{ji}(u))/2$ and $u_j - v(\{j\})$, and define u' as follows;

$$u'_i = u_i + \delta \tag{7.20}$$

$$u'_j = u_j - \delta \tag{7.21}$$

$$u'_k = u_k \text{ for } k \neq i, j \tag{7.22}$$

Let K^* be a coalition with $i \in K^*$, $j \notin K^*$, and having the largest excess among all coalitions that contain i and exclude j. Finally, in ordering the coordinates of the function $\theta(u)$, suppose that K^* is placed after any coalitions L for which $e(K^*, u) = e(L, u)$. The coordinate of $\theta(u)$ corresponding to K^* is k^*.[3]

Now compare $\theta(u)$ and $\theta(u')$: $\theta_k(u) \leq \theta_k(u')$ for $k < k^*$. The coalitions coming before K^* can fall into any of three categories: (a) they contain both i and j, (b) they contain neither i nor j, and (c) they contain i but not j. No such coalition can contain j and omit i because of the condition $s_{ij}(u) > s_{ji}(u)$. For any coalition falling into (a) or (b), $\theta_k(u) = \theta_k(u')$, and for any coalition in category, (c), $\theta_k(u) < \theta_k(u')$. Meanwhile, for $k > k^*$, $\theta_k(u') < \theta^*_k(u)$; therefore, $\theta(u') <_L \theta(u)$ and so u is not in the nucleolus. This establishes that u is not in the nucleolus if $(u, \mathcal{T})$ is not in the kernel, which implies that $(u, \mathcal{T})$ is in the kernel of u is in the nucleolus. QED

COROLLARY *Let (N, v) be a game in characteristic function form. (N, v) has a nonempty bargaining set, and, for each coalition structure $\mathcal{T}$, there is a payoff configuration in the bargaining set.*

Proof. This corollary follows from Theorems 7.1 to 7.3. By Theorem 7.2 $\text{nuc}[A_{\mathcal{T}}(N, v)]$ is not empty, because it is compact. Compactness follows from two things: First, the payoff space is bounded because the $v(K)$ are finite. Second, $I_{\mathcal{T}}(N, v)$ is, by definition, closed; hence, $A_{\mathcal{T}}(N, v)$ is also closed. By Theorem 7.3, $u \in \text{nuc}[A_{\mathcal{T}}(N, v)]$ implies that $(u, \mathcal{T})$ is in the kernel, and by Theorem 7.1 the kernel is in the bargaining set. Thus $(u, \mathcal{T})$ is in the bargaining set. That there is a point in the nucleolus for any coalition structure follows from $\text{nuc}[A_{\mathcal{T}}(N, v)]$ being nonempty. QED

5.3.4 *The nucleolus of S-equivalent games*

The nucleolus is defined with reference to both a game and a set of outcomes. Thus, $\text{nuc}(X)$ is the nucleolus over the set of outcomes X that are a subset of the outcomes that can be achieved in some particular game (N, v). In speaking of "the nucleolus of a game (N, v)" it would be natural to mean $\text{nuc}(X)$ for that game with X being the set of achievable outcomes. In fact, it is sufficient to examine $\text{nuc}[I(N, v)]$, the nucleolus over the set of imputations of the game, $I(N, v)$. Now let (N, v^*) be a zero–one normalized game and consider the *S-equivalence class of* (N, v^*). This class is the set of games that have (N, v^*) as their zero–one normalized form. The nucleolus of each of the members of this set of games are related by equation (7.13).

LEMMA 7.7 *Let (N, v) be S-equivalent to the zero–one normalized game (N, v^*). Then $u \in \text{nuc}[I(N, v)]$ if and only if $u^* \in \text{nuc}[I(N, v^*)]$, where u and u^* are related by*

$$u_i^* = \frac{u_i - v(\{i\})}{v(N) - \sum_{i \in N} v(\{i\})} \tag{7.23}$$

Proof. For any $u \in I(N, v)$, there is exactly one $u^* \in I(N, v^*)$ that is related to u by equation (7.23). Conversely, for each $u^* \in I(N, v^*)$, there is one corresponding $u \in I(N, v)$. These facts are clear from the definition of S-equivalence. For such a pair, $e(K, u^*)$ is given by

$$e(K, u^*) = \frac{v(K) - \sum_{i \in K} v(\{i\}) - \sum_{i \in K} [u_i - v(\{i\})]}{v(N) - \sum_{i \in N} v(\{i\})}$$

$$= \frac{v(K) - \sum_{i \in K} u_i}{v(N) - \sum_{i \in N} v(\{i\})} \tag{7.24}$$

Now let $u, w \in I(N, v)$ and let u^* and w^* be the corresponding elements of $I(N, v^*)$, related by equation (7.23). Then, from equation (7.24), it follows that $\theta(u) \leqslant_L \theta(w)$ if and only if $\theta(u^*) \leqslant_L \theta(w^*)$, which proves the lemma. QED

5.4 Examples

Two three-person games are described below for which the nucleolus over the set of imputations, and corresponding payoff configurations in the kernel and bargaining set are given. In the first game, $v(K)$ is defined by $v(\{i\})=0$ for all $i \in N$, $v(\{1,2\})=4$, $v(\{1,3\})=2$, $v(\{2,3\})=3$, and $v(N)=6$. The payoff vector $(2, 3, 1)$ is the only element of the nucleolus over $I(N, v)$, the set of imputations. The excesses of $u=(2, 3, 1)$, are $e(\{1\}, u)=-2$, $e(\{2\}, u)=-3$, $e(\{3\}, u)=-1$, $e(K, u)=-1$ for each two-person coalition, and the excess is 0 for both N and $\varnothing$. This gives $\theta(2, 3, 1)=(0, 0, -1, -1, -1, -1, -2, -3)$. For any other imputation, u', the excesses of the three two-person coalitions would still sum to -3, but at least one of the coalitions would have an excess greater than -1. This would make the third coordinate of $\theta(u')$ larger than -1 without changing the first two coordinates; therefore, $\theta(2, 3, 1) <_L \theta(u')$ for such a u'. Note, too, that if we consider the set of all individually rational payoffs, any such u' that is not an imputation would also have a vector of excesses that is lexicographically larger than $\theta(2, 3, 1)$, because $e(N, u')$ would be strictly positive. The payoff configuration $((2, 3, 1), \{1, 2, 3\})$ is in the kernel and the bargaining set.

For the second example, let $v(N)=4$ and let the characteristic function be otherwise unchanged from the preceding example. The payoff vector $(4/3, 7/3, 1/3)$ is in $\text{nuc}[I(N, v)]$. The argument to support this claim is similar to that used in the first example: The excesses are $1/3$ for each of the two-person coalitions, $-4/3$, $-7/3$, and $-1/3$ for coalitions $\{1\}$, $\{2\}$, and $\{3\}$, respectively, and 0 for coalitions $\varnothing$ and N. Thus, $\theta(4/3, 7/3, 1/3)=(1/3, 1/3, 1/3, 0, 0, -1/3, -4/3, -7/3)$. If $u \in I(N, v)$ is chosen with $u_1 \neq 1/3$ or $u_2 \neq 1/3$ or both, then one of the two-person coalitions will have an excess larger than $1/3$ and $\theta(u) >_L \theta(4/3, 7/3, 1/3)$; therefore, only $(4/3, 7/3, 1/3)$, is in the nucleolus. The payoff configuration $((4/3, 7/3, 1/3), \{1, 2\}, \{3\})$ is in the kernel and the bargaining set.

6 The Shapley value and the Banzhaf index

The Shapley (1953b) value can be calculated for any superadditive game (N, v) in characteristic function form having a finite number of players, and it has the further advantage of giving a unique outcome that satisfies both individual rationality and group rationality. The payoff to each player is a weighted average of the contributions that the player makes to each of the coalitions to which she belongs, with the weights depending on the number of players, n, and the number of members in each coalition. Another appealing aspect of the Shapley value is that it can be characterized by four easily understood conditions.

6.1 Description of the Shapley value

The Shapley value itself is denoted $\phi(v)$, where $\phi(v) \in R^n$ and $\phi_i(v)$ is the Shapley value payoff to the ith player. The formula is

$$\phi_i(v) = \sum_{K \subset N} [v(K) - v(K \backslash \{i\})] \frac{(k-1)!\,(n-k)!}{n!} \qquad (7.25)$$

The Shapley value is defined by four conditions, given in Defintion 7.25, and following the definition, it is proved that the conditions imply that the Shapley value is characterized by the formula in equation (7.25). The four conditions are (a) group rationality, $\sum_{i \in N} \phi_i(v) = v(N)$, (b) if a player, i, adds nothing more than $v(\{i\})$ to any coalition, then the player receives only $v(\{i\})$, (c) if two games are identical except for the order in which the players are listed, then the Shapley values for the players are the same, and (d) if a game is formed by adding two games together, then the Shapley value of the new game is the sum of the values of the two original games. To state condition (c) formally it is necessary to say what is meant by two games that differ only with respect to the order of the players. This is done in Definition 7.24.

DEFINITION 7.24 *Let i^* and j^* be a specific pair of players and suppose the games (N, v) and (N', v') are related in the following way: (a) $n = n'$, (b) $v(K) = v'(K)$ if $i^*, j^* \in K$ or if $i^*, j^* \notin K$, and (c) $v(K \cup \{i^*\}) = v'(K \cup \{j^*\})$ if $i^*, j^* \notin K$. Then (N, v) is a* **simple permutation** *of (N', v'). Players i^* and j^* are the* **permuted players.**

The conditions defining the Shapley value are enumerated below in Definition 7.25. These conditions are sometimes called the *Shapley axioms*.

DEFINITION 7.25 *For a game $\Gamma = (N, v)$, the* **Shapley value** *$\phi(v)$, is defined by:*

(a) $\sum_{i \in N} \phi_i(v) = v(N)$.

(b) If, for some $i \in N$, $v(K) = v(K \backslash \{i\}) + v(\{i\})$ for all $K \subset N$, $i \in K$ then $\phi_i(v) = v(\{i\})$.

(c) If (N, v) is a simple permutation of (N', v') with i^* and j^* being the permuted players, then $\phi_{i^*}(v) = \phi_{j^*}(v')$, $\phi_{i^*}(v') = \phi_{j^*}(v)$, and, for all $i \in N \backslash \{i^*, j^*\}$, $\phi_i(v) = \phi_i(v')$.

(d) If (N, v) and (N, w) are two games having the same player set N, and (N, z) is defined by $z(K) = v(K) + w(K)$ for all $K \subset N$, then $\phi(z) = \phi(v) + \phi(w)$.

It is surprising that the four conditions enumerated in Definition 7.25 are enough to determine a unique payoff vector. The first asserts group rationality. The second requires that a player i who contributes only $v(\{i\})$ to each coalition must get a payoff of $v(\{i\})$. The third is a symmetry condition in the sense that it forces the solution point payoffs to depend on the structure of $v(K)$, but not on how the individual players are numbered. If, for example, a game (N, v) is rearranged in the sense that Al, who is player 1, is to become player i_1, and Betty, who is player 2, is to become player i_2, and so forth, the solution point payoffs to the players in the new game are the same as in the old game. That is, the new game (N', v') is to have the feature that $v\{\text{Al, Betty, George}\}) = v'(\{\text{Al, Betty, George}\})$, and so on for all coalitions, and $\phi_{\text{Al}}(v) = \phi_{\text{Al}}(v')$, and so forth for all coalitions. The final condition requires a natural consistency among triplets of games that are related in the manner that the games (N, v), (N, w), and (N, z) are related. If the characteristic function z equals the sum of the characteristic functions v and w, then $\phi(z) = \phi(v) + \phi(w)$.

6.2 *Existence of the Shapley value*

Proof of existence and uniqueness of the Shapley value is done in several steps. In Section 6.2.1, existence is proved for simple games. This is useful because in Section 6.2.2 it is shown that any game can be treated like a sum of simple games, and the Shapley value for a game that is the sum of several games is the sum of their respective Shapley values.

6.2.1 *The Shapley value for simple games*

To prove the existence of a Shapley value and to derive its expression, a handy concept to use is the *marginal value of a coalition*, which is defined recursively, as follows:

$$c_{\{i\}}(v) = v(\{i\}) \qquad \text{for all } i \in N \tag{7.26}$$

$$c_K(v) = v(K) - \sum_{\substack{L \subset K \\ L \neq K}} c_L(v) \qquad \text{for all } K \subset N \text{ with } k \geqslant 2 \tag{7.27}$$

Thus, the marginal value of a coalition K is $v(K)$ minus the marginal values of all conditions that are both smaller than K and are subsets of K. Equation (7.27) can be rewritten as

$$c_K(v) = \sum_{L \subset K} (-1)^{k-l} v(L) \qquad \text{for } K \subset N \tag{7.28}$$

Deriving existence of the unique Shapley value is accomplished by showing that a game (N, v) can be regarded as a weighted sum of a number of simple games (N, v_K) defined as follows:

$$v_K(L) = 1 \qquad \text{if } K \subset L \qquad v_K(L) = 0 \text{ otherwise} \tag{7.29}$$

The game (N, cv), for a scalar c, has a characteristic function $cv(K)$.

LEMMA 7.8 *For the game (N, cv_K), the Shapley value is $\phi_i(cv_K) = c/k$ if $i \in K$ and $\phi_i(cv_K) = 0$ if $i \notin K$.*

Proof. If $i \notin K$, then by (b) of Definition 7.25, $\phi_i(cv_K) = 0$, and if $i \in K$, $\phi_i(cv_K) = c/k$ by condition (c). QED

6.2.2 *The Shapley value for superadditive games*

Extending Lemma 7.8 to general superadditive games is accomplished by showing that any characteristic function can be represented as a weighted sum of characteristic functions for simple games. This is done in Lemma 7.9. Next, in Lemma 7.10 it is shown that if a characteristic function equals one characteristic function minus another, then the Shapley value for this game is also the one characteristic function minus the other.

LEMMA 7.9 *The characteristic function, v, of a game (N, v) satisfies*

$$v = \sum_{\substack{K \subset N \\ L \neq \varnothing}} c_K(v) v_K \tag{7.30}$$

where the function $c_K(v)$ is defined by equation (7.28) *and the v_K are defined by equation* (7.29).

Proof. Proving the lemma requires showing that

$$v(K) = \sum_{\substack{L \subset K \\ L \neq \varnothing}} c_L(v) v_L(K) \qquad \text{for all } K \subset N \tag{7.31}$$

Using equation (7.28) in equation (7.31) gives

$$\begin{aligned} v(K) &= \sum_{L \subset K} \sum_{M \subset L} (-1)^{l-m} v(M) \\ &= \sum_{M \subset K} \left[\sum_{l=m}^{k} (-1)^{l-m} \frac{(k-m)!}{(l-m)!\,(k-l)!} \right] v(M) \end{aligned} \tag{7.32}$$

The term in brackets in equation (7.32) is always 0 when $m < k$; therefore, equation (7.32) reduces to $v(K) = v(K)$ which establishes the lemma. QED

LEMMA 7.10 *Let (N, v), (N, w), and (N, z) be games in which $z = v - w$. Then $\phi(z) = \phi(v) - \phi(w)$.*

Proof. This follows from Definition 7.25d. QED

THEOREM 7.4 *A superadditive game has a unique Shapley value given by*

$$\phi_i(v) = \sum_{K \subset N} [v(K) - v(K) \backslash \{i\})] \frac{(k-1)!\,(n-k)!}{n!}, \qquad i \in N \tag{7.33}$$

Proof. Using Lemmas 7.8 and 7.10 in equation (7.31) gives

$$\phi_i(v) = \sum_{\substack{K \subset N \\ i \in K}} \frac{c_K(v)}{k}, \qquad i \in N \tag{7.34}$$

And using equation (7.28) in equation (7.34), the latter becomes

$$\begin{aligned} \phi_i(v) &= \sum_{\substack{K \subset N \\ i \in K}} \frac{1}{k} \left[\sum_{K \supset L} (-1)^{k-1} v(L) \right] \\ &= \sum_{K \supset L} \sum_{K \subset N} \frac{(-1)^{k-1}}{k} [v(L) - v(L \backslash \{i\})] \\ &= \sum_{K \subset N} \sum_{l=0}^{n-k} \frac{(-1)^{l} (n-k)!}{(k+l)(n-k-l)!\,l!} [v(K) - v(K \backslash \{i\})] \end{aligned} \tag{7.35}$$

TABLE 7.1 A four-person game and its Shapley value

Coalition	$v(K)$	kC_k^n	$v(K)-v(K\backslash\{i\})$ for Player 1	2	3	4
{1}	0	4	0	0	0	0
{2}	0	4	0	0	0	0
{3}	0	4	0	0	0	0
{4}	0	4	0	0	0	0
{1, 2}	2	12	2	2	0	0
{1, 3}	5	12	5	0	5	0
{1, 4}	3	12	3	0	0	3
{2, 3}	6	12	0	6	6	0
{2, 4}	8	12	0	8	0	8
{3, 4}	5	12	0	0	5	5
{1, 2, 3}	7	12	1	2	5	0
{1, 2, 4}	11	12	3	8	0	9
{1, 3, 4}	9	12	4	0	6	4
{2, 3, 4}	10	12	0	5	2	4
{1, 2, 3, 4}	15	4	5	6	4	8

Shapley value: $\phi_1 = \frac{33}{12}$, $\phi_2 = \frac{49}{2}$, $\phi_3 = \frac{41}{12}$, $\phi_4 = \frac{57}{12}$

To evaluate the coefficient of $[v(K) - v(K\backslash\{i\})]$, denote $(n-k)!/[(n-k-l)!\,l!]$ by C_l^{n-k} and note

$$\sum_{l=0}^{n-k} \frac{(-1)^l}{k+l} C_l^{n-k} = \sum_{l=0}^{n-k} (-1)^l C_l^{n-k} \int_0^1 x^{k+l-1}\, dx$$

$$= \int_0^1 \sum_{l=0}^{n} (-1)^l C_l^{n-1} x^{k+l-1}\, dx$$

$$= \int_0^1 x^{k-1} \sum_{l=0}^{n} (-1)^l C_l^{n-k} x^l\, dx$$

$$= \int_0^1 x^{k-1}(1-x)^{n-k}\, dx = \frac{(k-1)!\,(n-k)!}{n!} \tag{7.36}$$

Using equation (7.36) in equation (7.35) completes the proof. QED

As an exercise, the reader might wish to prove that if (N, v) has the zero–one normalized form (N, v^*), and u is the Shapley value of (N, v), then the Shapley value of (N, v^*) is given by equation (7.13). Table 7.1 contains a game in characteristic function form, along with information useful in computing the Shapley value for the game.

6.3 *The Banzhaf power index*

The Banzhaf (1965) index is defined for simple games and is based on counting, for each player, the number of coalitions to which the player is crucial to winning. Let (N, v) be a zero–one normalized simple game and recall from Definition 7.11 that a winning coalition is one for which $v(K) = 1$ and a losing coalition is one for which $v(K) = 0$. Each coalition K that wins when $K\backslash\{i\}$ loses is called a *swing* for player i, because the membership of player i in the coalition is crucial to the coalition winning. For a game (N, v) let $\sigma_i(N, v)$ be the number of swings for i, and let $\sigma_0(N, v) = \sum_{i\in N} \sigma_i(N, v)$ be the total number of swings of all players in the game. Then the normalized Banzhaf index is $b_i(N, v) = \sigma_i(V, v)/\sigma_0(N, v)$.

The Banzhaf index can be generalized to nonsimple games. (See Owen (1978), who also provides axioms for the index that are along the lines of the conditions defining the Shapley value.) The generalization is made by using the formula $b_i(N, v) = \sum_{K\subset N} [v(K) - v(K\backslash\{i\})]/(2^{n-1})$. This formula has in common with the Shapley value that a player's value is computed as a weighted sum of his marginal contributions to all the coalitions of which he is a member; however, the Banzhaf weights are different from the Shapley weights. For Banzhaf, all coalitions are weighted equally, no matter what their size, while the Shapley value weights vary with the size of the coalition.

7 An application to power in government

Table 7.2 shows the Shapley and Banzhaf values for several related games that have interesting interpretations relating to the U.S. government. To pass a law at the federal level, it is necessary to have a bare majority in the House of Representatives, a bare majority in the Senate, and the agreement of the president. A second way to pass a law requires a two-thirds majority in both the House and Senate, but does not require the agreement of the president. Suppose that a coalition K having the power to pass a law has a characteristic function value of $v(K) = 1$ and that any other coalition L has a value $v(L) = 0$.

Table 7.2 contains the Shapley and Banzhaf values for three sets of rules. The rules differ only with respect to the votes needed to override a presidential veto. With the president's agreement, only a bare majority of both houses is needed in each case. The Shapley and Banzhaf values for the situation approximating the U.S. government rules are in the table as Shapley 2 and Banzhaf 2. These calculations are done on the basis of a 40-member House and 15-member Senate, which, of course, are not the actual sizes of these two bodies. Thus, $v(K) = 1$ if K includes 21 or more House members plus 8 or more Senate members plus the president, and $v(K) = 1$ if K includes 27 or more House members plus 10 or more senators. Otherwise, $v(K) = 0$.

Note that the Banzhaf value accords much less weight to the president—.1345 for Shapley versus .0972 for Banzhaf—because the Banzhaf value weights each coalition equally, whereas, the Shapley value weights a coalition inversely to the number of coalitions that exist of the given size. Because the president is crucial, on the whole, in smaller winning coalitions

TABLE 7.2 The Shapley and Banzhaf power indexes for an election game

Shapley 1	With president	Without president	Single value	Total value
House	21	22	.0018	.4704
Senate	8	9	.0324	.4857
President	—	—	.0439	—
Shapley 2				
House	21	27	.0019	.4741
Senate	8	10	.0261	.3914
President	—	—	.1345	—
Shapley 3				
House	21	40	.0076	.3022
Senate	8	15	.0187	.2797
President	—	—	.4181	—
Banzhaf 1	With president	Without president	Single value	Total value
House	21	22	.0152	.6092
Senate	8	9	.0235	.3525
President	—	—	.0383	—
Banzhaf 2				
House	21	27	.0146	.5852
Senate	8	10	.0212	.3176
President	—	—	.0972	—
Banzhaf 3				
House	21	40	.0145	.5806
Senate	8	15	.0212	.3182
President	—	—	.1012	—

(because with the president, many fewer House and Senate members are needed), the Shapley value will favor him relative to the Banzhaf value.

Shapley 1 and Banzhaf 1 are based on different rules for overriding a veto. The president's power is greatly reduced by allowing only 22 House members and 9 senators to override a presidential veto. Correspondingly, the Shapley value for the president is .0439, roughly a third of the Shapley 2 value, and the Banzhaf value is .0383, also greatly reduced and, of course, less than the corresponding Shapley value. Curiously, in both cases, most of the power lost by the president is gained by the Senate. The power of the House is changed very little.

This may be contrasted with Shapley 3 and Banzhaf 3 where, without the president, all members of both House and Senate are required to pass a law. The president's power, of course, rises, but the two indices behave very differently. The Shapley value of the president rises from .1345 to .4181, that is, by a factor of 3, but the Banzhaf value goes up only slightly.

In circumstances such as those of the example, the two indices are often called power indices, the idea being that the sum of the values across players is 1, and the value of a particular player or coalition gives the slice of total power wielded by that player or coalition. Individuals must decide for themselves whether they find such an interpretation congenial.

8 Concluding comments

Various cooperative game solutions have been examined for transferable utility games in characteristic function form. The solutions fall into several interesting categories. First, the core is based on criteria with which it is difficult to quarrel. A point is in the core if it gives to each player and coalition at least as much as that player or coalition could guarantee for itself. These minimal criteria are both too much and too little; they are too much because the core in many games can be very large, and they are too little because there are many games that have no core. Also, the core leaves one with little sense of how bargaining processes might proceed and affect the outcome.

The stable set remedies some of the undesirable aspects of the core. It seems a bit more connected with the bargaining process, and, for many more games than for the core, the stable set is not empty. But, as with the core, the stable set can include very many points, and, additionally, a game can have more than one stable set.

The bargaining set gets closer to the bargaining process than the stable set. As with all solution concepts in Chapters 7 and 8, the bargaining process is not formally modeled, but the definition of a solution can be suggestive as to the kind of bargaining that might be expected in the background. With the bargaining set, players are conceived as being divided into coalitions, and whether a particular payoff vector is a solution payoff depends on the coalition structure, $\mathcal{T}$, that accompanies it. Thus, pairs, $(u, \mathcal{T})$, are considered and, to some extent players are locked into the coalition structure $\mathcal{T}$. The bargaining set gets closer to the bargaining process and it exists for a large class of games, but, as in the core and the stable set, there can be many possible outcomes in the solution.

The kernel and the nucleolus are related to the bargaining set, and the kernel, in particular, shares much with it. The nucleolus is unique when the coalition structure is given and the set of payoff vectors being considered is convex. In essence, the nucleolus gives a method for picking one element from the bargaining set among the elements in the bargaining set that are associated with a given coalition structure $\mathcal{T}$. One must decide whether the criterion used for choosing this element is appealing.

The Shapley value goes in a quite different direction. It has little to say about bargaining processes, but is, instead, based on conditions that are, and are meant to be, reasonable. The value exists and is unique for a large class of games, but, again, one must decide whether the conditions that define it are ultimately acceptable. Like the core, the Shapley value outcome seems based on an action by the coalition of the whole, with lesser coalitions never forming, but, nonetheless, affecting the outcome. Nothing has the status quo nature that the coalition structure has for the bargaining set.

In sum, then, these cooperative game solutions are diverse and no single solution dominates the field.

Exercises

1. Suppose a three-person game with the characteristic function $v(\{1\}) = 5$, $v(\{2\}) = 8$, $v(\{3\}) = 4$ $v(\{1, 2\}) = 15$, $v(\{1, 3\}) = 20$, $v(\{2, 3\}) = 15$, $v(\{1, 2, 3\}) = 30$.
   a. Find a point in the core of this game.
   b. What is the zero–one normalization of this game?
   c. Assuming that superadditivity must hold and that the $v(K)$ are fixed for all $K \neq N$, what is the lowest value to which $v(N)$ can be changed? If $v(N)$ is changed to this value, is the core empty or nonempty?
2. Calculate the Shapley value for the game in part (a) of problem 1. Is the Shapley value in the core?
3. Is superadditivity either necessary or sufficient for a nonempty core?
4. Prove that the following is a necessary condition for a nonempty core. Let $(T_1, \ldots, T_m)$ be any partition of N. Then $v(N) \geqslant \sum_{i=1}^{m} v(K)$.

Notes

1. In some of the earlier literature, coalition K is called a *blocking coalition* and is said to *block x*.
2. Let K and L be coalitions, each of which contains at least one player who is not in the other coalition. Lemmas 7.5 and 7.6 can be easily extended to any simple game in which both K and L are winning.
3. These specifications on placing K^* are not restrictions on the model; they make the remainder of the proof easier to state without affecting the substance.

8

n-Person cooperative games without transferable utility

This chapter generalizes Chapters 6 and 7 in the sense that the models studied here are n person and assume nontransferable utility. Just as some of the models in Chapter 6 do not have counterparts in Chapter 7, not all of the models from either of the preceding chapters have counterparts in this chapter. The major topics presented are on the core and on a generalized Shapley value for nontransferable utility games.

1 Introduction to nontransferable utility games

1.1 A brief comparison between transferable and nontransferable utility

Transferable utility is an undesirable assumption to make in many economic situations. For example, where two or more consumers are involved, transferable utility requires that there must be a commodity with respect to which the utility of each consumer is linear and that enters into each consumer's utility function separately from other commodities. Where x_i, y_i, and z_i are quantities of three goods, X, Y, and Z, the utility function of each consumer, i, must be of the form $u_i(x_i, y_i, z_i) = x_i + w_i(y_i, z_i)$. Transferable utility is built into this formulation because all players' utility functions are separably linear in the same good X; thus, one person can increase the utility of any other person by one unit at a cost of just one unit to himself. Sometimes the commodity X that plays this special role is called *money*, and it is said that a unit of money is equivalent to a unit of utility for each person. While this might be acceptable in some circumstances, one would be reluctant to have all results in economics based on such a condition. Fortunately, some of the cooperative game solutions examined in Chapter 7 can be restated for models lacking the transferable utility restrictions, and others can be generalized.

1.2 Overview of the chapter

This chapter explores some nontransferable utility models. Section 2 deals with the reformulation of the characteristic function; for, when transferable utility is given up, it is no longer possible to describe the payoffs available to a particular coalition as a sum of utility that the group can guarantee. Instead, each coalition K has a set of k-dimensional payoff vectors in R^k that it can achieve. It is easily seen in Section 2 that the characteristic function for a transferable utility game is a special case of the characteristic function for a nontransferable utility game. The remaining sections look at the solutions that were examined in Chapter 7; not all of which will carry over. In Section 3, the core is redefined in terms of the generalized characteristic function and a theorem on existence of a nonempty core is stated and proved. Redefining the core is easy and natural, but the existence proof becomes much more formidable than was the case for transferable utility. In Section 4 the stable set is examined and recast. In Section 5, the bargaining set, kernel, and nucleolus are briefly discussed. In Section 6 the Shapley value is generalized to the nontransferable utility case. Section 7 includes applications of the core, and Section 8 includes concluding comments.

2 The characteristic function and other basic tools

2.1 Imputations, domination, and the characteristic function

The characteristic function for a nontransferable utility game must specify all obtainable payoff vectors for each coalition, and this must be done by actually specifying a set of payoff vectors. Note, for example, that the ability of the coalition $\{1, 2, 3\}$ to attain the payoff $(4, 8, 12)$ does not, in the absence of transferable utility, impart any information on that coalition's ability to achieve, say, $(5, 9, 10)$, a payoff vector whose component payoffs also sum to 24. On the one hand, this second payoff vector could be beyond the set of payoffs that $\{1,2, 3\}$ can reach; on the other hand, it may be interior to the coalition's attainable set of payoffs even if $(4, 8, 12)$ is on the frontier of that set. Letting $V(K)$ denote the characteristic function, $V(K)$ is the *set* of all k-dimensional vectors of payoffs that the coalition K can guarantee to itself. As with the games of Chapter 7, there is a payoff that each player, i, can guarantee himself. As before, it is denoted u_i, and $V(\{i\}) = \{\underline{u}_i\}$ for all $i \in N$.

Suppose, for example, that the three-player game in which any two or three players can dictate the division of \$100 is being examined, and assume that the utility for money of each of the three players is $U_1(x_1) = \ln(x_1 + 1)$ for player 1, $U_2(x_2) = \sqrt{x_2}$ for player 2, and $U_3(x_3) = -1/(x_3 + 2)$ for player 3. Then $V(\{1\}) = \{0\}$, $V(\{2\}) = \{0\}$, and $V(\{3\}) =$

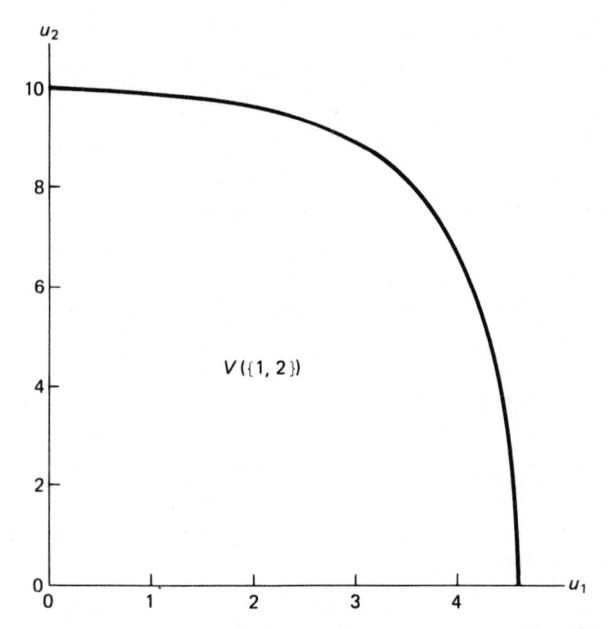

FIGURE 8.1 The attainable payoffs for a two-person coalition.

$\{-.5\}$. For the coalition $K = \{1, 2\}$, $V(K)$ consists of all utility pairs that can be obtained by distributing between 0 and \$100 between the two players. That is,

$$V(\{1,2\}) = \{u_1, u_2) \mid u_1 = \ln(x_1 + 1),\ u_2 = \sqrt{x_2},\ 0 \leqslant x_1,\ x_2, \text{ and } x_1 + x_2 \leqslant 100\} \quad (8.1)$$

This set is depicted in Figure 8.1 and its upper right boundary is given by the equation $u_2 = (101 - e^{u_1})^{.5}$.

If x is a payoff vector in R^n, then x^K denotes the k-dimensional payoff vector for coalition K implied by x. Thus, $x^K = (x_{i_1}, x_{i_2}, \ldots, x_{i_k})$, where $K = \{i_1, i_2, \ldots, i_k\}$. A transferable utility game can have its characteristic function represented by a set function $V(K)$; however, such a set function is related to $v(K)$ by the rule

$$V(K) = \left\{x^K \in R^k \mid x_i \geqslant v(\{i\}),\ i \in K,\ \sum_{i \in K} x_i \leqslant v(K)\right\} \quad (8.2)$$

Thus, for example, $V(\{1,2\})$ in a transferable utility game is always a right triangle with vertices at $(\underline{u}_1, \underline{u}_2)$, $(\underline{u}_1 + v(\{1,2\}),\ \underline{u}_2)$, and $(\underline{u}_1, \underline{u}_2 + v(\{1,2\}))$. Parallel to the practice in Chapter 7, a game in characteristic function form will be denoted $\Gamma = (N, V)$.

DEFINITION 8.1 *The* **characteristic function** $V(K)$ *for a nontransferable utility game* (N, V) *is a set-valued function where, for each coalition* K, $V(K)$ *is the set of all payoff vectors* u^K *that the coalition* K *can achieve and that satisfies the condition* $u_i^K \geqslant \underline{u}_i$ *for all* $i \in K$.

Imputations are defined essentially as in Chapter 7. They consist of all payoff vectors in $V(N)$ that are not strongly dominated. That is, $u \in I(N, V)$ if $u \in V(N)$ and there is no $u' \in V(N)$ for which $u \ll u'$.

DEFINITION 8.2 *A payoff vector* $u \in V(N)$ *is an* **imputation** *in the game* $\Gamma = (N, V)$ *if there is no* $u' \in V(N)$ *such that* $u_i < u'_i$ *for all* $i \in N$. *The set of imputations is denoted* $I(N, V)$.

Domination can also be defined in the present setting. The idea remains the same: One payoff vector dominates another via the coalition K if the first vector gives more to each player in K than does the second, and if the first vector is attainable by K.

DEFINITION 8.3 *For* $x, y \in I(N, V)$, y **dominates** x **via** K *if* $y^K \gg x^K$ *and* $y^K \in V(K)$.

DEFINITION 8.4 y **dominates** x *if, for some coalition* K, y *dominates* x *via* K.

2.2 Balanced games

Balanced games are important because such games can be proved to have nonempty cores. This is done in Section 3. Balanced games are defined in relation to *balanced collections of coalitions*. In a game (N, V) there are $2^n - 1$ possible coalitions, not counting the null coalition, $\varnothing$. A collection of coalitions is a subset of this family of coalitions. For example, in a game with four players, the set of (nonempty) coalitions is $W = \{\{1\}, \{2\}, \{3\}, \{4\}, \{1,2\}, \{1,3\}, \{1,4\}, \{2,3\}, \{2,4\}, \{3,4\}, \{1,2,3\}, \{1,2,4\}, \{1,3,4\}, \{2,3,4\}, \{1,2,3,4\}\}$, and a collection of coalitions is a subset of W, such as $W_1 = \{\{1\}, \{4\}, \{1,3\}, \{2,3,4\}, \{2,3\}\}$ or $W_2 = \{\{2\}, \{3\}, \{1,4\}, \{2,4\}, \{3,4\}, \{1,2,3\}\}$. A collection of coalitions is balanced if it is possible to assign a positive number to each coalition in the collection so that the assigned numbers add to unity when summed over the coalitions to which any one player belongs.

DEFINITION 8.5 *Let* W *be the set of coalitions in the game* $\Gamma = (N, V)$, *and let* W_1 *be a subset of* W. W_1 *is a* **balanced collection of coalitions** *if there are numbers* $\delta_K > 0$, $K \in W_1$ *such that* $\sum_{\{K \in W_1 \mid i \in K\}} \delta_K = 1$ *for each* $i \in N$.

For example, $W_1 = \{\{1\}, \{4\}, \{1,3\}, \{2,3,4\}, \{2,3\}\}$ is not a balanced collection, but W_2 is. to see that $W_2 = \{\{2\}, \{3\}, \{1,4\}, \{2,4\}, \{3,4\}, \{1,2,3\}\}$ is balanced, let

$$\delta_{\{2\}} = \frac{1}{6} \qquad \delta_{\{3\}} = \frac{1}{12} \qquad \delta_{\{1,4\}} = \frac{5}{12}$$

$$\delta_{\{2,4\}} = \frac{1}{4} \qquad \delta_{\{3,4\}} = \frac{1}{3} \qquad \delta_{\{1,2,3\}} = \frac{7}{12}$$

Summing the weights for the coalitions containing player 1 yields $\frac{5}{12} + \frac{7}{12} = 1$. Doing the same for player 2 yields $\frac{1}{6} + \frac{1}{4} + \frac{7}{12} = 1$, $\frac{1}{12} + \frac{1}{3} + \frac{7}{12} = 1$

for player 3, and $\frac{5}{12}+\frac{1}{4}+\frac{1}{3}=1$ for player 4. To see that W_1 is not balanced, note that positive numbers must be found so that $\delta_{\{2,3\}}+\delta_{\{2,3,4\}}=1$ (this sum is over the coalitions containing player 2) and $\delta_{\{1,3\}}+\delta_{\{2,3\}}+\delta_{\{2,3,4\}}=1$ (this sum is over the coalitions containing player 3). It is impossible to find three such positive numbers, which shows that W_1 is not a balanced collection.

A *balanced game* is a game in which a special relationship exists between $V(N)$ and the $V(K)$; however, this special relationship needs to hold only with respect to balanced collections of coalitions. To get the gist of this relationship, suppose $u \in R^n$ is a payoff vector and W_1 represents any balanced collection of coalitions. Then if $u^K \in V(K)$ for all K in any balanced collection W_1, it is required that $u \in V(N)$.

DEFINITION 8.6 *A game $\Gamma=(N, V)$ is* **balanced** *if $u \in V(N)$ whenever, for any balanced collection W_1 and any $K \in W_1$, $u^K \in V(K)$.*

3 The core

3.1 The assumptions and the definition of the core

The class of games to be considered here is described by Assumptions (8.1) to (8.4), which specify that (a) there be a finite number of players, (b) the sets $V(K)$ be nonempty and compact, (c) that any payoff vector in $V(K)$ can be altered by reducing a player's payoff, and the resulting payoff vector is still in $V(K)$ if the player's payoff is not reduced below $\underline{u}_i$, and (d) the game is balanced.

ASSUMPTION 8.1 *In the game (N, V), the number of players is finite.*

ASSUMPTION 8.2 *In the game (N, V), for each $K \subset N$, $V(K)$ is nonempty and compact. For each $i \in N$, $V(\{i\})=\{\underline{u}_i\}$, and, for any $u^K \in V(K)$, $u^K \geqslant \underline{u}^K$.*

ASSUMPTION 8.3 *In the game (N, V), if $u^K \in V(K)$ and $\underline{u}^K \leqslant u'^K \leqslant u^K$, then $u'^K \in V(K)$.*

ASSUMPTION 8.4 *The game (N, V) is balanced.*

Figure 8.2 illustrates Assumption 8.3 for a two-player coalition. The assumption would be violated if any of the three shaded regions were not part of $V(\{1,2\})$. For this two-player coalition, Assumption 8.3 implies that $V(\{1,2\})$ is bounded below by a horizontal line through $\underline{u}_2$, on the left by a vertical line through $\underline{u}_1$, and above and to the right by a curve from $(\underline{u}_1, u_2^*)$ to $(u_1^*, \underline{u}_2)$ that never slopes upward. Assumption 8.3 is sometimes called a *free disposal* assumption, because it is analogous to economic axioms of that name. The core is defined essentially as in Chapter 7:

DEFINITION 8.7 *The* **core** *of the game (N, V), denoted $C(N, V)$, is the subset of the set of imputations consisting of imputations that are not dominated.*

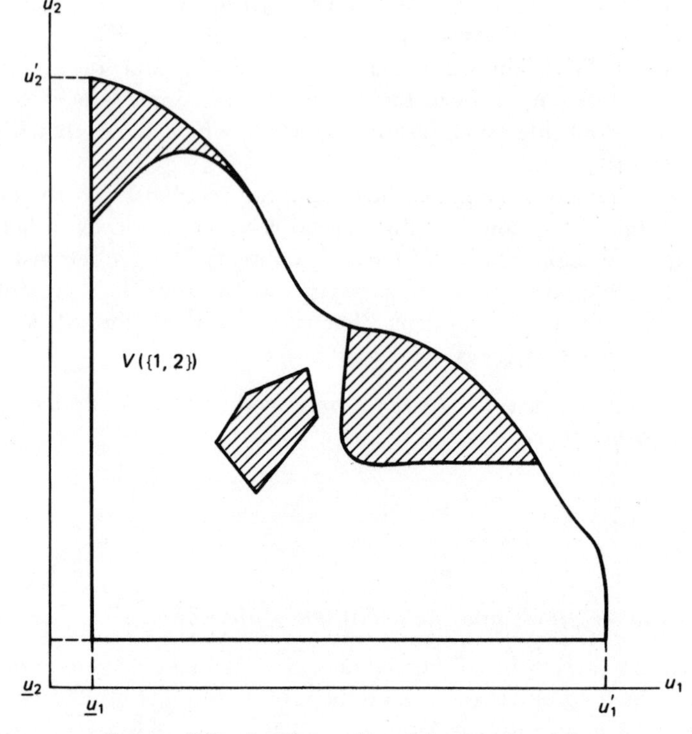

FIGURE 8.2 An illustration of Assumption 8.3.

As in Chapter 7, u is the core of the game (N, V) if it is both group- and individually rational (i.e., $u \in I(N, V)$), and if it gives to each coalition at least as much as the coalition can ensure itself (i.e., there is no $K \subset N$ and $x^K \in V(K)$ such that $x^K \gg u^K$).

3.2 The core for a finite-cornered game

The nonemptiness of the core is proved following Scarf (1967, 1973) using an algorithm that relies on there being only a finite number of points to check. This algorithm is an extension of the algorithm developed by Lemke and Howson (1964) for bimatrix games. It cannot work directly on all games satisfying Assumptions 8.1 to 8.4; however, the larger class of games is dealt with in Section 3.3 via a limiting argument. A set $V(K)$ that satisfies Assumptions 8.1 to 8.4 can be approximated using a finite number of points in R^k. This is illustrated in Figure 8.3. For an arbitrary coalition K, a set $V(K)$, analogous to Figure 8.3 can be defined by means of $\underline{u}$ and m_K payoff vectors, $u^{K,1}, \ldots, u^{K,m_K}$. In Figure 8.3, $m_K = 3$, and $u^K \in V(K)$ if (a) $\underline{u}^K \leqslant u^K \leqslant u^{K,1}$ or (b) $\underline{u}^K \leqslant u^K \leqslant u^{K,2}$ or (c) $\underline{u}^K \leqslant u^K \leqslant u^{K,3}$. The part of $V(K)$ satisfying (a) is shaded with diagonal lines, that satisfying (b) with horizontal lines, and that satisfying (c) with vertical lines. A characteristic

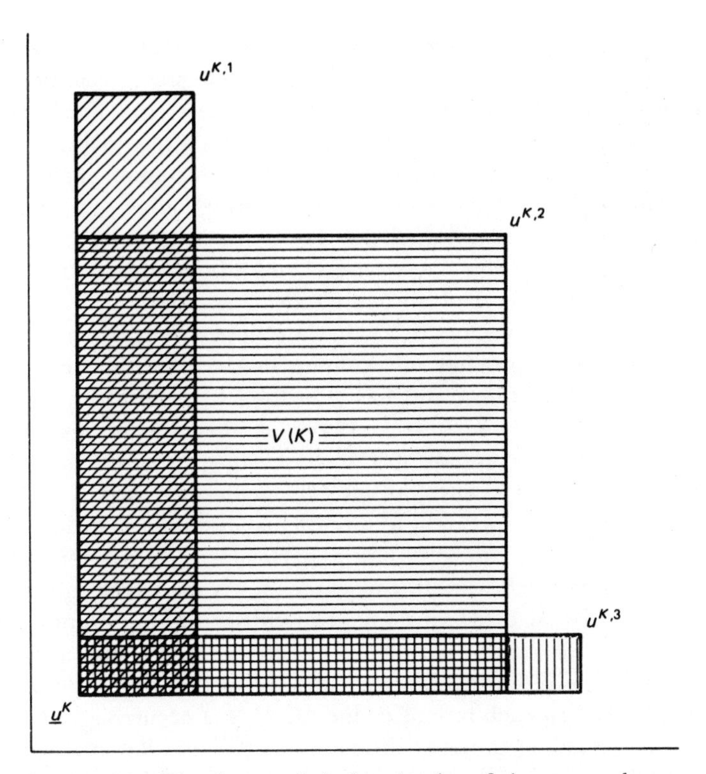

FIGURE 8.3 The characteristic function in a finite-cornered game.

function based on such points is called a *finite-cornered characteristic function*, and a *finite-cornered game* is a game whose characteristic function is finite cornered.

DEFINITION 8.8 *V is a* **finite-cornered characteristic function** *for the game (N, V) if, for each $K \subset N$, $k \geq 2$, there exist $u^{K,j} \in R^k$, $j = 1, \ldots, m_K$, such that $\underline{u}^K \leq u^{K,j}$ for all j and*

$$V(K) = \bigcup_{j=1}^{m^K} \{x^K \in R^k \mid \underline{u}^K \leq x^K \leq u^{K,j}\} \tag{8.3}$$

DEFINITION 8.9 *(N, V) is a* **finite-cornered game** *if V is a finite-cornered characteristic function for the game.*

3.2.1 A two-matrix representation of finite-cornered games

A finite-cornered game can be represented using two matrices, each having n rows and $n^* = \sum_{K \subset N} m_K$ columns, where $m_{\{i\}} = 1$ for all i and $m_\varnothing = 0$. This representation will be used to introduce Scarf's algorithm and to prove existence of a nonempty core for a finite-cornered game satisfying Assumptions 8.1 and 8.4. Note, by the way, that a finite-cornered game automatically satisfies Assumptions 8.2 and 8.3.

One matrix, called the *incidence matrix*, I, identifies all the coalitions. The first n columns each correspond to one of the single-player coalitions, with column i having a 1 in the ith row and 0 everywhere else. The next $m_{\{1,2\}}$ columns correspond to the coalition $\{1, 2\}$, the next $m_{\{1,3\}}$ to $\{1, 3\}$, and so forth through all coalitions of two or more players. A column corresponding to a particular coalition has a 1 in any row corresponding to a member of the coalition and a 0 in each other row. Thus, in a column for $\{1, 2, 4\}$, the first, second, and fourth row entries are 1, and all the other entries are 0.

The second matrix, called the *payoff characterization matrix*, B, has a column for each of the vectors, $u^{K,j}$, that are upper right corners of $V(K)$. Each one-player coalition has one column in the B matrix. For $\{i\}$ the assigned column is column i and has $\underline{u}_i$ in the ith row and M_{jk} in the jth row. The remaining columns are ordered as they are in the incidence matrix. After the first n columns of B, the next $m_{\{1,2\}}$ columns correspond to the coalition $\{1, 2\}$, and so forth, as with the incidence matrix I. Thus, each of the m_K columns corresponding to a particular coalition K is devoted to a different vector among the $u^{K,j}$. For a particular $u^{K,j}$, corresponding to column r of the matrix B, the lth row entry (b_{lr}) is the payoff to player l in $u^{K,j}$ if $l \in K$ and equal to M_{lr} if $l \notin K$. The entries M_{ij} are dummy entries whose values are all larger than any of the other entries, and, in addition, chosen so that no two of the M_{ij} are equal. One further restriction is placed on the entries M_{ij}: In each row of B, the M_{ij} $(j \leq n)$ occurring in the first n columns, corresponding to the coalitions $\{i\}$, $i \in N$, are larger than are the M_{il} $(l > 0)$ for entries in other columns.

TABLE 8.1 The basic data for a finite-cornered game

Payoff vectors defining the $V(K)$

	Coalition															
Player	1	2	3	1, 2				1, 3		2, 3			1, 2, 3		
1	5	*	*	15	13	9	7	10	8	*	*	*	20	16	11
2	*	2	*	3	5	6	9	*	*	10	7	4	8	11	12
3	*	*	4	*	*	*	*	10	12	5	6	13	16	14	8

Incidence matrix

	Coalition														
Player	1	2	3	1, 2				1, 3		2, 3			1, 2, 3		
1	1	0	0	1	1	1	1	1	1	0	0	0	1	1	1
2	0	1	0	1	1	1	1	0	0	1	1	1	1	1	1
3	0	0	1	0	0	0	0	1	1	1	1	1	1	1	1

Payoff characterization matrix

	Coalition														
Player	1	2	3	1, 2				1, 3		2, 3			1, 2, 3		
1	5	97	95	15	13	9	7	10	8	87	86	85	20	16	11
2	99	2	94	3	5	6	9	89	88	10	7	4	8	11	12
3	98	96	4	93	92	91	90	10	12	5	6	13	16	14	8

An illustration is presented in Table 8.1. The first part of the table shows the payoff vectors that define the characteristic function. The vector $\underline{u} = (5, 2, 4)$, and, for player 1, $V(\{1\}) = \{5\}$. This set is defined by the vector (5) in R^1. The first column, headed by $\{1\}$, has * in rows 2 and 3 to denote that there are no entries corresponding to players 2 and 3. The upper right boundary of $V(\{1, 2\})$ is defined by the corners (15, 3), (13, 5), (9, 6), and (7, 9), which are vectors in R^2, the two-dimensional space with coordinates corresponding to payoffs to players 1 and 2. The information in this part of the table provides the material to form the incidence matrix I and the payoff characterization matrix B. For I, there is a column corresponding to each column at the top of the table. For each column, the entry in a row is 1 if the corresponding player is in the coalition to which the column belongs, and is 0 otherwise. Thus * in the top matrix of the table corresponds to 0 in the incidence matrix. The matrix B repeats all the numbers from the top matrix of the table, but where * appears at the top matrix of the table, the corresponding entry in B is a large number. Each of these large numbers must be bigger than any number appearing in the payoffs in the top matrix of the table in the payoff vectors defining the $V(K)$, and the large entries in the first three columns must be larger than all the rest of these special entries.

3.2.2 *Defining a basis for each matrix*

The way that the matrices I and B are used is that each matrix has associated with it a *basis,* which is a set of n columns from a matrix; a basis for I is a set of n columns of I and similarly for B. If the same set of columns is a basis for both, then it defines a core point. Specifically, let $\delta \geqslant 0$ be a vector with exactly as many components as there are columns of I (i.e., $\delta \in R^{n^*}_+$), and let $e \in R^n$ be a vector whose components are all equal to 1. Then the n columns of I numbered $j_1, \ldots, j_n$ comprse a *feasible basis* for I if there exists δ such that $I\delta = e$, $\delta_j > 0$ for $j = j_i$, $i \in N$, and $\delta_i = 0$ otherwise. Recall that in I a column really denotes a coalition, so a collection of columns $\{j_1, \ldots, j_n\}$ corresponds to a collection of coalitions. Thus a feasible basis is a balanced collection of coalitions. Note that $\sum_{j=1}^{n^*} b_{ij}\delta_j$ is the sum of the weights assigned to the coalitions in the collection $\{j_1, \ldots, j_n\}$ that correspond to those coalitions to which player i belongs.

Now turning to B, a collection of columns of B, $W = \{j_1, \ldots, j_n\}$, is an *ordinal basis* of B if a vector β defined using the columns in W satisfies a special condition. Let

$$\beta_i = \min\{b_{ij_1}, \ldots, b_{ij_n}\}, \qquad i \in N \tag{8.4}$$

and define $\beta = (\beta_1, \ldots, \beta_n)$. Denote the lth column of B by b_l. If $\beta \ll b_l$ does not hold for any l, then the columns $j_1, \ldots, j_n$ are an ordinal basis. In other words, for β associated with an ordinal basis, there is no column of B that is strictly larger, component by component. If β were attainable (i.e., if $\beta \in V(N)$), then β would be in the core.

Thus, if W is a feasible basis, it is a balanced collection of coalitions, and if W is an ordinal basis, the associated β would, if attainable, be in the core. However, if W is *both a feasible and an ordinal basis,* then $\beta \in V(N)$ (i.e., $\beta \in C(N, V)$). This follows from the game being balanced.

DEFINITION 8.10 *Let $W = \{j_1, \ldots, j_n\}$ be a collection of columns of the incidence matrix I. W is a* **feasible basis** *for I if there exists $\delta \in R_+^{n^*}$ such that $I\delta = e$.*

DEFINITION 8.11 *Let $W = \{j_1, \ldots, j_n\}$ be a collection of columns of the characterization matrix, B, and let β be defined by $\beta_i = \min_{i \in N} b_{ij}$, $i \in N$. Then W is an* **ordinal basis** *for B if, for each column j of B, there is at least one index i such that $\beta_i \geq b_{ij}$.*

3.2.3 *Scarf's algorithm*

The purpose of the algorithm, then, is to find a collection of columns, W, that is simultaneously a feasible and an ordinal basis. The method of the algorithm is to start with two collections of columns, W_f^0 and W_o^0, the former being a feasible basis and the latter an ordinal basis, with the two sets having $n-1$ of their members in common. As will be seen, it is easy to start in this way. In a sense, one is close to finding a core point at this beginning place. The algorithm proceeds by removing that member of W_o^0 that is not found in W_f^0. This is called a *pivot* on the ordinal basis. The rules of proceeding are such that there is exactly one possible column to add to the ordinal basis when the mismatched column is removed. This results in a new ordinal basis, W_o^1. One of two things must be true about W_o^1: (a) the newly added column corresponds to the (formerly) mismatched column in W_f^0, or (b) the newly added column is not one of the columns found in W_f^0. If (a) holds, then the search is over, and the corresponding β^1, defined by equation (8.4) using W_o^1, is an element of the core. If (b) holds, then the next step is to *pivot* on the feasible basis. This is done by adding to W_f^0 the mismatched column of W_0^1. As with the pivot on the ordinal basis, there is only one possible column to remove, resulting in either a complete matching between W_o^1 and W_f^1 or a mismatch in one column. Again, in the former case β^1 is a core point, and in the latter, another pivot step must be taken. This time the pivot step is on the ordinal basis. This process continues, with ordinal and feasible pivots alternating. The process must end in a finite number of pivot steps with a core point being found, because there are only a finite number of ordinal and feasible bases and the pivoting process can never cycle back and repeat some pair of bases for a second time.

3.2.4 *Preliminary results*

To simplify the process of proof, the games that are first examined are finite-cornered games in which elements of B in any given row are distinct and in which the δ_j associated with the n columns of a feasible basis are all strictly positive and different from one another. These extra conditions are later removed, but they simplify proofs and make it easier to see why the results hold. To summarize the results proved in this section, Lemma 8.1 establishes that when a new column is added to a feasible basis, there is only one column that can possibly be removed. Then, in Lemma 8.2, it is shown that when a column is removed from an ordinal basis, there is only one possible column that can replace it to form a new ordinal basis.

LEMMA 8.1 *Let (N, V) be a finite-cornered game that satisfies Assumptions* 8.1 *and* 8.4, *and let I and B be the incidence and payoff characterization matrices. Assume that no two elements of the ith row of B have the same value $(i \in N)$ and that, for any feasible basis $W = \{j_1, \ldots, j_n\}$, the weights $\delta(W)$ associated with W corresponding to the columns in W are positive and unique. For any columns $j^* \notin W$, there is a unique feasible basis consisting of j^* and $n-1$ of the columns in W if the convex set $\Delta = \{\delta \in R^{n^*} \mid \delta \geqslant 0, I\delta = e\}$ is bounded.*

Proof. The introduction into the basis of j^* and the adjustment of the positive δ_j to keep $I\delta = e$ satisfied is precisely a linear programming pivot step. The condition that the positive elements of $\delta(W)$ also be distinct from one another guarantees that, as the weight on j^* is increased and the other weights are appropriately altered, two weights cannot reach 0 simultaneously. The condition that Δ is bounded implies that, as the weight on j^* increases, the weight on at least one column in the original basis, W, must decline. This, in turn, means that the weight on j^* must rise to a point at which one of the previous columns drops out of the basis. QED

LEMMA 8.2 *Under the conditions of Lemma* 8.1, *let $W = \{j_1, \ldots, j_n\}$ be an ordinal basis of B. For an arbitrary column of the basis, say j_1, suppose that the remaining columns are not all contained in $\{1, \ldots, n\}$. Then there is a unique column $j^* \neq j_1$ such that $W^* = \{j^*, j_2, \ldots, j_n\}$ is an ordinal basis.*

Proof. Three things are done in this proof. The first is to define an ordinal pivot step. After that, it is shown that such a step is unique and must lead to a new ordinal basis. Finally, it is shown that nothing but an ordinal pivot step can lead to a new ordinal basis.

From the definition of an ordinal basis, there must be exactly one column among $j_2, \ldots, j_n$ that contains two entries that are the smallest entries in their respective rows. Although one of these row minimizers, occurring in row i^*, is a row minimizer in W, the other, occurring in row i', is not. Denote by j_l the column in which these two row minimizers are found, and let β be the vector of row minimizers for W. Thus, the new row minimizer in row i', after removal of j_1, is $b_{i'j_l}$. Next, look among all the columns *not* in W and, among them, find all the columns for which $b_{ij} > \beta_i$ $(i \neq i^*)$ and $b_{ij} > b_{ij_l}$. From these columns, select the one in which b_{i^*j} is maximized and denote the column j^*. This completes the specification of the ordinal pivot step.

Now it is shown that the ordinal pivot step described above leads to a unique new basis. Letting β' denote the vector of row minimizers for $W' = \{j^*, j_2, \ldots, j_n\}$, note that for $i \neq i^*$, i^*, we have $\beta'_i = \beta_i$, and that $\beta'_{i'} = b_{i'j_1}$ and $\beta'_{i^*} = b_{i^*j^*}$. The columns not included in W can be divided into two groups; those selected for examination in the search for j^* and those not selected. For those not selected, each one has $b_{ij} < \beta'_i$ for at least one $i \neq i^*$; therefore, introducing one of these would force out some member of W other than j_1. Among those selected, if any column apart from j^* had been introduced, $b_{ij^*} > \beta'_i$, $i \in N$, which would violate the definition of ordinal basis. Thus, the pivot step is unique and leads to a new ordinal basis.

It remains to see that a new column other than j^* could not have been chosen with its row minimizer in row i'. Supposing there is such a row, then the resulting β' would be identical to the old β in all entries except i'. Therefore, $b_{ij^*} \geqslant b_{ij_1}$, which means that column j_1 could not improve on β'. On the other hand if j^* is introduced, $b_{i'j^*} \leqslant \beta_{i'}$ which implies that $b_{i'j^*} = b_{i'j_1}$; thus j^* could only be j_1. Therefore, a new column $j^* \neq j_1$ cannot be chosen with its row minimizers in i^*. QED

3.2.5 *Existence of a nonempty core*

Putting Lemmas 8.1 and 8.2 together, suppose there is a W^0_o (ordinal basis) and W^0_f (feasible basis) that form a starting point for the present discussion, and assume that the two bases have $n-1$ columns in common. Two methods are allowed by the algorithm for pivoting to a new basis: (1) add to the feasible basis the mismatched column from the ordinal basis or (2) remove the mismatched column from the ordinal basis. Thus, either the mismatched column from W_o becomes matched by putting it into W_f or it ceases to be mismatched by removing it from W_o. Both of these steps are unique. When the algorithm is applied, there is only one step to take at each point, because one of the two possible pivots moves the algorithm *backwards* to the previous pair of bases. Thus, the only way to attain a new pair of bases is to take the one step that does not reverse the previous pivot.

Assume that the two special conditions of the lemmas are met (that the set Δ is bounded and, if a column is removed from W^0_o, that at least one remaining column is not among columns $1, \ldots, n$). Denote by j^0_o the column in W^0_o that is not also found in W^0_f (the mismatched column in W^0_o) and by j^0_f the mismatched column from W^0_f. From the lemmas, there are only two possible pivot steps to take. One step is to drop j^0_o from W^0_o, the other is to add j^0_o to W^0_f. Supposing that the next pivot step does not result in a core point, dropping j^0_o causes (a unique) j^1_o to be added and it is a new mismatch in the new ordinal basis W^1_o. Adding j^0_o to W^0_f causes some other (unique) member to drop out. If that member is not j^0_f, then j^0_f $(=j^1_f)$ remains the mismatch in the new W^1_f, and some new j^1_o (corresponding to the column that was removed from W^0_f) is the mismatch in W^0_o.

Consider the possibilities if the first pivot were taken. After the pivot step, j^1_o is the mismatch in W^1_o and the two possible pivots are (i) drop j^1_o

from W_o^1 or (ii) add j_o^1 to W_f^0. Taking step (i) merely puts the bases back to their previous position; therefore, the only *new* alternative is (ii). Similarly, if the second pivot were taken, W_o^0 acquires a new mismatched column, j_o^1. After this pivot, two steps can be taken: j_o^1 can be dropped from W_o^0 or j_o^1 can be added to W_f^0. The latter simply reverses what was just done, so the first step is the only new alternative.

It is now possible to prove that a game satisfying the extra restrictions assumed in Lemmas 8.1 and 8.2 has a nonempty core. The technique of proof is to show that one can always find a special pair of bases W_f^0 and W_o^0, having $n-1$ columns in common, from which only one pivot step is possible. This starting point ensures that the pivoting process must always move forward to bases that have not already been encountered, and the finite corneredness of the game guarantees that pivoting cannot go on forever. The process must reach a termination, but it can only terminate at a core point.

LEMMA 8.3 *Let* (N, V) *be a finite-cornered game satisfying Assumptions* 8.1 *and* 8.4, *in which no two elements of any row of* B *have the same value and in which the weights* $\delta(W)$ *associated with any column in a feasible basis* W *are positive and distinct. Then* $C(N, V)$ *is not empty.*

Proof. Let $W_f^0 = \{1, \ldots, n\}$ and $W_o^0 = \{2, \ldots, n, j_o^0\}$, where j_o^0 is chosen to be that (unique) column among columns $n+1, \ldots, n^*$ having the largest first-row entry. Thus, $\beta = (b_{1j_o^0}, \underline{u}_2, \ldots, \underline{u}_n)$, and this is clearly the only ordinal basis that can have columns $2, \ldots, n$ as members. From this starting point, only one pivot is possible; removing j_o^0 from W_o^0 is ruled out, because only j_o^0 can be put with columns $2, \ldots, n$ to form an ordinal basis. Any other column used in place of j_o^0 will result in an associated β vector that is strongly dominated by columns j_o^0. This violates the definition of an ordinal basis.

Note, too, that Δ must be bounded, which means that Lemma 8.1 can be applied. Next note that in taking the other pivot step, adding j_o^0 to W_f^0, one of the columns $2, \ldots, n$ becomes the mistmatch in W_o^0, therefore, the special condition in Lemma 8.2 is met for the next pivot step. With Lemmas 8.1 and 8.2 applying, there is, after the first pivot step (which is unique) only one possible pivot step that does not backtrack. Because there can be only a finite number of possible bases of either type, and, aside from the special starting point for the algorithm and a core point, there must always be two possible pivots, there must be a collection of n columns that is simultaneously a feasible and an ordinal basis. QED

Moving from Lemma 8.3 to a theorem establishing that $C(N, V)$ is not empty for any game satisfying Assumptions 8.1 to 8.4 is easy and rests on using Lemma 8.3 with sequences of games. This is carried out in two steps. For the first step, taken in Lemma 8.4, finite-cornered games are retained with the special conditions mentioned in Lemma 8.1 being dropped. Then Theorem 8.1 makes the final extension.

LEMMA 8.4 *Any finite-cornered game* (N, V) *satisfying Assumptions* 8.1 *and* 8.4 *has a nonempty core.*

Proof. Let I^l and B^l, $l = 1, 2, \ldots$, be a sequence of incidence and payoff characterization matrices that converge to I and B, respectively. The members of the sequence are chosen to satisfy all the conditions of Lemma 8.1 and I and B represent a finite-cornered game and satisfy Assumptions 8.1 and 8.4. For the game (N^l, V^l), represented by I^l and B^l, let β^l be the core point found by the algorithm and let δ^l be the corresponding set of weights. A cluster point of the sequence $\{\delta^l\}$ is a solution for $I\delta = e$. Let the sequence $\{\beta^l\}$ be the companion sequence to the convergent subsequence of $\{\delta^l\}$. This subsequence of core points has a convergent subsequence whose limit may be denoted β. The payoff vector β is a core point of the game represented by I and B. QED

3.3 The core of a balanced game

Lemma 8.4 provides a foundation upon which the final result can easily be built. Essentially, the class of games covered by Lemma 8.4 can be used to approximate any game satisfying Assumptions 8.1 to 8.4. For an arbitrary game (N, V) satisfying these assumptions, it is possible to find a convergent sequence of games satisfying the conditions of the lemma that has the game (N, V) as its limit. Then it is proved that the limiting game has a nonempty core if the members of the convergent sequence have nonempty cores.

THEOREM 8.1 *A game* (N, V) *satisfying Assumptions* 8.1 *to* 8.4 *has a nonempty core.*

Proof. Let $\{(N^l, V^l)\}$ be a sequence of finite-cornered games satisfying Assumptions 8.1 and 8.4, the members of which are represented by I^l and B^l, respectively, and suppose that the members of this sequence are chosen to converge to a game (N, V) that satisfies Assumptions 8.1 to 8.4. This sequence can be selected to be dense in the limit. Letting β^l be a core point of (N^l, V^l), a cluster point of the sequence $\{\beta^l\}$ is a core point of (N, V). QED

The algorithm can be illustrated using the game in Table 8.1. The initial feasible basis is $W_f^0 = \{1, 2, 3\}$ and the initial ordinal basis is $W_o^0 = \{2, 3, 10\}$. In choosing this ordinal basis, any two of the first three columns can be used, but once the two columns are selected, there is only one possible choice from the remaining 12 columns. Were any column other than 10 used with 2 and 3, the resulting β vector would be strictly dominated by column 10. From this starting point, the only possible pivot step is to put column 10 into the feasible basis, which results in the removal of column 2 and $W_f^1 = \{1, 3, 10\}$.[1] Now the mismatched column in the ordinal basis is column 2, which is removed. Column 14 is added, resulting in $W_o^1 = \{3, 10, 14\}$. Next, column 14 is added to the feasible basis and column 1 is dropped, resulting in $W_f^2 = \{3, 10, 14\}$. At this point, the two bases are the same, so the β vector associated with the ordinal basis (16, 10, 4), is a core point. Note that a point is in the core if it is not strongly dominated, and this core point is, in fact, weakly dominated by column 14.

4 The stable set

It was shown that the definition of the core is essentially the same for nontransferable and transferable utility games. In both situations, the core is based on domination. The same holds for the stable set. Just as in Chapter 7, a stable set for a game (N, V) consists of a subset B of the set of imputations in B that dominates x. Second, if an imputation y is not in B, then there is some imputation x in B that dominates y.

5 The bargaining set, the kernel, and the nucleolus

Nearly all of the definitions for the bargaining set can be carried over with no alteration; however, it is necessary to restate the definition of individually rational payoff configurations to put it in terms of the characteristic function V.

DEFINITION 8.12 *An* **individually rational payoff configuration** *is a pair* $(x, \mathcal{T})$ *where* $x \in R^n$, $\mathcal{T}$ *is a coalition structure,* $x^K \in V(K)$ *and* $y^K \not> x^K$ *for all* $y^K \in V(K)$ *and all* $K \in \mathcal{T}$. *The set of all individually rational payoff configurations for the game* (N, V) *relative to the coalition structure* $\mathcal{T}$ *is denoted* $I_{\mathcal{T}}(N, V)$.

From here, the definitions of *objection, counterobjection,* and the *bargaining set* from Section 6 of Chapter 7 can be used with $I_{\mathcal{T}}(N, V)$ substituted for $I_{\mathcal{T}}(N, v)$.

For the kernel and the nucleolus, it is less obvious how to generalize. Both are based on the concept of *excess,* which is intimately bound to the transferable utility characteristic function.

6 Extending the Shapley value to games without transferable utility

The extension of the Shapley value presented here is due to Shapley (1969) and uses a concept from Harsanyi (1959). Harsanyi found a generalization of the Shapley value to nontransferable utility games that also generalizes the Nash bargaining solution to n-person games. This was later refined in Harsanyi (1963), and his formulation has considerable appeal for the way in which it balances off all the two-person games that can be found embedded in the original game by (a) taking each coalition K with $1 \leq k < n$ and considering a two-person game in which the players are K and $N \backslash K$, and (b) for each pair of players, i and j, holding fixed the circumstances of all but these two players and looking at a two-person game between i and j. The details of Harsanyi's model are formidable, which accounts for some of the appeal of Shapley's much simpler formulation. In addition to Harsanyi's original articles, expositions can be found in Friedman (1977; Chapter 11) and Harsanyi (1977, Chapter 12). The latter treatment is exceptionally fine. Another extension of the Shapley value is due to Owen (1972).

The concept borrowed from Harsanyi is called the *λ-transfer value,* and the two approaches differ with respect to the rules defining how the value is characterized. The assumptions of the model and Shapley's definition are found in Section 6.1; then, in Section 6.2 existence of a λ-transfer value is proved. The model is illustrated in Section 6.3 by a simple example.

6.1 The assumptions of the model and the λ-transfer value

Shapley's own version is extremely simple and is, perhaps, as easily justified as any of the others. His basic device is to define a family of transferable utility games that can be associated with the original game. Each such game is based on giving weights to the players. Each game has a Shapley value; however, for most of these games, the Shapley value cannot be achieved in the (suitably weighted version of the) original game. A λ-transfer value for the original game is, essentially, the Shapley value of a member of this family that can be attained in the original game. Assumptions 8.1 and 8.2 are retained and supplemented by Assumptions 8.5 and 8.6. Both concern $V(N)$, the former being Assumption 8.3 applied only to $V(N)$ and the latter being convexity.

ASSUMPTION 8.5 *If $u' \in V(N)$ and $\underline{u} \leqslant u \leqslant u'$, then $u \in V(N)$.*

ASSUMPTION 8.6 *$V(N)$ is convex.*

To define the family of transferable utility games that are associated with (N, V), let Λ denote the unit simplex in R^n. That is, $\Lambda = \{\lambda \in R^n \mid \lambda \geqslant 0, \sum_{i \in N} \lambda_i = 1\}$. For $\lambda \in \Lambda$, (N, v_λ) is a transferable utility game with a characteristic function defined by

$$v_\lambda(K) = \max_{u^K \in V(K)} \sum_{l \in K} \lambda_l u_l^K, \qquad K \subset N \tag{8.5}$$

The weights λ assign a relative importance to each of the players and equality would mean $\lambda_i = 1/n$ for each player. Denote the Shapley value of the game (N, v_λ) by $\phi(N, v_\lambda)$, and define for $\lambda' \in \Lambda$ the payoff vector

$$u' = (\phi_1(N, v_{\lambda'})/\lambda'_1, \ldots, \phi_n(N, v_{\lambda'})/\lambda'_n) \tag{8.6}$$

Then if $u' \in V(N)$ it is the *λ-transfer value* of the game (N, V) and is Shapley's proposed value solution for nontransferable utility games.

DEFINITION 8.13 *For a nontransferable utility game (N, V) and for $\lambda \in \Lambda$, (N, v_λ) is a companion transferable utility game where v_λ is defined by equation* (8.5).

If the Shapley value of (N, v_λ) *satisfies*

$$u' = (\phi_1(N, v_\lambda)/\lambda_1, \ldots, \phi_n(V, v_\lambda)/\lambda_n) \in V(N) \tag{8.7}$$

then u' *is the* **λ-transfer value** *of the game.*

Figure 8.4 illustrates the relationship between the game (N, V) and the game (N, v_λ). For the coalition $\{1, 2\}$ the region bounded above by the curve BAC is $V(\{1, 2\})$. The point A is where $\lambda_1 u_1 + \lambda_2 u_2 = v_\lambda(\{1, 2\})$ is maximized on $V(\{1, 2\})$, and the straight line DAE satisfies $\lambda_1 u_1 + \lambda_2 u_2 = v_\lambda(\{1, 2\})$. Thus, the set with upper right boundary DAE is the set of attainable payoff vectors for $\{1, 2\}$ in the game (N, v_λ). What is done here for the coalition $\{1, 2\}$ must be repeated for all coalitions K, including the coalition of the whole, N, in order to form v_λ. For the sake of illustration, it is supposed that the Shapley value for the game (N, v_λ) awards u^λ to $\{1, 2\}$ in the diagram.

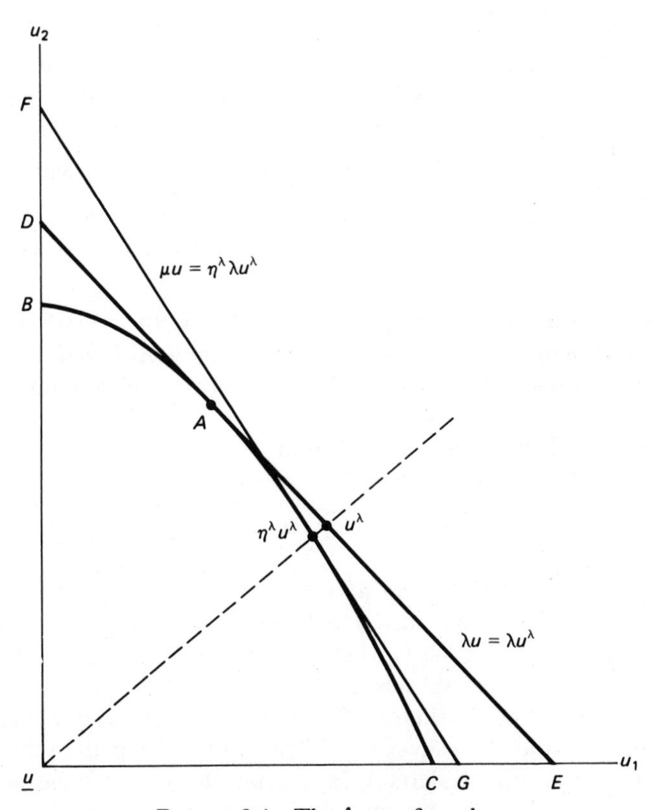

FIGURE 8.4 The λ-transfer value.

6.2 Existence of the λ-transfer value

The existence of a λ-transfer value for a game satisfying Assumptions 8.1, 8.2, 8.5, and 8.6 is proved by defining a correspondence that takes elements of Λ into subsets of Λ, and showing that this correspondence has a fixed point that must be the weights associated with the λ-transfer value. This is carried out in two steps, the first of which is Lemma 8.5, where it is proved that $\phi(N, v_\lambda)$ is continuous in λ. When the Shapley value of (N, v_λ) is outside $V(N)$, it is possible to give a scalar measure of how far outside it is. A measure is defined in Lemma 8.6 and shown to be continuous in λ. Then, in Theorem 8.2 an appropriate correspondence is defined, using the measure from Lemma 8.6, and shown to be upper semicontinuous. From this the existence of the λ-transfer value is established.

LEMMA 8.5 *Under Assumptions* 8.1, 8.2, 8.5, *and* 8.6, $\phi(N, v_\lambda)$ *is a continuous function of* λ *for all* $\lambda \in \Lambda$.

Proof. The characteristic function, v_λ is continuous in λ, because $\max_{u^K \in V(K)} \sum_{l \in K} \lambda_l u_l^K$, which defines the characteristic function, is continuous in λ. From equation (7.25), giving the Shapley value, it is clear that $\phi(N, v_\lambda)$ is continuous in each $v(K)$; therefore, $\phi(N, v_\lambda)$ is continuous in λ, which establishes the lemma. QED

LEMMA 8.6 *Let* $u^\lambda = (\phi_1(N, v_\lambda)/\lambda_1, \ldots, \phi_n(N, v_\lambda)/\lambda_n)$. *For* $\lambda \in \Lambda$ *define* $\eta^\lambda \in R_+$ *by the condition that* $\eta^\lambda(u^\lambda - \underline{u}) + \underline{u}$ *is on the upper right boundary of* $V(N)$. *That is,* $\eta(u^\lambda - \underline{u}) + \underline{u} \in V(N)$ *for* η *equal to* η^λ *and* $\eta(u^\lambda - \underline{u}) \notin V(N)$ *for* $\eta > \eta^\lambda$. *Then,* η^λ *is continuous in* λ.

Proof. The lemma follows from the continuity of $\phi(N, v_\lambda)$ with respect to λ. QED

THEOREM 8.2 *A game* (N, V) *satisfying Assumptions* 8.1, 8.2, 8.5, *and* 8.6 *has a* λ-*transfer value.*

Proof. Note first that $u^\lambda = (\phi_1(N, v_\lambda)/\lambda_1, \ldots, \phi_n(N, v_\lambda)/\lambda_n)$ is a continuous function of λ and that either u^λ lies on the upper right boundary of $V(N)$ or it is above and to the right of $V(N)$. With that in mind, let

$$T(\lambda) = \left\{ \mu \in \Lambda \,\middle|\, \sum_{i \in N} \mu_i u_i^\lambda = \sum_{i \in N} \lambda_i u_i^\lambda \quad \text{and} \quad \sum_{i \in N} \mu_i u_i \leq \eta^\lambda \sum_{i \in N} \mu_i u_i^\lambda, u \in V(N) \right\} \tag{8.8}$$

If $\lambda \notin T(\lambda)$, then $u^\lambda \notin V(N)$ and u^λ cannot be the λ-transfer value of the game; however, if $\lambda \in T(\lambda)$, then $u^\lambda \in V(N)$ and u^λ is the λ-transfer value. The only remaining question is whether $T(\lambda)$ has a fixed point. Clearly, it is a mapping from Λ to subsets of Λ, because $\eta^\lambda u^\lambda$ is a Pareto optimal element of $V(N)$. The continuity of $\eta^\lambda u^\lambda$ in λ ensures that $T(\lambda)$ is upper semicontinuous, and the convexity of $V(N)$ implies that the sets $T(\lambda)$ are convex. Therefore, the Kakutani fixed-point theorem can be applied to ensure that $T(\lambda)$ has a fixed point, which completes the proof. QED

The proof of Theorem 8.2 is illustrated in Figure 8.4. For λ, recall that the point A in the figure is where the weighted sum of the players' payoffs, using the weights λ, is maximized. The payoff vector u^{λ}, which is a weighted transformation of $\phi(N, v_{\lambda})$ (again, using the weights λ) lies on the line DAE that is tangent to $V(N)$ (i.e., the supporting hyperplane through A), and it is used to obtain the weights μ into which λ maps by drawing a straight line from $\underline{u}$ to u^{λ}. The point on this line that intersects the upper right boundary of $V(N)$, $\eta^{\lambda}u^{\lambda}$, is used to obtain the new weights μ. A tangent to $V(N)$ at $\eta^{\lambda}u^{\lambda}$ (the line FG) gives the new weights. Figure 8.5 illustrates a situation in which these weights would not be unique; however, they are nonnegative by Assumption 8.5.

Note that a coalition K, smaller than N, can receive a payoff vector lying outside of $V(K)$. It is only for $K = N$ that the coalition's payoff must lie on the upper right boundary of $V(K)$. It is tempting to interpret the weights λ as the relative importance of the players. Certainly within the confines of the model, such a procedure may be useful; however, establishing the "worth" of a player within a game context and transferring that "worth" to other situations may be unwarranted.

6.3 An example

An example is easily constructed for three players. Suppose a characteristic function where $\underline{u} = (0, 0, 0)$ and the upper right boundary of the sets $V(K)$ for the two- and three-player coalitions are $u_2 = 16 - u_1^2/6$ for $V(\{1, 2\})$,

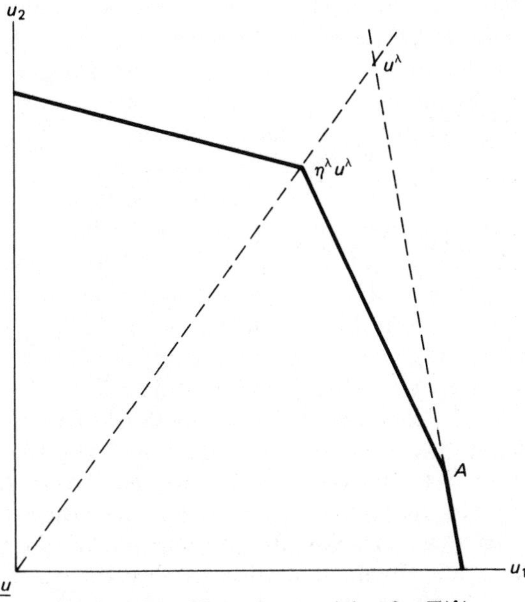

FIGURE 8.5 Nonunique weights for $T(\lambda)$.

$u_3 = 14 - .02u_1^2$ for $V(\{1,3\})$, $u_3 = 10 - .045u_2^2$ for $V(\{2,3\})$, and $u_3 = 14 - u_1^2/87.5 - .036u_2^2$ for $V(\{1,2,3\})$. For weights $\lambda = (.2, .3, .5)$, the λ-transfer value of the game is $(35/2, 25/3, 8)$. This can be seen by first finding the characteristic function for the game (N, v_λ), then obtaining the Shapley value, $\phi(N, v_\lambda)$, for this game, and, finally, noting that $(35/2, 25/3, 8) = (\phi_1(N, v_\lambda)\lambda_1, \phi_2/(N, v_\lambda)/\lambda_2, \phi_3(N, v_\lambda)/\lambda_3)$.

When $.2u_1 + .3u_2$ is maximized on $V(\{1,2\})$, the resulting maximum is $v_\lambda(\{1,2\}) = 5$. Doing the same for the remaining coalitions of two and three players yields $v_\lambda(\{1,3\}) = 8$, $v_\lambda(\{2,3\}) = 6$, and $v_\lambda(\{1,2,3\}) = 10$. The Shapley value for this game is $\phi(N, v_\lambda) = (3.5, 2.5, 4)$ and the λ-transfer value is $(3.5/.2, 2.5/.3, 4/.5) = (17\frac{1}{2}, 8\frac{1}{3}, 8)$.

7 Applications of the core

Two examples are discussed in this section. The first is an old and well-known core application: Edgeworth's model of general equilibrium without production. This is the classic case, mentioned earlier in this volume, of an economy in which there are consumers who can trade with one another, but there are no producers. The second example is from the political science literature and relates to decision making by majority vote in a committee or a society.

7.1 A model of general economic equilibrium with trade

The first example is based on Edgeworth's (1881) general equilibrium model of pure trade, within which framework Edgeworth invented the core. His approach was to postulate a two-person economy, depicted in the Edgeworth box in Figure 8.6, and to ask which trades ought one expect the traders to make. In the figure, E is the endowment point of the traders, indicating that Ann has 50 apples and 100 bananas, and Ben has 80 apples and 60 bananas. For Ann, the origin is the lower left corner of the box, and quantities rise for her as one moves up and to the right. The reverse holds for Ben, for whom the upper right corner of the box is the origin and quantities rise as one moves down and to the left. A point in the box represents an allocation of the total resources of the community (130 apples and 160 bananas), between the two traders.

Edgeworth postulated two criteria for the acceptability of an allocation. First, a final allocation should yield to each trader at least as much utility as he obtains from his endowment. Second, it should be impossible to find a trade that Pareto dominates a final allocation. The shaded region in Figure 8.6 satisfies the first criterion and the curve $ABCD$ satisfies the second. Thus, the curve BC satisfies both. For the two-person game, the segment BC is the set of commodity allocations corresponding to payoff vectors in the core.

Note that the discussion has been carried out without reference to

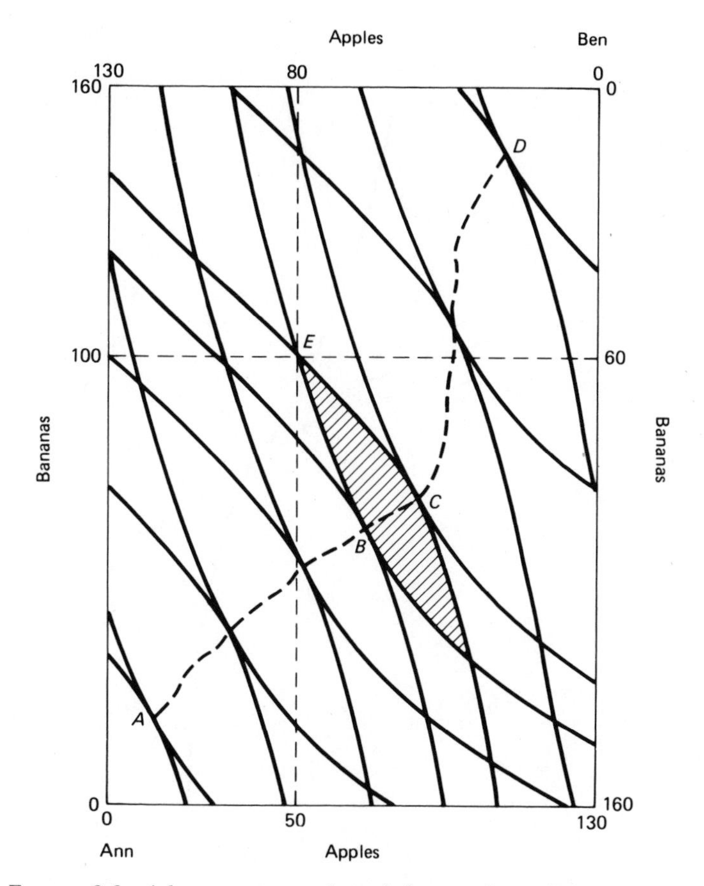

FIGURE 8.6 Advantageous trade and the core in an Edgeworth box.

markets or prices. This is natural in the sense that markets in which prices are used as a mechanism of trade ought to be competitive, or at least competitive on one side (i.e., on the buyers' side or on the sellers' side). In the two-person setting, it is not natural for one trader to inform another that he will trade at a predetermined ratio any quantities the other person wishes. This could happen, but traders need not be constrained to such behavior.

Edgeworth then considered an economy with $2k$ traders. Half of them have both the same endowment and the same preferences as Ann, the other half share these characteristics with Ben. Edgeworth conjectured, and others later proved (see Debreu and Scarf (1963, 1972)), that, as k goes to infinity the core commodity allocations converge to the set of commodity allocations associated with competitive equilibria. In other words, this remarkable result says that when the number of traders is extremely large, the set of "acceptable" trades consists of those trades that are achievable as equilibrium outcomes in competitive economic markets with prices.

The Edgeworth model with two types of traders and k of each type can

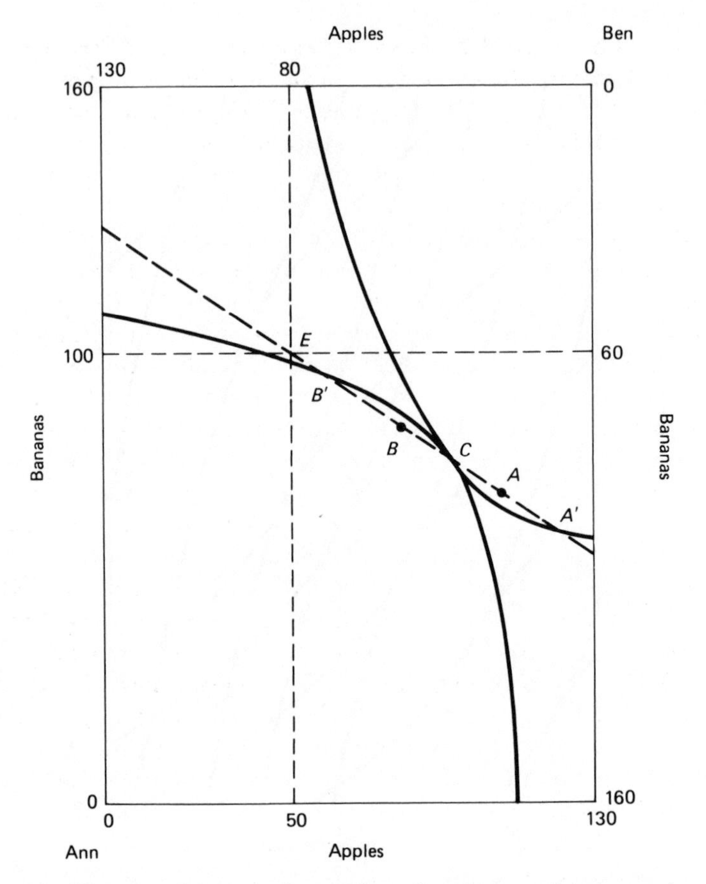

FIGURE 8.7 The core shrinks as the number of consumers of each type increases.

be adequately represented in the Edgeworth box in Figures 8.6 and 8.7, because all core allocations must give identical commodity bundles to all traders of a single type. The importance of this fact is that the commodity allocations in the core can be fully represented in a space of unchanging dimension as k increases. Similarly, the payoff vectors in the core give identical payoffs to all traders of a single type; therefore, at a core allocation one need know only the utility received by a representative member of each type. More generally, if there are n types of traders, m commodities, and k traders of each type, then the core can be represented in a space of dimension $n \cdot m$, no matter what the value of k. Thus, it is meaningful to talk of the core becoming smaller as k increases.

To see why all traders of a given type must receive identical commodity bundles at a Pareto optimal allocation, suppose an allocation z that does not have this feature. Let z^{ih} denote the commodity bundle assigned to player h of type i, where the index h can go from 1 to k. From each i form the commodity bundle $\bar{z}^i = \sum_{h=1}^{k} z^{ih}/k$. Thus, $\bar{z}^i$ is the mean bundle received by the players of type i. Now form a coalition L having n members, one of each type, with the member of each type i being that one receiving the least desirable commodity bundle among all the bundles of his type. Note that the allocation $(\bar{z}^1, \ldots, \bar{z}^n)$ is achievable by L and each member i finds $\bar{z}^i$ at least as desirable as his original assignment. Furthermore, if member's original assignment was different from $\bar{z}^i$, then $\bar{z}^i$ is superior, due to convexity of preferences.

To see how the core shrinks as k increases and why any outcome that is not a competitive equilibrium will eventually be dropped from the core as k rises, consider the commodity allocation C in Figure 8.7. This point is not a competitive equilibrium, as can be seen from the price line through C and E. (Recall that E is the endowment point.) On that price line, the point A is preferred to C by the players of type Ann and the point B is preferred to C by the players of type Ben. Now measure the distances EA and EB, and suppose a coalition L made up of k_A type-Ann- players and k_B type-Ben players. If $k_A \cdot EB = k_B \cdot EA$, then a trade among the members of L is feasible that places the type-Ann players at A and the type-Ben players at B. If this coalition can be formed, then the point C cannot be a core commodity allocation, but EA and EB need not stand in a ratio that allows this coalition. This problem is remedied by noting that A could be placed anywhere in the (open) interval from C to A', and B could be placed anywhere in the interval from C to B'. Therefore, k must be large enough that two integers k_A and k_B can be chosen to satisfy $1 < k_A/k_B < EA'/EB'$. Clearly, as k increases, a threshold value $k(C)$ is reached such that, for all $k > k(C)$, the point C cannot be in a core commodity allocation. A commodity allocation corresponding to a competitive equilibrium cannot be eliminated from the core in this way, no matter how many consumers there are of each type.

7.2 Group decision and the core

The second example is from political science and deals with characteristics of group decision practices. A democratic political organization ought to make decisions that are best for the group that the organization represents; however, there are at least two obstacles to this end. One is that those who actually make decisions for the organization may have objectives that are at variance with the best interests of the group. Another is that it may be very difficult to determine what is, in fact, in the best interests of the group. Where the group is democratic, finding the "best" policy must come down to aggregating the preferences of the group members in some fashion. Suppose there are three people, Ann, Ben, and Carrie, and three alternatives among which they must choose, I, II, and III. Suppose that these three alternatives are mutually exclusive and are joint in nature, as would be the case if the three people were the citizens of a medieval town and the issue were the sort of wall to be built around their village for their defence. I might be *no wall*, II might be a *wooden wall*, and III might be a *stone wall*. If, for instance, all three ranked the alternatives in order, from best to worst, III, II, I, then decision would be trivial. It is when everyone is not in agreement that an interesting group decision issue arises.

Voting is a common method for group decision in democratic organizations. If the three-person village uses majority voting as a means of ranking each pair of alternatives, it is well known that no clearcut social ordering need emerge. If Ann's preferences are I, II, III, Ben's are II, III, I, and Carrie's are III, I, II, then, in pairwise votes, I beats II, II beats III, and III beats I. This, of course, is the *Condorcet paradox* (see Black (1958)), and is intimately related to Arrow's (1951) general possibility theorem. The Condorcet paradox can be stated in game theoretic terms by noting that I dominates II, II dominates III, and III dominates I; thus, the core of the game is empty. If, on the other hand, the decision criterion were unanimity instead of majority vote, then the paradox would dissolve. No alternative would beat any other, which would leave all three tied for "best." Put in terms of the core, no alternative would dominate another; hence, all would be in the core.

The model elaborated here is a voting game with n voters. It is possible for the weights of the voters to differ; however, the decision structure is similar to that of a *simple game*. Any coalition is either *winning* or *not winning*. A winning coalition has the power to select any policy from the set of policies, X. The set $X \subset R^m$ is called the *policy space* and should be pictured in the following terms: Imagine that the group has m issues that it must decide. Each issue is quantifiable. For example, one might be the height of the town wall, another could be the budget of the fire department, another could be the surface area of the town swimming pool, and so forth. Each voter i has preferences over the policy space that are represented by a utility function $u_i(x)$. Following Schofield (1978), conditions are given below that guarantee that the core is empty.

CONDITION 8.1 *The policy space $X \subset R^m$ is convex.*

CONDITION 8.2 *Each player i has preferences over X that are represented by a continuously differentiable utility function $u_i(x)$. Each function u_i achieves a maximum and a minimum on X. The minimum is $\underline{u}_i$ The first partial derivatives of u_i are 0 only at extreme points.*

CONDITION 8.3 *For a coalition K, $V(K) = \{u^K(x) \in R^k \mid x \in X\}$ if K is winning and $V(K) = \{\underline{u}^K\}$ otherwise. N is winning. For at least one player, i $\{i\}$ is not winning. If K is winning and $K \subset L$, then L is winning. If K is winning, then $N \backslash K$ is not winning.*

In Condition 8.2 it is specified that each $u_i(x)$ achieves a maximum on X. This states that some policy in X is preferred by player i to all others. Figure 8.8 illustrates the preferences of a player in a two-dimensional issue space with the player's most preferred policy vector at B. The structure set up in Condition 8.3 affords a very natural way to determine whether policy x^1 wins over policy x^2. One policy wins over another if there is a winning

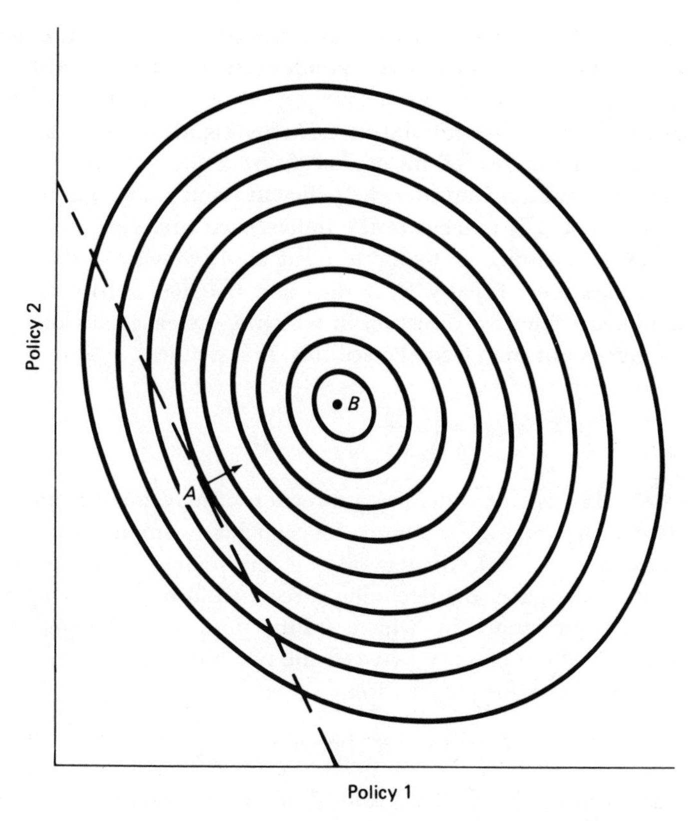

FIGURE 8.8 The preference contours of one player.

coalition that prefers one policy to the other. Let

$$K(x^1/x^2) = \{i \in N \mid u_i(x^1) > u_i(x^2\} \tag{8.9}$$

Then x^1 wins over x^2 if and only if $K(x^1/x^2)$ is a winning coalition. Note that $K(x^1/x^2)$ and $K(x^2/x^1)$ cannot both be winning, and that neither need be winning. In the political context, policy x^1 wins over x^2 when x^1 would win an election against x^2. In terms of cooperative game theory, winning is the same as domination. x^1 defeating x^2 is the same as x^1 dominating x^2 (via $K(x^1/x^2)$).

Now suppose that there is a finite sequence of policies $x^0, x^1, \ldots, x^r$, with $x^r = x^0$. If x^{j-1} wins over x^j for $j = 1, \ldots, r$, then *cyclic group preferences* occur. This is analogous to the situation in the Condorcet paradox, and none of the policies in the list is a group winner, because each is beaten by some other policy. Any policy that can be made part of a preference cycle, as in this example, cannot be in the core, and cannot be a "best" choice for the group. An important concern in political theory is over characterization of conditions under which "best" outcomes must be (or cannot be) available. Schofield (1978) gives conditions under which local and global cyclic group preferences will occur. These results are based on examination of the directions of movement from a given policy that increase the utility of a player.

The continuity and differentiability conditions assumed on the u_i make it easy to tell the directions of movement from a given policy that will increase a player's utility. Consider the policy at point A in Figure 8.8. The dashed line through A is tangent to the indifference curve on which A lies, and the arrow, perpendicular to the tangent at A, shows the direction in which u^i increases most rapidly. Any direction that lies above and to the right of the dashed line is a direction in which u_i increases, as long as the size of the move is not too large. To see this, first define

$$u_i'(x) = \left(\frac{\partial u_i(x)}{\partial x_1}, \ldots, \frac{\partial u_i(x)}{\partial x_m}\right) \tag{8.10}$$

and let $\delta \in R^m$. The vector δ defines a direction of movement from x. This direction is one in which u_i increases, for sufficiently small change, if the inner product $\delta \cdot u'(x) > 0$. In general, the improving directions from x define an open half space, the declining directions form another open half space, and the directions in which utility does not change form a hyperplane that is the boundary between the two open half spaces. The set of improving directions for player i from x is

$$C_i(x) = \{\delta \in R^m \mid \delta \cdot u_i'(x) > 0\} \tag{8.11}$$

for any x for which $u_i'(x) \neq 0$. If x is a local minimum, then $C_i(x) = X$ and if it is a local maximum, then $C_i(x) = \varnothing$. In the former case, all directions are improving, and in the latter, none is.

The set of improving directions from x for a coalition K is the intersection of the improving directions for the members of $K: C_K(x) = \bigcap_{i \in K} C_i(x)$. Such sets are illustrated in Figure 8.9 for the two-player coalitions in a three-player game. The indifference contours are not drawn, because they would add unnecessary confusion to the diagram; however, the tangent to each indifference curve at x is drawn, as is the gradient. For player 1, the dashed line $11'$ is the tangent and the arrow labeled u_1' is the gradient. Parallel notation applies to the other two players. The shaded region labeled $C_{(1,2)}$ is the region of improving directions from x for the coalition $\{1, 2\}$. Improving directions for $\{1, 3\}$ and $\{2, 3\}$ are similarly shown.

The set of directions that is of particular interest is the union of all the $C_K(x)$, taken over winning coalitions, in other words, the set of directions that is improving for at least one winning coalition. Denote this set $C_W(x)$, and denote the convex hull of $C_W(x)$ by $\bar{C}_W(x)$. A principal result in Schofield (1978) is that if $\bar{C}_W(x) = X$ then x is part of a preference cycle. The result is proved by showing that it is possible to find a (continuous) path that begins and ends at x, along which at least one coalition is improving continually. As a rule, the path is divided into several segments, the union of which is the whole path. Each segment is identified with a

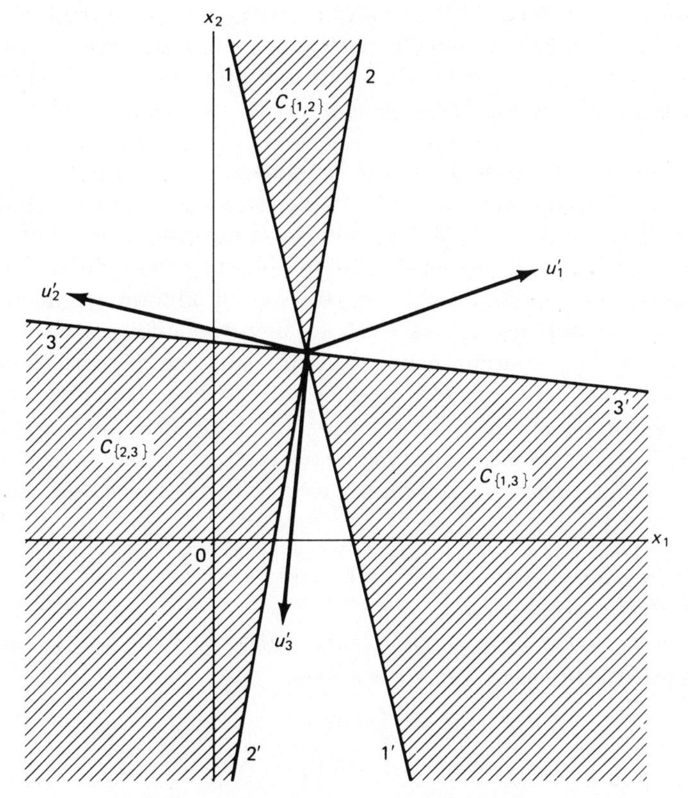

FIGURE 8.9 Sets of improving directions for various coalitions.

particular coalition, and that coalition has its members' utility increasing as one moves along that segment of the path. What makes it possible to show that such a path exists is that there must be a (nonempty) neighborhood of x such that $\bar{C}_W(y) = X$ for all y in the neighborhood. In Figure 8.9, $\bar{C}_W(y) = X$.

As the result stands, it is not clear whether preference cycles must be samll and close to x, or whether they may range quite far. Suppose, for example, that $X = R^m$ and that $\bar{C}_W(x) = R^m$ for all $x \in R^m$. Is there a way to know if a cycle can be found that arbitrarily includes any pair of policies x and y? The answer is that a cycle can be found among any such pair. In general, where C^* is the set of all elements of X for which $\bar{C}_W(x) = X$, if two points of C^* can be connected by a continuous curve that lies entirely in C^*, then those two points can be part of a preference cycle. Finally, Schofield gives conditions under which $C^* = X$.

8 Concluding comments

The cooperative game solutions reviewed in this chapter for nontransferable utility games are, I believe, interesting and useful for applications to economics and politics. The examples presented in this and other chapters are far from exhaustive, but do represent a few interesting illustrations. Moreover, they firmly make the case that game theory has contributed greatly to the clarification and development of important problems in both disciplines.

The variety of cooperative game solutions in Chapters 6 to 8, is noteworthy. On the positive side, several important solutions have been successfully adapted from transferable to nontransferable utility models. On the negative side, no single solution concept comes close to attaining acceptance as *the* solution. Of course, the wall of uniform acceptance of the noncooperative equilibrium of Nash for non-cooperative games contains some cracks. These pertain mainly to issues surrounding perfection and to the extreme and superb rationality that must be assumed of the players.

Still, looking back over the 45 years since the first publication of *The Theory of Games and Economic Behavior,* I think it is fair to say that the promise von Neumann and Morgenstern held out is well on the way to being fulfilled.

Exercises

1. Let there be a three-person nontransferable utility game in which the characteristic function is defined as follows:

$$V(\{i\}) = \{0\} \text{ for } i \in N$$

$$V(\{1,2\}) = \{(u_1, u_2) \in R^2_+ \mid u_2 \leq 48 + u_1 - \tfrac{1}{8}u_1^2\}$$

$$V(\{1,3\}) = \{(u_1, u_3) \in R^2_+ \mid u_3 \leq 10 + \tfrac{2}{3}u_1 - \tfrac{2}{27}u_1^2\}$$

$$V(\{2,3\}) = \{(u_2, u_3) \in R^2_+ \mid u_3 \leq 10 + u_2 - \tfrac{1}{45}u_2^2\}$$

$$V(N) = \{(u_1, u_2, u_3) \in R^3_+ \mid u_3 \leq 33 + \tfrac{1}{3}u_1 + \tfrac{2}{3}u_2 - \tfrac{1}{28}u_1^2 - \tfrac{1}{49}u_2^2\}$$

   Find a λ-transfer value for this game. Is the solution you have found in the core of the game?

2. For the game in problem 1, what is the transferable utility game, (N, v_λ) associated with the solution you found? What is the Shapley value for (N, v_λ)?

3. Suppose an economy comprised of four individuals. Two type 1 individuals have identical utility functions, $u_1 = x_1 y_1$, and two type 2 individuals have identical utility functions, $u_2 = x_2 y_2^2$. The type 1 individuals are each endowed with 40 units of X and 60 units of y; the type 2 individuals are each endowed with 60 units of X and 40 units of y. These people can trade with one another.
   a. Show that a trade that leaves the type 1 individuals with (60, 42.86) each and the type 2 individuals with (40, 57.14) is an imputation in the trading game.
   b. Show that the trade specified in part a is not in the core.

Note

1. Some of the special restrictions in Lemma 8.1 are intended to rule out various forms of degeneracy. These restrictions are that all row entries are distinct and that the weights associated with a feasible basis are strictly positive. The latter

condition is actually violated at the first pivot step, and inevitably so. The pivot adds a two-player coalition, $\{2, 3\}$, and must remove one of the one-player coalitions. It turns out that the weights on $\{2\}$ and $\{3\}$ reach 0 simultaneously; therefore, it cannot happen that one falls out while the other remains with a positive weight. To deal with this, it is alright to use an arbitrary rule to decide which to drop, and then to leave the other in the basis with a weight of 0.

APPENDIX

Mathematical notation, definitions, and theorems

This appendix includes some mathematical notation and definitions, and a few theorems that are used throughout the text. All notation, definitions, and theorems are stated for points, sets, functions, and so forth, defined for finite dimensional Euclidean spaces, unless the contrary is explicitly noted.

Notation

R	The set of finite real numbers.
R^n	The set of vectors that have n real-valued components. $R^n = \{x = (x_1, \ldots, x_n) \mid x_i \in R,\ i = 1, \ldots, n\}$. The n-dimensional Euclidean space.
R^n_+	The set of vectors in R^n that have nonnegative components. $R^n_+ = \{x \in R^n \mid x \geqslant 0\}$
R^n_{++}	The set of vectors in R^n that have strictly positive components. $R^n_{++} = \{x \in R^n \mid x \gg 0\}$.
$\{\ \}$	A set of points. $\{a, b, c\}$ is the set of points consisting of a and b and c. $\{a \in R \mid a < 4\}$ is the set of real numbers less than 4.
$\in$	Membership in a set. $a \in A$ means that a is an element of the set A.
$\subset$	*Set inclusion.* $A \subset B$ means that the set A is contained in the set B.
$\cup$	The union of two sets. $A \cup B$ is the set of all points that are in A, B, or both.
$\cap$	The intersection of two sets. $A \cap B$ is the set of all points contained in both A and B.
$x\backslash u_i$	Where x is a vector, $x\backslash u_i$ denotes the vector $(x_1, x_2, \ldots, x_{i-1}, u_i x_{i+1}, \ldots, x_n)$.
$K\backslash L$	Where K and L are sets, $K\backslash L$ denotes the set of points that are members of K but are not members of L. $K\backslash L$ is the intersection of K and the complement of L.
$\varnothing$	The empty set.
$\tilde{K}$	The complement of K, that is, the set of points that are not in K.
$\times$	Cartesian product. $A \times B$ denotes the set of points (a, b), where $a \in A$ and $b \in B$.
$\leqslant$	Applied to vectors, $x \leqslant y$ means $x_i \leqslant y_i$, $i = 1, \ldots, n$.
$<$	Applied to vectors, $x < y$ means $x \leqslant y$ and $x \neq y$.
$\ll$	Applied to vectors, $x \ll y$ means $x_i < y_i$, $i = 1, \ldots, n$.

$\mid$	Used with other symbols to reverse the meaning of the accompanying symbol. For example, $\neq$ means "is not equal to," $\notin$ means "is not an element of," $\nless$ means "is not smaller than."
$[a, b]$	The closed interval. $[a, b] = \{x \in R \mid a \leqslant x \leqslant b\}$.
$[a, b)$	The half-open interval. $[a, b) = \{x \in R \mid a \leqslant x < b\}$.
$(a, b]$	The half-open interval. $(a, b] = \{x \in R \mid a < x \leqslant b\}$.
(a, b)	The open interval. $(a, b) = \{x \in R \mid a < x < b\}$.
$f^i(x)$	The partial derivative of $f(x)$ with respect to the ith argument of the function.
f^{ij}	The second partial derivative of $f(x)$ with respect to the ith and jth arguments of the function.
$\max_{i \in A} f(x)$	The largest value achieved by the function f on the set A. That is, $\max_{x \in A} f(x) = f(x^*)$ if and only if $x^* \in A$ and $f(x^*) \geqslant f(x)$ for all $x \in A$.

Definitions

affine function Let $x \in R^n$, y, $a \in R^m$, and let A be a $m \times n$ matrix. Then $y = a + AX$ is an affine function.

bounded set A set is bounded if there is a finite upper bound on the distance between any two points in the set. The set A is bounded if, for all a, $b \in A$, $d(a, b) \leqslant M < \infty$.

Cartesian product The Cartesian product of two sets, A and B, consists of all ordered pairs (a, b) such that $a \in A$ and $b \in B$. If $A \subset R^m$ and $B \subset R^n$, then $A \times B \subset R^{m+n}$.

closed set A set is closed if, for any convergent sequence of points in the set, the limit is in the set.

cluster point A cluster point of a sequence is the limit point of a subsequence. Suppose that $\{x^i\}$ is a sequence of points. If a subsequence of $\{x'\}$ converges to a limit then that limit is a cluster point of the original sequence.

compact set A set in R^n is compact if and only if it is closed and bounded.

concave function Let f be a function with domain $A \subset R^n$ and range contained in R. The function is concave if, for any $x,y \in A$ and any $\lambda \in [0, 1]$, $f[\lambda x + (1 - \lambda)y] \geqslant \lambda f(x) + (1 - \lambda)f(y)$. The function is strictly concave if, for any $x,y \in A$ and any $\lambda \in (0, 1)$, $f[\lambda x + (1 - \lambda)y] > \lambda f(x) + (1 - \lambda)f(y)$.

contraction Let f be a function with domain $A \subset R^n$ and range contained in R^m. The function is a contraction if, for any distinct $x,y \in A$, $d(f(x), f(y)) < d(x,y)$. A contraction is a function that obeys a Lipschitz condition with ratio $k < 1$.

convex combination A convex combination of the points $x^1, \ldots, s^n$ is $\lambda_1 x^1 + \cdots + \lambda_n x^n$, where $\lambda_i \geqslant 0$, $i = 1, \ldots, n$, and $\sum_i \lambda_i = 1$.

convex function Let f be a function with domain $A \subset R^n$ and range contained in R. The function is convex if, for any $x,y \in A$ and any $\lambda \in [0, 1]]$, $f[\lambda x + (1 - \lambda)y] \leqslant \lambda f(x) + (1 - \lambda)f(y)$. The function is strictly convex if, for any $x,y \in A$ and any $\lambda \in (0, 1)$, $f[\lambda x + (1 - \lambda)y] < \lambda f(x) + (1 - \lambda)f(y)$.

convex hull The convex hull of a set A is the smallest convex set that contains A. The convex hull of A consists of all points that are convex combinations of members of A.

convex set A set is convex if any convex combination of points in the set is also in the set. A convex set is the convex hull of itself.

correspondence A correspondence is a mapping that associates a subset of R^m with each point in its domain.

disjoint sets Two sets are disjoint if their intersection is empty.

distance The distance between two points in a set, $d(x,y)$, is defined by a norm: $d(x,y) = \|x - y\|$. Distance must satisfy three properties: $d(x,y) \geqslant 0$, $d(x,x) = 0$, and $d(x,z) \leqslant d(x,y) + d(y,z)$.

domain The domain of a function or a correspondence $f(x)$ is the set of values of x for which the function or correspondence is defined.

equivalence relation An equivalence relation is a binary relation that is reflexive, symmetric, and transitive. For example, "$=$" is an equivalence relation. Reflexivity means that $a = a$. Symmetry means that, if $a = b$, then $b = a$. Transitivity means that if $a = b$ and $b = c$, then $a = c$.

function A function is a mapping that associates a point in R^n with a point in R^m.

infimum The infimum of a set of points is the greatest lower bound of the set.

interior The interior of a set A is the largest open set contained in A. A point a is in the interior of A, denoted $\mathring{A}$, if, for some small positive ε, all points b such that $d(a,b) < \varepsilon$ are in A.

limit The point x^0 is the limit of the sequence of points $\{x^i\}$ if, for any positive ε, there is a finite k such that $d(x^0, x^i) < \varepsilon$ for all $i > k$.

linear function A linear function is an affine function for which $a = 0$.

Lipschitz condition Let f be a function with domain $A \subset R^n$ and range contained in R^m. The function obeys a Lipschitz condition with ratio k if, for any $x, y \in A$, $d(f(x), f(y)) \leqslant kd(x,y)$.

norm A norm, denoted $\|x\|$, is a function defined on the elements of a set A that satisfies $\|x\| \geqslant 0$ for all $x \in A$, $\|0\| = 0$, and $\|x + y\| \leqslant \|x\| + \|y\|$.

quasiconcave function Let f be a function with domain $A \subset R^n$ and range contained in R^m. The function is quasiconcave if, for any $x, y \in A$ and any $\lambda \in [0,1]$, $f[\lambda x + (1-\lambda)y] \geqslant \min\{f(x), f(y)\}$. The function is strictly quasiconcave if, for any $x, y \in A$ and any $\lambda \in (0,1)$, $f[\lambda x + (1-\lambda)y] > \min\{f(x), f(y)\}$.

partition A partition of a set A is a division of A into pairwise disjoint subsets, $B_1, \ldots, B_n$, whose union equals A. That is, $\bigcup_i B_i = A$, and for all i, j with $i \neq j$, $B_I \cap B_j = \varnothing$.

range Let f be a function with domain $A \subset R^n$ and range contained in R^m. The range of $f(x)$ is $\{y \in R^m \mid y = f(x), x \in A\}$.

supremum The supremum of a set of points is the least upper bound of the set.

unit simplex The unit simplex in R^n is $\{x \in R^n \mid x \geqslant 0, \sum_i x_i = 1\}$.

Theorem 1 Let f be a continuous function with compact domain $A \subset R^n$ and range contained in R^m. Then there is $x^* \in A$ such that $f(x^*) \geqslant f(x)$ for all $x \in A$.

Theorem 2 Let $\{x^i\}$ be an infinite sequence of points contained in a compact subset of R^n. Then $\{x^i\}$ has a cluster point.

Theorem 3 Let f and g be concave functions with domain $A \subset R^n$ and range contained in R^m. Then for $a, b \in R_+$, $h = af + bg$ is concave on A.

Answers to exercises

Chapter 2

1. Whether these are games of complete information depends, in part, on whether each player knows the payoff functions of the other players. One player can easily rank outcomes in order of another player's preference, but the other player's utility assignment need not be known. Tennis, chess, and monopoly are games of perfect information: each move by any player is seen by all players. Blackjack is not a game of perfect information, because nature makes moves that are not seen by all players. For example, the dealer does not see all the cards held by other players. In football, a player would presumably be a single team. Both teams are frequently doing many things simultaneously and a player is not going to be in a single node information set at all times. Tennis, blackjack, chess, and monopoly are games of perfect recall. Football is probably not, because there are times when a football player makes a move in ignorance of some past activities of his teammates.
3. Player 1 has eight pure strategies. Each player specifies his first move unconditionally, then specifies his second move conditional on his own first

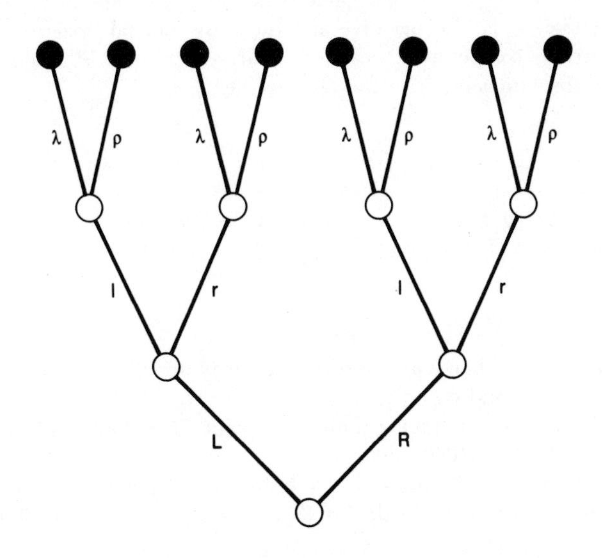

move. For example, $R\rho\lambda$ means "choose R on the first move, then choose ρ if R was first chosen and choose λ if L was first chosen." The eight strategies are (1) $R\rho\rho$, (2) $R\rho\lambda$, (3) $R\lambda\rho$, (4) $R\lambda\lambda$, (5) $L\rho\rho$, (6) $L\rho\lambda$, (7) $L\lambda\rho$, and (8) $L\lambda\lambda$. Player 2 has four pure strategies, each specifying what to do given the move of player 1. For example, lr is choose l if R and choose r if L. The four are (1) rr, (2) rl, (3) lr, and (4) ll. The probability of λ on the second move of player 1, given L, is $(s_{17}+s_{18})/(s_{15}+s_{16}+s_{17}+s_{18})$.

4. The strategies of player 1 may be written $R\rho\lambda$, interpreted as "choose R on the first move, then choose ρ if the moves match and λ if they do not." As above, the eight strategies are (1) $R\rho\rho$, (2) $R\rho\lambda$, (3) $R\lambda\rho$, (4) $R\lambda\lambda$, (5) $L\rho\rho$, (6) $L\rho\lambda$, (7) $L\lambda\rho$, and (8) $L\lambda\lambda$. The probability of the two players matching is $p = (s_{11}+s_{12}+s_{13}+s_{14})(s_{21}+s_{22}) + (s_{15}+s_{16}+s_{17}+s_{18})(s_{23}+s_{24})$. The probability of λ on the second move of player 1, given matching, is $[(s_{13}+s_{14})(s_{21}+s_{22}) + (s_{17}+s_{18})(s_{23}+s_{24})]/p$.
5. The game in problem 3 is one of perfect recall, so a mixed strategy induces a unique behavior strategy. This is not the case in problem 4, which describes a game of imperfect recall.

Chapter 3

1. a. $v = \frac{20}{7}$, $s_1^* = (\frac{2}{7}, \frac{5}{7})$, $s_2^* = (\frac{4}{7}, \frac{3}{7})$
   b. $v = \frac{20}{7}$, $s_1^* = (\frac{2}{7}, \frac{5}{7})$, $s_2^* = (\frac{4}{7}, \frac{3}{7}, 0)$
   c. $v = \frac{11}{4}$, $s_1^* = (\frac{1}{4}, \frac{3}{4})$, $s_2^* = (\frac{1}{4}, 0, \frac{3}{4})$
3. $s_1^* = (\frac{1}{9}, \frac{8}{9})$, $s_2^* = (\frac{4}{7}, \frac{3}{7})$, $P(s^*) = (\frac{20}{7}, \frac{16}{5})$
4. $r_1(s) = 3.125 + 1.875s_2$, $r_2(s) = 50 - .5s_1$, $s^* = (50, 25)$
5. $r_1(s) = 3.125 + 1.875s_2$, $r_2(s) = 50 - .5s_1$, $s^* = (30, 35)$
6. $r_1(s) = 10 + 1.25s_2$, $r_2(s) = 25 - 1.5s_1$, $s^* = (\frac{330}{23}, \frac{80}{23})$
7. The Rosen theorem can be applied to the game in problem 3, because the Jacobian of the best-reply mapping is negative quasidefinite.
8. With both versions of the strategy space for player 2, equilibrium is at (0, 10). To determine whether one of the games is strictly competitive, it helps to know that $P_2(s) = -[P_1(s)]^2$. The first game is not strictly competitive, because, when P_1 is negative, P_2 is also negative and they rise and fall together. The second game is strictly competitive, because P_1 is always positive, P_2 is always negative, and their absolute values rise and fall together.

Chapter 4

1. a. (1, 1), (3, 2), (4, 3)
   b. Yes, with $s^0 = (4, 3)$, $s^1 = (1, 1)$, $s^2(3, 2)$
   c. 6 periods
   d. .95, .805
2. If $s^* = (2, 4)$ and $s^c = (4, 3)$, then $\alpha_1 > \frac{5}{14}$ and $\alpha_2 > \frac{3}{16}$. If $s^* = (2, 4)$, but (1, 1) is chosen if player 1 abandons the trigger strategy and (3, 2) is chosen if player 2 does, then $\alpha_1 > \frac{1}{3}$ and $\alpha_2 > \frac{3}{17}$.
3. a. $s^* = (10, 10)$ is a unique equilibrium point. The best-reply mapping is a contraction. $P(s^*) = (1000, 1500)$
   b. For $\alpha_1 > \frac{3}{4}$ and $\alpha_2 > \frac{49}{85}$ a subgame perfect trigger strategy equilibrium can be maintained. $s^* = (35, 30)$ and $P(s^*) = (1750, 3000)$

c. $v_1 = 250$, $v_2 = 2000/3$. Player 1 must choose 0 to minimax the other player. His own payoff is then 0.

4. $P_1(6, 12) = 3{,}150$ $P_2(6, 12) = 2{,}928$. This payoff can be supported for periods $1, \ldots, T-1$ by a trigger strategy under which the players revert to (5, 10) if anyone deviates from the path μ^0. in deviating from (6, 12), player 1 can achieve a payoff of 3,163.2 for one period. Player 2 can achieve 2,928.6 by deviating. The extra gain either player would receive is easily offset by one period at (5, 10) rather than at (8, 15).
5. The single-shot noncooperative equilibrium is at $q_i = 90/(n+1)$ for all i. The corresponding payoffs are $[90/(n+1)]^2$ for each firm. For any n and any i, $v_i = 0$. To realize a payoff of 22 per firm per period, set $q_i = 11$. To sustain this, any deviator is subjected to three periods during which the deviator is minimaxed. The deviator produces 0 and the other firms produce 90/7 each. The play reverts to the deviator choosing $q^d = 10.55$ and the remaining firms choosing $q^p = 11.15$.

Chapter 5

1. $$s_{1t} = \frac{5 - 5s_{1,t-1} + s_{2,t-1} + 4s_{1,t+1}}{28}$$

   $$s_{2t} = \frac{10 - 6s_{1,t-1} + 2s_{1t}}{16}$$

   $s^* = (.15, .55)$

2. a. If $p > .577$, the best-reply of player 1 is to choose NC in the first iteration and C in the second. Specifically, the criterion is $-1 + 3p + 3p^2 > 3p$. The left-hand side of the inequality is the expected payoff to playing NC, then C, and the right-hand side is the payoff to choosing C both times. If both players have the same beliefs, with $p > .577$, they will both play NC followed by C. Their realized payoffs will be 2 each and their expected payoffs will be $-1 + 3p + 3p^2$ each.
   b. The criterion for choosing between NC followed by C and C both times remains the same. Realized payoffs will be $2(T-1)$ and expected payoffs will be $[3p - 1 + p^{T-1}(1 - 3p^2)]/(1-p)$.
   c. Realized payoffs are 198 for all three cases. Expected payoffs are 17 for $p = .9$, 125 for $p = .99$, and 191 for $p = .999$.

Chapter 6

1. a. The threat point is $d = (10, 15)$. The attainable set H is the convex hull of the utility pairs associated with the points in the set of possible trades. These are $u^1 = (0, 37)$, $u^2 = (4, 36)$, $u^3 = (6, 35)$, $u^4 = (10, 15)$, $u^5 = (12, 11)$, $u^6 = (14, 25)$, $u^7 = (20, 10)$, $u^8 = (16, 8)$, and $u^9 = (21, 5)$; therefore, $H = \{u \in R^2 \mid u = \sum_{i=1}^{9} \lambda_i u^i$, $\sum_{i=1}^{9} \lambda_i = 1$, $\lambda \geq 0$, $i = 1, \ldots, 9)$. The subset of H containing payoffs weakly dominating the threat point, $H^* \in \{u \in H \mid u \geq d\}$, is the convx hull of (10, 15), (10, 30), (14, 25), and (18, 15).
   b. (14, 25)
2. a. $m = (10, 15)$, $M = (18, 30)$
   b. $(\frac{102}{7}, \frac{165}{7})$
3. a. Reference point: (10.5, 21); solution: (13.85, 25.1875)
   b. Reference point: (14, 22.5); solution: (14.5, 23.75)

Solution	Reference point	Solution payoffs
Nash	(2, 2)	(3.4142, 3.4142)
R–K–S	(2, 2), (4, 4)	(3.4142, 3.4142)
Min expectations	(2, 2)	(3.4142, 3.4142)
Min compromise	(3, 3)	(3.4142, 3.4142)
Smallest rectangle	(2, 2)	(3.4142, 3.4142)

5.

Solution	Reference point	Solution payoffs
Nash	(0, 2)	(3.49, 3.334)
R–K–S	(0, 2), (4, 4)	(3.6, 3.2)
Min expectations	(2, 2)	(3.4142, 3.4142)
Min compromise	(3, 3)	(3.4142, 3.4142)
Smallest rectangle	(2, 2)	(3.4142, 3.4142)

Chapter 7

1. a. (10, 10, 10)
   b. $v(\{i\}) = 0$ for $i = 1, 2, 3$, $v(\{1, 2\}) = \frac{2}{13}$, $v(\{1, 3\}) = \frac{11}{13}$, $v(\{2, 3\}) = \frac{3}{13}$, $v(N) = 1$
   c. $v(N) = 28$ is the minimum. (8, 8, 10) is in the core.
2. $\phi(v) = (10.5, 9.5, 10)$. The Shapley value of this game is in the core.
3. Superadditivity is neither necessary nor sufficient. The following three-person game violates superadditivity and has (2, 2, 2) in the core: $v(\{i\}) = 2$ for all i, $v(K) = 3$ for all two-player coalitions, and $v(N) = 6$. On the other hand, simple games are superadditive and generally have empty cores. For example, the three-person game in which $v(\{i\}) = 0$ for all i and $v(K) = 1$ for all two- and three-player coalitions has no core.
4. If this inequality is violated for a partition, then no imputation can be found that gives at least $v(T_i)$ to each coalition in the partition.

Chapter 8

1. The weights are $(\frac{1}{3}, \frac{1}{6}, \frac{1}{2})$ and the associated payoffs are $u = (14, 49, \frac{43}{3})$. This payoff vector cannot be improved on by any coalition; hence, it is in the core.
2. For $\lambda = (\frac{1}{3}, \frac{1}{6}, \frac{1}{2})$ $v_\lambda(\{i\}) = 0$ for $i \in N$, $v_\lambda(\{1, 2\}) = 10$, $v_\lambda(\{1, 3\}) = 8$, $v_\lambda(\{2, 3\}) = 15$, and $v_\lambda(N) = 20$. The Shapley value is $\phi(v_\lambda) = (\frac{28}{6}, \frac{49}{6}, \frac{43}{6})$.
3. a. Calculate the marginal rate of substitution for each type at the indicated point. Their respective marginal rates of substitution are y_1/x_1 and $y_2/2x_2$. Therefore, the trade is group rational. Each types achieves higher utility than at its original points, before trade, so the trade is individually rational. The respective pretrade and posttrade utilities for the two types are 2,400 and 2,571.6 for type 1 and 96,000 and 130,599 for type 2.
   b. To see that this trade is not in the core, consider a trade involving two type 1 individuals and one type 2 individual under which the type 1 individuals end up with (55, 47) and the type 2 individual ends up with (30, 66). This trade is within the power of the three people involved. Their achieved utilities are 2,585 for the type 1 individuals and 130,680 for the type 2 individual.

References

Abreu, Dilip. 1986. "Extremal Equilibria of Oligopolistic Supergames." *Journal of Economic Theory* 39:191–225.

———. 1988. "*On the Theory of Infinitely Repeated Games with Discounting.*" *Econometrica* 56: 383–96.

Arrow, Kenneth, J. 1951. *Social Choice and Individual Values*. New York: Wiley.

———. 1971. *Essays in the Theory of Risk Bearing*. Chicago: Markham.

Arrow, Kenneth J., and Gerard Debreu. 1954. "Existence of an Equilibrium for a Competitive Economy." *Econometrica* 22:265–90.

Aumann, Robert J. 1959. "Acceptable Points in General Cooperative *n*-person Games." In A. W. Tucker and R. D. Luce, eds., *Contributions to the Theory of Games,* vol. IV. Princeton: Princeton University Press.

———. 1961. "The Core of a Cooperative Game without Side Payments." *Transactions of the American Mathematics Society* 98:539–52.

———. 1976. Agreeing to Disagree. *Annals of Statistics* 4:1236–9.

———. 1981. "Survey of Repeated Games." In R. J. Aumann et al., *Essays in Game Theory*. Mannheim: Bibliographisches Institut.

Aumann, Robert, J. and Lloyd Shapley. 1976. "Long Term Competition—A Game Theoretic Analysis." Mimeograph.

Banzhaf, J. F., III. 1965. "Weighted Voting Doesn't Work: A Mathematical Analysis." *Rutgers Law Review* 19:317–43.

Bartle, Robert G. 1964. *The Elements of Real Analysis*. New York: Wiley.

Benoit, Jean-Pierre, and Vijay Krishna. 1985. "Finitely Repeated Games." *Econometrica* 53:905–22.

Berge, Claude. 1957. *Théorie Générale des Jeux à n Personnes*. Paris: Gauthier-Villars.

———. 1963. *Topological Spaces,* translated by E. M. Patterson, New York: Macmillan.

Bernheim, B. Douglas. 1984. "Rationalizable Strategic Behavior." *Econometrica* 52:1007–28.

Bernheim, B. Douglas, Bezalel Peleg, and Michael D. Whinston. 1987. "Coalition-proof Nash Equilibria I. Concepts." *Journal of Economic Theory* 42:1–12.

Bernheim, B. Douglas, and Michael D. Whinston. 1987. "Coalition-proof Nash Equilibria II. Applications." *Journal of Economic Theory* 42: 13–29.

Binmore, Ken, and Partha Dasgupta, eds. 1987. *The Economics of Bargaining*. Oxford: Blackwell.

Black, Duncan. 1958. *The Theory of Committees and Elections*. Cambridge: Cambridge University Press.

Blackwell, David. 1965. "Discounted Dynamic Programming." *Annals of Mathematical Statistics* 36:226–35.

Brito, D. L., A. M. Buoncristiani, and M. D. Intriligator. 1977. "A New Approach to Nash's Bargaining Problem." *Econometrica* 45:1163–72.

Cournot, Augustin. 1838. *Recherches sur les Principes Mathématiques de la Théorie des Richesses*. Paris: Hachette.

———. 1960. *Researches into the Mathematical Principles of the Theory of Wealth*. English edition of Cournot (1838), translated by Nathaniel T. Bacon. New York: Kelley.

Crawford, Vincent P. 1979. "A Procedure for Generating Pareto-Efficient Egalitarian–equivalent Allocations." *Econometrica* 47:49–60.

———. 1980. "A Self-administered Solution of the Bargaining Problem." *Review of Economic Studies* 47:385–92.

Debreu, Gerard. 1952. "A Social Equilibrium Existence Theorem." *Proceedings of the National Academy of Science* 38:886–93.

———. 1959. *Theory of Value*. New York: Wiley.

Debreu, Gerard, and Herbert E. Scarf. 1963. "A Limit Theorem on the Core of an Economy." *International Economic Review* 4:235–46.

———. 1972. "*The Limit of the Core of an Economy*." In C. B. McGuire and Roy Radner, eds., *Decision and Organization*. Amsterdam: North-Holland.

Denardo, Eric V. 1967. "Contraction Mappings in the Theory Underlying Dynamic Programming." *SIAM Review* 9:165–77.

Dieudonné Jean 1960. *Foundations of Modern Analysis*. New York: Academic Press.

Edgeworth, Francis Y. 1881. *Mathematical Psychics*. London: Kegan Paul.

Fan, Ky. 1952. "Fixed-point and Minimax Theorems in Locally Convex Topological Linear Spaces." *Proceedings of the National Academy of Sciences* 38:121–26.

Farrell, Joseph, and Eric Maskin. 1987. "Renegotiation in Repeated Games." Harvard University.

Fellner, William J. 1949. *Competition Among the Few*. New York: Knopf.

Flaherty, M. Thérèse. 1980. "Industry Structure and Cost Reducing Investment." *Econometrica* 48:1187–1209.

Friedman, James W. 1971. "A Non-cooperative Equilibrium for Supergames." *Review of Economic Studies* 38:1–12.

———. 1974. "Non-cooperative Equilibria in Time-dependent Supergames." *Econometrica* 42:221–37.

———. 1977. *Oligopoly and the Theory of Games*. Amsterdam: North-Holland.

———. 1981. "A Note on the Turnpike Properties of Time Dependent Super games." *Econometrica* 49:1087–88.

———. 1983. *Oligopoly Theory*. New York: Cambridge University Press.

———. 1984. "On Characterizing Equilibrium Points in Two Person Strictly Competitive Games." *International Journal of Game Theory* 12:245–47.

———. 1985. "Cooperative Equilibria in Finite Horizon Noncooperative Supergames." *Journal of Economic Theory* 35:390–98.

———. 1990. "A Modification of the Folk Theorem to Apply to Time-Dependent Supergames." *Oxford Economic Papers*.

Fudenberg, Drew, and Eric Maskin. 1986. "The Folk Theorem in Repeated Games with Discounting and with Incomplete Information." *Econometrica* 54:533–54.

Gale, David, and Hukukane Nikaido. 1965. "The Jacobian Matrix and the Global Univalence of Mappings." *Mathematische Annalen* 159:81–93.

Gillies, D. B. 1953. *Some Theorems on n-Person Games*. Ph.D. dissertation, Department of Mathematics, Princeton University.

Green, Edward J. 1980. "Noncooperative Price Taking in Large Dynamic Markets." *Journal of Economic Theory* 22:155–82.

Green, H. A. John. 1976. *Consumer Theory*. Rev. ed. London: Macmillan.

Green, Jerry, and Jean-Jacques Laffont. 1979. *Incentives in Public Decision-Making*. Amsterdam: North-Holland.

Groves, Theodore, and John Ledyard. 1977. "Optimal Allocation of Public Goods: A Solution to the 'Free Rider' Problem." *Econometrica* 45:783–809.

Harsanyi, John C. 1956. "Approaches to the Bargaining Problem Before and After the Theory of Games." *Econometrica* 24:144–56.

———. 1959. "A Bargaining Model for the Cooperative *n*-Person Game." In A. W. Tucker and R. D. Luce, eds., *Contributions to the Theory of Games*, vol. 4. Princeton: Princeton University Press.

———. 1963. "A Simplified Bargaining Model for the n-person Cooperative Games." *International Economic Review* 4:194–220.

———. 1967. "Games with Incomplete Information Played by 'Bayesian' Players, Part I: The Basic Model," *Management Science* 14:159–82.

———. 1968a. "Games with Incomplete Information Played by 'Bayesian' Physics, Part II." *Management Science* 14:320–334.

———. 1968b. "Games with Incomplete Information Played by 'Bayesian' Players, Part III." *Management Science* 14:486–502.

———. 1977. *Rational Behavior and Bargaining Equilibrium in Games and Social Situations*. Cambridge: Cambridge University Press.

Harsanyi, John C., and Reinhard Selten. 1988. *A General Theory of Equilibrium Selection in Games*. Cambridge: MIT Press.

Hildenbrand, Werner, and Alan Kirman. 1976. *Introduction to General Equilibrium Analysis*. Amsterdam: North-Holland.

Hillier, Frederick S., and Gerald J. Lieberman. 1974. *Operations Research*. San Francisco: Holden-Day.

Hume, David. 1888. *A Treatise of Human Nature*. Edited by L. A. Selby-Bigge. Oxford: Clarendon Press.

Kakutani, Shizuo. 1941. "A Generalization of Brouwer's Fixed Point Theorem." *Duke Mathematical Journal* 8:457–59.

Kalai, Ehud, and Dov Samet. 1982. "Persistent Equilibria in Strategic Games." Unpublished paper, Northwestern University.

Kalai, Ehud, and Meir Smorodinsky. 1975. "Other Solutions to Nash's Bargaining Problem." *Econometrica* 43:513–18.

Kaneko, Mamoru. 1982. "Some Remarks on the Folk Theorem in Game Theory." *Mathematical Social Sciences* 3:281–90.

Kelley, John L. 1955. *General Topology*. New York: Van Nostrand.

Kirman, Alan P., and Matthew J. Sobel. 1974. "Dynamic Oligopoly with Inventories." *Econometrica* 42:279–87.

Kohlberg, Elon, and Jean-François Mertens. 1986. "On the Strategic Stability of Equilibria." *Econometrica* 54:1003–39.

Kreps, David M., and Robert Wilson. 1982a. "Reputation and Imperfect Information." *Journal of Economic Theory* 27:253–79.

———. 1982b. "Sequential Equilibrium." *Econometrica* 50:863–94.

Kuhn, Harold W. 1953. "Extensive Games and the Problem of Information." In H. W. Kuhn and A. W. Tucker, eds., *Contributions to the Theory of Games*, Vol. 2. Princeton: Princeton University Press.

Kurz, Mordecai. 1976. "Altruistic Equilibrium." In Bela Balassa and Richard Nelson, eds., *Economic Progress, Private Values and Public Policy.* Amsterdam: North-Holland.

Lambson, Val E. 1984. "Self-Enforcing Collusion in Large Dynamic Markets." *Journal of Economic Theory* 34:282–91.

Leininger, Wolfgang, Lars Thorlund-Petersen, and Jörgen Weibull. 1988. "A Note on Strictly Competitive and Zero-Sum Games." Universität Bonn.

Lemke, C. E., and J. T. Howson. 1964. "Equilibrium Points of Bimatrix Games." *SIAM Journal of Applied Mathematics* 12:413–23.

Lucas, William F. 1969. "The Proof that a Game May Not Have a Solution." *Transactions of the American Mathematical Society* 136:219–229.

Luce, R. Duncan, and Howard Raffia. 1957. *Games and Decisions.* New York: Wiley.

McKenzie, Lionel. 1960. "Matrices with Dominant Diagonals and Economic Theory." In Kenneth J. Arrow, Samuel Karlin, and Patrick Suppes, eds., *Mathematical Methods in the Social Sciences, 1959.* Stanford: Stanford University Press.

Maschler, Michael, Bezalel Peleg, and Lloyd Shapley. 1979. "Geometric Properties of the Kernel, Nucleolus, and Related Concepts." *Mathematics of Operations Research* 4:303–38.

Milgrom, Paul. 1981. "An Axiomatic Characterization of Common Knowledge." *Econometrica* 49:219–22.

Miller, Charles L. 1982. "A Reaction Function Equilibrium for a Simple Dynamic duopoly Model with Random Demand." Unpublished paper, Department of Economics, Johns Hopkins University.

Moulin, Hervé. 1979. "Dominance Solvable Voting Schemes." *Econometrica* 47:1337–51.

———. 1988. *Axioms of Cooperative Decision Making.* Cambridge: Cambridge University Press.

Myerson, Roger B. 1978. "Refinements of the Nash Equilibrium Concept." *International Journal of Game Theory* 7:73–80.

Nash, John F., Jr. 1950. "The Bargaining Problem." *Econometrica* 18:155–62.

———. 1951. "Non-Cooperative Games." *Annals of Mathematics* 54:286–95.

———. 1953. "Two-Person Cooperative Games." *Econometrica* 21:128–40.

Nikaido, Hukukane, and Kazuo Isoda. 1955. "Note on Noncooperative Convex Games." *Pacific Journal of Mathematics* 5:807–15.

Nishimura, Kazuo, and James Friedman. 1981. "Existence of Nash Equilibrium in n Person Games without Quasiconcavity." *International Economic Review* 22:637–48.

Owen, Guillermo. 1967. "An Elementary Proof of the Minimax Theorem." *Management Science* 13:765.

———. 1972. "Values of Games Without Side Payments." *International Journal of Game Theory* 1:95–109.

———. 1978. "Characterization of the Banzhaf–Coleman Index." *SIAM Journal of Applied Mathematics* 35:315–27.

———. 1982. *Game Theory.* 2nd ed. New York: Academic Press.

Pearce, David G. 1984. "Rationalizable Strategic Behavior and the Problem of Perfection." *Econometrica* 52:1029–50.

———. 1987. "Renegotiation-proof Equilibria: Collective Rationality and Interpersonal Cooperation." Yale University.

Peleg, Bezalel, and Menahem Yaari. 1973. "On the Existence of A Consistent

Course of Action When Tastes are Changing." *Review of Economic Studies* 40:391–401.

Porter, Robert H. 1983. "Optimal Cartel Trigger Price Strategies." *Journal of Economic Theory* 29:313–38.

Prescott, Edward C. 1973. "Market Structure and Monopoly Profits: A Dynamic Theory." *Journal of Economic Theory* 6:546–57.

Radner, Roy. 1980. "Collusive Behavior in Noncooperative Epsilon-Equilibria of Oligopolies with Long but Finite Lives." *Journal of Economic Theory* 22:136–54.

Rawthorn, Bob. 1988. "Finite Equilibria in Finite Horizon Noncooperative Supergames: A correction and New Formulation." Cambridge University.

Riker, William H. 1962. *The Theory of Political Coalitions.* New Haven: Yale University Press.

Rives, Norfleet W., Jr. 1975. "On the History of the Mathematical Theory of Games." *History of Political Economy* 7:549–65.

Roberts, A. Wayne, and Dale E. Varberg. 1973. *Convex Functions.* New York: Academic Press.

Rogers, Philip D. 1969. "Nonzero-Sum Stochastic Games." ORC 69-8, Operations Research Center, University of California, Berkeley.

Rosen, J. B. 1965. "Existence and Uniqueness of Equilibrium Points for Concave n-person Games." *Econometrica* 33:520–34.

Rosenthal, R. W. 1979. "Sequences of Games with Varying Opponents." *econometrica* 47:1353–66.

———. 1981. "Games of Perfect Information, Predatory Pricing and the Chain-Store Paradox." *Journal of Economic Theory* 25:92–100.

Roth, Alvin E. 1977. "Individual Rationality and Nash's Solution to the Bargaining Problem." *Mathematics of Operations Research* 2:64–66.

———. 1979. *Axiomatic Models of Bargaining*. Berlin: Springer.

Rubinstein, Ariel. 1979. "Equilibrium in Supergames with the Overtaking Criterion." *Journal of Economic Theory* 21:1–9.

———. 1982. "Perfect Equilibrium in a Bargaining Model." *Econometrica* 50:97–109.

———. 1987. "A Sequential Strategic Theory of Bargaining." In Truman F. Bewley, ed., *Advances in Economic Theory, Fifth world congress, Econometric Society*: Cambridge: Cambridge University Press.

Scarf, Herbert E. 1967. "The Core of an n-Person Game." *Econometrica* 35:50–69.

———. 1979. *The Computation of Economic Equilibria.* New Haven: Yale University Press.

Schmeidler, David. 1973. "Equilibrium Points of Nonatomic Games." *Journal of Statistical Physics* 7:295–300.

Schofield, Norman. 1978. "Instability of Simple Dynamic Games." *Review of Economic Studies* 45:575–94.

Selten, Reinhard. 1965. "Spieltheoretische Behandlung eines Oligopolmodelles mit Nachfrageträgheit," *Zeitschrift für die Gesamte Staatswissenschaft* 12:301–24.

———. 1975. "Reexamination of the Perfectness Concept for Equilibrium Points in Extensive Games." *International Journal of Game Theory* 4:25–55.

———. 1978. "The Chain Store Paradox." *Theory and Decision* 9:127–59.

Shapley, Lloyd. 1953a. "Stochastic Games." *Proceedings of the National Academy of Science* 39:1095–1100.

———. 1953b. "A Value for n-Person Games." In H. W. Kuhn and A. W. Tucker, eds., *Contributions to the Theory of Games,* Vol. 2. Princeton: Princeton Universiy Press.

———. 1969. "Utility Comparison and the Theory of Games." In G. Th. Guilbaud, ed., *La Décision, Aggregation et Dynamique.* Colloques Internationaux du Centre de la Recherche Scientifique No. 171, Editions Centre National de la Recherche Scientifique.

Shapley, Lloyd, and Martin Shubik. 1966. "Quasi-cores in a Monetary Economy with Nonconvex Preferences." *Econometrica* 34:805–27.

Shubik, Martin. 1982. *Game Theory in the Social Sciences.* Cambridge: MIT Press.

———. 1984. *A Game-Theoretic Approach to Political Economy.* Cambridge: MIT Press.

Simon, Herbert A. 1957. *Models of Man.* New York: Wiley.

Sobel, Matthew J. 1971. "Noncooperative Stochastic Games." *Annals of Mathematical Statistics* 42:1930–35.

Stähl, Ingolf. 1972. *Bargaining Theory.* Stockholm: Stockholm School of Economics.

Sutton, John. 1986. "Non-cooperative Bargaining Theory: An Introduction." *Review of Economic Studies* 53:709–24.

Thomson, William. 1981. "A Class of Solutions to Bargaining Problems." *Journal of Economic Theory* 25:431–41.

Tideman, T. Nicolaus, and Gordon Tullock. 1976. "A New and Superior Process for Making Social Choices." *Journal of Political Economy* 84:1145–59.

van Damme, Eric. 1987. *Stability and Perfection of Nash Equilibria.* Berlin: Springer-Verlag.

Varian, Hal R. 1984. *Microeconomic Analysis.* Rev. ed. New York: Norton.

von Neumann, John. 1928. "Zur Theorie der Gesellschaftsspiele." *Mathematische Annalen* 100:295–320.

———. 1959. "On the Theory of Games of Strategy." English translation of von Neumann (1928), translated by Sonya Bargmann. In A. W. Tucker and R. D. Luce, ed., *Contributions to the Theory of Games,* vol. 4. Princeton: Princeton University Press.

von Neumann, John, and Oskar Morgenstern. 1944. *Theory of Games and Economic Behavior.* Princeton: Princeton University Press.

Zeuthen, Frederik. 1930. *Problems of Monopoly and Economic Warfare.* London: Routledge & Kegan Paul.

Author index

Subject index